Alexander Waibel · Rainer Stiefelhagen
Editors

Computers in the Human Interaction Loop

Springer

Editors
Alexander Waibel
Universität Karlsruhe (TH)
Germany

Rainer Stiefelhagen
Universität Karlsruhe (TH)
Germany

ISSN 1571-5035
ISBN 978-1-84996-819-5 e-ISBN 978-1-84882-054-8
DOI 10.1007/978-1-84882-054-8
Springer Dordrecht Heidelberg London New York

British Library Cataloguing in Publication Data
A catalogue record for this book is available from the British Library

Printed on acid-free paper

Springer is part of Springer Science+Business Media (www.springer.com)

Human–Computer Interaction Series

Human–Computer Interaction is a multidisciplinary field focused on human aspects of the development of computer technology. As computer-based technology becomes increasingly pervasive – not just in developed countries, but worldwide – the need to take a human-centered approach in the design and development of this technology becomes ever more important. For roughly 30 years now, researchers and practitioners in computational and behavioral sciences have worked to identify theory and practice that influences the direction of these technologies, and this diverse work makes up the field of human–computer interaction. Broadly speaking, it includes the study of what technology might be able to do for people and how people might interact with the technology.

In this series, we present work which advances the science and technology of developing systems which are both effective and satisfying for people in a wide variety of contexts. The human–computer interaction series will focus on theoretical perspectives (such as formal approaches drawn from a variety of behavioral sciences), practical approaches (such as the techniques for effectively integrating user needs in system development), and social issues (such as the determinants of utility, usability and acceptability).

Author guidelines: www.springer.com/authors/book+authors > Author Guidelines

For further volumes:
http://www.springer.com/series/6033

Preface

Considerable human attention is expended on operating and attending to computers, and humans are forced to spend precious time fighting technological artifacts rather than doing what they enjoy and do well: human-human interaction and communication.

Instead of forcing people to pay attention to the artifact, that artifact should pay attention to humans and their interaction. Rather than forcing humans to work within a loop of computers, we would like to see computers serving in a Human Interaction Loop (CHIL). A CHIL computing environment aims to radically change the way we use computers. Rather than expecting a human to attend to technology, CHIL attempts to develop computer assistants that attend to human activities, interactions, and intentions. Instead of reacting only to explicit user requests, such assistants proactively provide services by observing the implicit human request or need, much like a personal butler would.

In 2004, a consortium of 15 laboratories from nine countries constituted itself as the CHIL Consortium to study this CHIL computing paradigm. Project CHIL, "Computers in the Human Interaction Loop," was supported as an Integrated Project (IP 506909) of the European Union under its 6th Framework Program and represents one of Europe's largest concerted efforts in the area of advanced human interfaces. It began on January 1, 2004, and ended successfully in August 2007. It was coordinated administratively by the Fraunhofer Institut für Informations- und Datenverarbeitung (IITB) and scientifically by the Interactive Systems Labs (ISL) at the University of Karlsruhe. The CHIL team includes leading research laboratories in Europe and the United States who collaborate to bring friendlier and more helpful computing services to society.

Instead of requiring user attention to operate machines, CHIL services attempt to understand human activities and interactions to provide helpful services implicitly and unobtrusively. To achieve this goal, machines must understand the human context and activities better; they must adapt to and learn from the human's interests, activities, goals, and aspirations. This requires machines to better perceive and understand all the human communication signals including speech, facial expressions, attention, emotion, gestures, and many more. Based on the perception and understanding of human activities and social context, innovative context-aware and proactive computer services became possible and were explored. Several prototype services were assembled and tested for performance and effectiveness: (1) the Connector (a proactive phone/communication device), (2) the Memory Jog (for supportive information and reminders in meetings), (3) collaborative supportive workspaces and meeting monitoring.

To realize the CHIL vision, the key research activities concentrate on four central areas:

- Perceptual technologies: Proactive, implicit services require a good description of human interaction and activities. This implies a robust description of the perceptual context as it applies to human interaction: Who is doing What, to Whom, Where and How, and Why. Unfortunately, such technologies – in all of their required robustness – do not yet exist. The consortium identified a core set of needed technologies and set out to build and improve them for use in CHIL services.
- Data collection and evaluation: Due to the inherent challenges (open environments, free movement of people, open distant sensors, noise, etc.), technologies are advanced under an aggressive R&D regimen in a worldwide evaluation campaign. In support of these campaigns, meeting, lectures and seminar data have been collected at more than five different sites, and metrics for the technology evaluations have been defined.
- Software infrastructure: A common and defined software infra-architecture serves to improve interoperability among the partners and offers a market-driven exchange of modules for faster integration.
- Services: based on the emerging technologies developed at different labs, using a common architecture, and within a user-centered design framework, CHIL services are assembled and evaluated. In this fashion, first prototypes are continually being (re-)configured and the results of user studies effectively exploited.

This book provides an in-depth discussion of the main research work and main results from an almost four-year effort on CHIL technologies and CHIL services. The work and insights reported represent a major advance in the state-of-the-art of modern context-aware computing and interface technologies.

A description of the CHIL framework, however, would not be complete without mentioning the excellent and constructive cooperation within the CHIL Consortium. The spirits of competitive cooperation as well as cooperative competition turned out to be key factors for the scientific achievements of CHIL.

July 2008

Alex Waibel, Rainer Stiefelhagen,
and the CHIL Consortium

The CHIL Consortium

Financial Coordinator	Prof. Hartwig Steusloff and Dr. Kym Watson Fraunhofer Institute IITB, Fraunhoferstr. 1 76131 Karlsruhe, Germany Email: steusloff@iitb.fraunhofer.de
Scientific Coordinator	Prof. Alex Waibel and Dr. Rainer Stiefelhagen Universität Karlsruhe (TH), Fakultät Informatik Am Fasanengarten 5, 76131 Karlsruhe, Germany Email: waibel@ira.uka.de, stiefel@ira.uka.de

Table 1. Contact persons.

Participant name	**Short name**	**Country**
Fraunhofer-Gesellschaft zur Förderung der Angewandten Forschung e.V. through its Fraunhofer Institut für Informations- und Datenverabeitung, IITB	FhG/IITB	Germany
Universität Karlsruhe (TH) through its Interactive Systems Laboratories (ISL)	UKA-ISL	Germany
DaimlerChrysler AG	DC	Germany
Evaluations and Language Resources Distribution Agency	ELDA	France
IBM Ceska Republika	IBM CR	Czech Republic
Research and Education Society in Information Technologies	RESIT (AIT)	Greece
Institut National de Recherche en Informatique et en Automatique through GRAVIR-UMR5527	INRIA[1]	France
Foundation Bruno Kessler - irst	FBK-irst	Italy
Kungl Tekniska Högskolan	KTH	Sweden
Centre National de la Recherche Scientifique through its Laboratoire d'Informatique pour la Mécanique et les Sciences de l'Ingénieur (LIMSI)	CNRS-LIMSI	France
Technische Universiteit Eindhoven	TUE	Netherlands
Universität Karlsruhe (TH) through its Institute IPD	UKA-IPD	Germany
Universitat Politècnica de Catalunya	UPC	Spain
The Board of Trustees of the Leland Stanford Junior University	Stanford	USA
Carnegie Mellon University	CMU	USA

Table 2. Partners in the CHIL Consortium.

[1] Institut National de Recherche en Informatique et en Automatique (INRIA) participated through GRAVIR-UMR5527, a joint laboratory of INRIA, Centre National de la Recherche

Acknowledgments

We gratefully acknowledge the financial support of the European Commission under the 6th Framework Program. Without the support of the EC and the efforts of the two heads of Unit E1, Nino Varile and Bernhard Smith, work in this area and this Integrated Project would not have been possible.

We would like to express our sincerest gratitude to the project officers at the EC, Mats Ljungqvist and Anne Bajart. As responsible representatives of the European Commission, they have accompanied the work of this IP at all times with great enthusiasm, encouragement, and competent advice and provided invaluable guidance and feedback, scientifically as well as administratively. From their guidance, a highly focused and disciplined, yet highly creative and innovative program resulted.

We would also like to thank the expert review panel for its valuable feedback and advice throughout the project: Josef Kittler, University of Cambridge (UK), Jean Vanderdonckt, Université Catholique de Louvain (Belgium), Michael Wooldridge, University of Liverpool (UK), and Hans Gellersen, Lancaster University (UK).

We also thank the members of the international advisory board of the CHIL project who provided numerous valuable comments and feedback early on, which streamlined the process and focused the Consortium's efforts. The board included Sadaoki Furui, Tokyo Institute of Technology (Japan), Martin Herman, NIST (USA), Joseph Mariani, Ministère délégué à la Recherche (France) at the time of CHIL, Matthew Turk, University of California, Santa Barbara (USA), Ipke Wachsmuth, University Bielefeld (Germany), and Seiichi Yamamoto, Advanced Telecommunications Research Institute International, Kyoto (Japan).

We would like to thank NIST, the National Insititute of Standards and Technology (USA), for its excellent cooperation. NIST included and supported CHIL's participation in its annual Rich Transcription (RT) evaluations during the entire duration of the project. Together with the NIST team, a new evaluation platform for the evaluation of multimodal perception technologies, CLEAR – Classification of Events, Activities and Relationships – was founded in 2006 and jointly organized in 2006 and 2007. We thank the NIST speech group for its cooperation, advice, and friendly support, especially John Garofolo, Rachel Bowers, Jonathan Fiscus, Audrey Le, Travis Rose, Martial Michel, Jerome Ajot, and Kathy Gallo. We also thank Vincent Stanford for his support in using NIST's Marc III microphone array in our smart rooms.

Last, but no means least, all credit is due to the efforts of the CHIL team. A select group from the best laboratories in Europe and the United States, the CHIL Consortium has brought together a remarkable group of research talent that in many ways achieved the enormous feat of demonstrating successful CHIL computing in a very short time. Driven by a joint vision, the CHIL researchers, students, and staff members worked way beyond the call of duty to make the program a success.

Only some of these appear as chapter authors of this book. However, there were more people, including administrative staff members, researchers, and students, who

Scientifique (CNRS), Institut National Polytechnique de Grenoble (INPG), and Université Joseph Fourier (UJF)

provided invaluable contributions to this effort. We would thus like to acknowledge their help here:

The Interactive Systems Labs (ISL) of Universität Karlsruhe (TH) would gratefully like to acknowledge the help of the following persons: Annette Römer and Silke Dannenmaier did a terrific job in managing the bookkeeping of our lab in this large research project, making countless travel arrangements, and organizing various meetings, workshops and events. Margit Rödder did a wonderful job in organizing the dissemination activities of the project, including the organization of two technology days, the project's presentation at the IST fair in Helsinki (for which we won a best exhibition prize!), and numerous meetings and project workshops. She also handled the production of the project's "technology catalog," various presentations and publications, and all interactions with the press. The ISL computer vision team was greatly supported by a number of student members. These include Johannes Stallkamp (face recognition from video), Christian Wojek (audiovisual activity classification and tracking), Julius Ziegler (articulated body tracking), Hua Gao (3D face recognition), Mika Fischer (face recognition), Lorant Toth (open set face recognition), Florian Van De Camp (tracking with active cameras) and Alexander Elbs (multiperson tracking). Gopi Flaherty provided great help in programming smart telephones. Tobias Kluge helped evaluate the Connector service. We also would like to thank the following members of the speech recognition team: Florian Kraft, Sebastian Stüker, Christian Fügen, Shajit Ikhbal and Florian Metze. Furthermore, Kenichi Kumatani contributed to work on acoustic source separation and audiovisual speech recognition. The following students contributed to speech recognition and acoustic source localization: Tobias Gehrig, Uwe Maier, Emilian Stoimenov, Ulrich Klee and Friedrich Faubel (feature enhancement). We also want to thank Cedrick Rochet for his help with data collection, quality control, and distribution of data for the CLEAR workshops. We would also like to thank Kornel Laskowski for his contributions to acoustic emotion recognition and Dan Valsan for his technical support. Finally, we would like to thank Andrea Krönert and Anja Knauer for their help in incorporating the valuable suggestions of the copy editor into the final manuscript of this book.

Fraunhofer IITB and Universität Karlsruhe – ISL would not have been able to coordinate the large integrated R&D CHIL project without a professional consortium management team. We especially thank Tanja Schöpke (Fraunhofer contracts department), Henning Metzlaff (Fraunhofer international project finances), and Marika Hofmann (project management assistant) for their excellent contributions to the project. Special thanks are also due to Fernando Chaves, Falk Walter, and Sibylle Wirth for their creative ideas in establishing the CHIL project server and supporting the dissemination of information to the public. Fraunhofer IITB gratefully acknowledges the significant R&D work on the CHIL software architecture achieved through the contributions of (in alphabetical order) Stefan Barfuß, Thomas Batz, Laurenz Berger, Tino Bischoff, Gottfried Bonn, Ingo Braun, Ulrich Bügel, Axel Bürkle, Fernando Chaves, Daniel Dahlmeier, Niels Dennert, Rusi Filipov, Walter Gäb, Torsten Geiger, Torsten Großkopf, Rainer Helfinger, Jörg Henß, Alice Hertel, Reinhard Herzog, Fritz Kaiser, Ruth Karl, Carmen Kölbl, Thomas Kresken, Pascal Lang, Jürgen Moßgraber, Wilmuth Müller (leader of the IITB R&D work), Michael Okon, Dirk

Pallmer, Uwe Pfirrmann, Manfred Schenk, Gerd Schneider, Michael Stühler, Gerhard Sutschet (leader of the CHIL workpackage on architecture), Daniel Szentes, Martin Thomas, Christoph Vedder, Katrin Vogelbacher, Gero Volke, Patrick Waldschmitt, Heiko Wanning, Martin Wieser, Boxun Xi, Mustafa Yilmaz, and Tianle Zhang.

IBM would like to thank all CHIL partners for their enthusiasm and excellent collaboration during the entire project. Special thanks go to the architecture team (WP2) for its cooperation and many valuable comments during the development of the middleware tools SitCom and CHiLiX, as well as the perception technologies teams under the WP4 and WP5 workpackages of the project. IBM would also like to acknowledge students of the Czech Technical University (CTU) involved in student projects within IBM-CTU cooperation: Petr Slivoň, Tomáš Solár, and Karel Pochop (3D visualization in SitCom), Michal Pavelka (visualization of history in SitCom), and Robert Kessl (support for CHiLiX and SitCom). In addition, IBM will like to acknowledge the work of colleagues Vit Libal and Stephen Chu in data collection and speech technology development efforts, as well as Andrew Senior (formerly with IBM), Zhenqiu Zhang (formerly with the University of Illinois, Urbana-Champaign), and Ambrish Tyagi (formerly with the Ohio State University) for work on visual tracking technology. In particular, Ambrish Tyagi has been the main contributor to the IBM data collection efforts. Furthermore, Patrick Lucey (Queensland University of Technology) contributed to multipose AVASR work on CHIL data. Last, but not least, we would like to thank our management team, in particular David Nahamoo, Chalapathy Neti, Michael Picheny, and Roberto Sicconi for their continued support during various stages of the project.

The AIT CHIL team would like to acknowledge Fotios Talanztis, who has played a leading role in AIT's audio research, as well as Christos Boukis for his valuable contributions in the same field. Special thanks to our students Siamak Azodolmolky, Ghassan Karame, Panos Kasianidis, Nikos Katsarakis, Vassilis Mylonakis, Ippokratis Pandis, Elias Retzeperis, George Souretis, Kostas Stamatis, and Andreas Stergiou, who have undertaken the (real-time) implementation of AIT's audio and visual tracking technologies, as well as of several middleware components of the CHIL architecture and applications.

INRIA would like to gratefully thank the following people for their participation and substantial contributions to the CHIL project: Patrick Reignier (situation modeling, situation implementation), Dominique Vaufreydaz (audio processing, speech detection and analysis), Jerome Maisonnasse (human interaction analysis, user studies and experiments), Sebastien Pesnel (video tracking system, perceptual components and system architecture), Alba Ferrer-Biosca (3D video tracking system, perceptual components), Alban Caporossi (video tracking system), and Augustin Lux (image processing, perceptual components).

FBK-irst would like to thank the following people for their substantial contributions to the project: Koray Balci and Nadia Mana (embodied agents), Alberto Battocchi, Ilenia Graziola, Vera Falcon (user studies for the Social Supportive Workspace and the Relational Report), Luca Brayda and Claudio Bertotti (to Mark III bugs, design of the new array), Luca Cristoforetti (audio acquisition system, data acquisition),

Massimiliano Ruocco (model-based head orientation from optical flow), Piergiorgio Svaizer (advice on signal processing), Francesco Tobia (low-level vision software architecture, data acquisition, participation in the CLEAR evaluation campaigns), and Stefano Ziller and Sebastian Rizzo (vision-based motion capture).

KTH would like to thank the CHIL Consortium for an excellent joint effort and a well-organized and smoothly run project. The multimodal team was considerably strengthened by Magnus Nordstrand, Gunilla Svanfeldt, and Mikael Nordenberg. Inger Karlsson and Kåre Sjölander dedicated themselves to the emotion recognition task. Caroline Bergling provided excellent and timely administrative support during the whole project.

Many researchers in the Spoken Language Processing and some in Language, Information and Representation Group at LIMSI contributed to the CHIL project The main activities at LIMSI concerned the activities of "Who and Where" and "What", to which the following persons directly contributed: Gilles Adda, Martine Adda-Decker, Claude Barras, Eric Bilinski, Olivier Galibert, Jean-Jacques Gangolf, Jean-Luc Gauvain, Lori Lamel, Cheung-Chi Leung, Sophie Rosset, Holger Schwenk, Delphine Tribout, and Xuan Zhu. Early in the project, some initial work was carried out for emotion recognition, involving the contributions of Laurence Devillers, Laurence Vidrascu, Martine L'Esprit, and Jang Martel. We also would like to thank Alexandre Allauzen, Hélène Bonneau-Maynard, Michèle Jardino, Patrick Paroubek, and Ioana Vasilescu for their support and fruitful discussions, as well as Bianca Dimulescu-Vieru and Cecile Woehrling for their help when called upon. We also would like to thank the people in the background who contributed to the smooth functioning of the CHIL project: the administrative and technical staff at the laboratory and the CNRS regional headquarters. In particular, we would like to mention Martine Charrue, Béatrice Vérin-Chalumeau, Karine Bassoulet-Thomazeau, Joëlle Ragideau, Nadine Pain, Bernard Merienne, Elisabeth Piotelat, Annick Choisier, Pascal Desroches, and Daniel Lerin for their help throughout the project. We also acknowledge the support from the laboratory director, Patrick Le Quéré, and co-director Philippe Tarroux.

The Department of Industrial Design of the Technische Universiteit Eindhoven wishes to thank the following people: Adam Drazin, Jimmy Wang, Olga Kulyk, Rahat Iqbal, Anke Eyck, Kelvin Geerlings, Ellen Dreezens, and Olga van Herwijnen, all of whom were involved in the research for some time, Martin Boschman, Leon Kandeleers, and Eugen Schindler for technical support, and Richard Foolen and Nora van den Berg for administrative support. Furthermore, we would like to thank Herman Meines of the Governmental Department of VROM in The Hague for allowing us to make recordings of strategic meetings, and Alan Kentrow, Ernst Teunissen, Shani Harmon, and Neil Pearse of the Monitor Group for spending their time in numerous face-to-face and telephone meetings to discuss data analysis and to explore opportunities for applying CHIL technologies to the coaching of teamwork skills.

The Institute for Program Structures and Data Organization of Universität Karlsruhe (TH) appreciates the contributions of Yue Ying Zhi to the ontology visualization front end of the CHIL Knowledge Base Server. Thanks are also due to Stefan Mirevski and Ivan Popov for laying out the foundations of the formal specification of the CHIL OWL API. We wish to express our gratitude to Tuan Kiet Bui, who devised the database back end of the CHIL Knowledge Base Server, implemented code generators for the CHIL OWL API, and contributed significantly to our site visit demonstrations. We are grateful to Guido Malpohl for his valuable advice and support in resolving networking issues during the development of the ChilFlow communication middleware. Special thanks go to the AIT CHIL team for their efforts in early adopting the CHIL Ontology.

TALP Research Center and the Image Processing Group of Universitat Politècnica de Catalunya (UPC) would like to acknowledge all groups and researchers who have contributed to the project activities. The authors are grateful to the CHIL partners who provided the evaluation seminar databases and also to all participants in the evaluations for their interest and contributions. We would like to thank the following individuals for their contributions to the success of the project: Joachim Neumann and Shadi El-Hajj for their contributions to the demos and the CHIL UPC movie; Albert Gil for the multicamera recording system in the UPC's smart room; Javier Ruiz and Daniel Almendro for the module wrappers and the solutions to real-time intermodule communication; Joan-Isaac Biel and Carlos Segura for the acoustic event detection system and the demo GUI. Carlos Segura is also thanked for helping to develop the isolated acoustic event database. We would like to acknowledge Jan Anguita and Jordi Luque for their work on speaker identification task, Pascual Ejarque and Ramon Morros for their contributions to multimodal person identification, Alberto Abad and Carlos Segura for their efforts on speaker tracking and orientation tasks, José Luis Landabaso and Jordi Salvador for their contributions to person tracking, Christian Ferran and Xavi Giró for their efforts in object detection, and Mihai Surdeanu for his contribution to question answering.

Stanford University would like to acknowledge the contributions from the following people: Anna Ho, Shailendra Rao, Robby Ratan, Erica Robles, Abhay Sukumaran, Leila Takayama, and Qianying Wang.

The Interactive Systems Laboratories at CMU would like to greatfully mention several lab members who helped and supported our work during the CHIL project: Jie Yang led efforts on video analysis and multimodal interaction; Datong Cheng assisted in collecting and analyzing video data; Tanja Schultz advised our developments in automatic recognition of meeting speech data at CMU as well as at Universität Karlsruhe (TH); Qin Jin worked on detecting of persons' identities from acoustic data; our data collection team supported the CHIL project in building a meeting room scenario, collecting meeting data, transcribing data, and labeling emotionally relevant behavior and instances of laughter. We would like to thank Nolan Pflug, Spike Katora, Joerg Brunstein (technical support and data collection), Timothy Notari, Julie Whitehead, Zachary Sloane, Matthew Bell, and Joseph P. Fridy (transcription and labeling) for their diligent work and great support.

Contents

List of Figures

List of Tables

Part I

The CHIL Vision and Framework

1

Computers in the Human Interaction Loop

Alex Waibel[1], Hartwig Steusloff[2], Rainer Stiefelhagen[1], Kym Watson[2]

[1] Universität Karlsruhe (TH), Interactive Systems Labs, Fakultät für Informatik, Karlsruhe, Germany
[2] Fraunhofer Institute IITB, Karlsruhe, Germany

It is a common experience in our modern world for humans to be overwhelmed by the complexities of technological artifacts around us and by the attention they demand. While technology provides wonderful support and helpful assistance, it also gives rise to an increased preoccupation with technology itself and with a related fragmentation of attention. But, as humans, we would rather attend to a meaningful dialog and interaction with other humans than to control the operations of machines that serve us. The cause for such complexity and distraction, however, is a natural consequence of the flexibility and choices of functions and features that the technology has to offer. Thus, flexibility of choice and the availability of desirable functions are in conflict with ease of use and our very ability to enjoy their benefits. The artifact cannot yet perform autonomously and therefore requires precise specification of every aspect of its behavior under a variety of different user interface conventions. Standardization and better graphical user interfaces and multimodal human-machine dialog systems including speech, pointing,and mousing have all contributed to improve this situation. Yet, they have improved only the input mode, and have not gone far enough in providing proactive, autonomous *assistance*, and still force the user into a human machine interaction loop at the exclusion of other human human interaction.

To change the limitations of present-day technology, machines must be engaged implicitly and indirectly in a world of humans, or: We must put Computers in the Human Interaction Loop (CHIL), rather than the other way round. Computers should proactively and implicitly attempt to take care of human needs without the necessity of explicit human specification. If technology could be CHIL-enabled, a host of useful services could be possible. Could two people get in touch with each other at the best moment over the most convenient and best media, without phone tag, embarrassing ring tones, and interruptions? Could an attendee in a meeting be reminded of participants' names and affiliations at the right moment without messing with a contact directory? Could computers offer simultaneous translation to attendees who speak different languages and do so unobtrusively just as needed? Can meetings be supported, moderated, and coached without technology getting in the way? Human assistants often work out such logistical support, reminders, and helpful assistance, and they do so tactfully, sensitively, and sometimes diplomatically. Why not machines?

Of course, the answers to these lofty goals and dreams are multifaceted and contain room for a long research agenda along multiple dimensions. It was this overarching vision and dream of CHIL computing that posed specific questions singled out as suitable research problems for a three-year program funded under an Integrated Project of the 6th Framework Program of the European Commission, the *project* CHIL. While originally conceived as a six- (or more) year vision, the CHIL program began with a three-year approved program support.[1] During such a three-and-a-half-year phase, all aspects of the vision were to be addressed, but emphasis was to be given to the perceptual problems needed to derive a better recognition and understanding of the human context, without which all attempts at autonomy would be futile.

A consortium of 15 laboratories in nine countries teamed to explore what is needed to build such advanced perceptual components along with the development of concepts of operations for preliminary CHIL computing services. The CHIL Consortium is one of the largest consortia tackling this problem to date. It has resulted in a research and development infrastructure by which different CHIL services can be quickly assembled, proposed, and evaluated, exploiting a user-centered design to ensure that the developed services will aim at real users' needs and demands. Examples of prototype concepts that have been explored in the project include:

- The Connector: A connector service attempts to connect people at the best time via the best media, whenever it is most opportune to connect them. In lieu of leaving streams of voice messages and playing phone tag, the Connector attempts to track and assess its masters' activities, preoccupations, and their relative social relationships and mediate a proper connection at the right time between them.
- The Memory Jog: This personal assistant helps its human user remember and retrieve needed facts about the world and people around him or her. By recognizing people, spaces, and activities around its master, the Memory Jog can retrieve the names and affiliations of other members in a group. It provides past records of previous encounters and interactions and retrieves information relevant to the meeting.
- Socially supportive workspaces: This service supports human gathering. It offers meeting assistants that track and summarize human interactions in lectures, meetings, and office interactions, provide automatic minutes, and create browsable records of past events.

While CHIL experimented with these concepts, it was not intended as a project to develop three specific application systems, nor was it to deploy any commercial services in their space. It was intended to explore them only as exemplary placeholders for the more general class of systems and services that put human-human interaction at the center and computing services in a supporting function on the periphery, rather

[1] In the third year of the project, a project extension of eight months was approved by the European Commision, resulting in an overall project duration of 44 months.

than the other way around. The prototypes and concepts were also to steer the technology development, so that the most urgently needed technology components could be advanced in a targeted manner.

During the early stages of such development, it was clear that dramatic advances in perceptual technologies were needed: perceptual technologies that can operate in realistic uncontrived environments, without idealizations and simplifications; technologies that observe and understand human activities in an unobtrusive and opportunistic way without depending on unrealistic recording conditions or assumptions. The complementarity of modalities (vision, speech, etc.) would also have to be considered and new ways of describing human activities defined. Short of describing every aspect of a perceptual scene, however, the CHIL service concepts helped prioritize potential choices to the most important descriptions of human interactions ("Who? What? Where? To Whom? How?") and guided the perceptual processing research of the project.

For a project of the magnitude of CHIL and for a bold new vision of computing to be explored in the short timeframe of three years, it must be supported by management and software tools that allow for the rapid integration of various service concepts, without lengthy and unwieldy system development efforts. Such tools must provide for playful "plug-and-play" and "compare-and-contrast" that inspire new concepts and permit their rapid realization.

These considerations led to the organization of CHIL "the project" in terms of four major areas:

- Perceptual technologies: Proactive, implicit services require a good description of human interaction and activities. This implies a robust description of the perceptual context as it applies to human interaction: Who is doing What, to Whom, Where,How, and Why. Unfortunately, such technologies – in all of their required robustness – did not yet exist, and the Consortium identified a core set of needed technologies and set out to build and improve them for use in CHIL services.
- Data collection and evaluation: Due to the inherent challenges (open environments, free movement of people, open distant sensors, noise, etc.), technologies are advanced under an aggressive research and development regimen in worldwide evaluation campaigns. In support of these campaigns, data from real meetings, lectures, and seminars in open office and lecture rooms have been collected at more than five different sites, and metrics for technology evaluation defined. The scenes and scenarios were not staged or scripted, but represent natural, realistic human office-seminar interaction scenes.
- Software infrastructure: A common and defined software infra-architecture serves to improve interoperability among partners and offer a market-driven exchange of modules for faster integration.
- Services: Based on the emerging technologies developed at different labs, using a common architecture, CHIL service concepts are then explored and evaluated. In this fashion, system prototypes are continuously developed and (re-)configured.

The present book is intended to review the major achievements, insights, and conclusions derived from the CHIL effort. It loosely follows the organization of the project. Following this introduction, the book is organized in three main parts:

Part II covers perceptual technologies. The book discusses the perceptual component technologies that matter for the description of human activity, interaction, and intent, including person tracking, person identification, head pose estimation, automatic speech recognition, and acoustic scene analysis. Also, some higher-level analysis tasks such as question answering, the extraction and use of human interaction cues, including emotion recognition, as well the classification of activities and modeling of situations are discussed. This part concludes with a chapter summarizing the yearly technology evaluations that have been conducted to assess the progress of such technologies, as well as the rich multimodal data sets that have been produced by the project.

Part III provides an overview of specific computing services that were developed as prototypical instances of the more general class of CHIL services. After a general introduction to the approaches and methodologies used to develop these services, the individual service prototypes are described: The Collaborative Workspace, the Memory Jog Service, the Connector Service, the Relational Cockpit, and the Relational Report Service are described in individual chapters.

Part IV then discusses issues related to a suitable software architecture for the development of perceptually aware multimodal CHIL services. Here details about the general software architecture, the CHILFlow middleware for distributed low-level data transfer, the defined data models, APIs, ontologies, and the software agent framework are presented.

Part V summarizes the achievements and discusses open problems and directions for further research.

Part II

Perceptual Technologies

2

Perceptual Technologies: Analyzing the Who, What, Where of Human Interaction

Rainer Stiefelhagen

Universität Karlsruhe (TH), Interactive Systems Labs, Fakultät für Informatik, Karlsruhe, Germany

To enable the vision of user-centered context-aware and proactive smart environments, such environments need to be able to automatically acquire information about the room, the people in it, and their interactions. Thus, a wide range of audiovisual perceptual technologies is needed that can deliver the relevant information about what is going on in the scene based on the analysis of acoustic and visual sensors. This includes information about the number of people, their identities, locations, postures, gestures, body and head orientations, and the words they utter, among other things. The availability of such cues can then provide the necessary answers to the basic questions of "who," "what," "where," "when," "how" and "why," which are needed to eventually enable computers to engage and interact with humans in a human-like manner.

In the CHIL project, considerable effort was thus spent in order to build novel techniques to sense who is doing what, where, with whom, and how in these environments: One of the first questions to answer is, *who* is the person or the group of people involved in the event(s)? And, second, *where* is this-and-that happening in the given environment? Answers to the "where" question can be provided by components for the detection and localization of people. Such components constitute the most basic but also the most important building blocks of perception, since much of the higher-level analysis requires the detection and localization of persons in the first place. The "who" question can be answered by components for identification of speakers and faces. In addition, components to detect people's head orientation and direction of attention, can give answers to the "to whom" question, for example, answering "who was talking to whom" or "who was attending to whom."

Another group of component technologies addresses the detection of *what* is happening, understanding what is being said, what a conversation is about, and what the decisions, conclusions, and action points were. Answers to these question can, for example, be delivered by components for automatic speech recognition and analysis of language. Further important related information can be provided by components for recognizing actions, activities, or acoustic events.

Of particular interest for CHIL environments is the analysis of *how* humans interact. This includes perceiving interaction cues such as human gesture, body pose, and head orientation, as well as analyzing people's attitude, affect, and emotion. Finally, in order to provide useful service to people in a socially appropriate, human-centered way, adequate multimodal presentation and interaction components are needed.

This part of the book presents an overview of the perceptual components developed in the CHIL project. In the various chapters, different approaches for the individual perceptual components will be summarized and compared, and lessons learned will be discussed.

An important aspect for the development of such perceptual components is the availability of realistic training and evaluation data. In the CHIL project, large audio-visual corpora have been collected and annotated to facilitate the development and evaluation of the envisioned perceptual components. To this effect, and to speed up overall progress in this research area, the project has organized a series of perceptual technology evaluations. These were first conducted internally within the project and then took place in the newly created open international evaluation workshop called CLEAR ("Classification of Events, Activities and Relationships"), which was successfully held with broad international participation in 2006 and 2007.

The remainder of this part is organized as follows: Chapters 3 to 5 discuss various perceptual technologies addressing the *who* and *where* questions, including chapters about person tracking (Chapter 3), person identification (Chapter 4), and head pose estimation (Chapter 5). Chapters 6 to 8 then address technologies related to answering the *what* question. These include a chapter on acoustic speech recognition (Chapter 6), a chapter on acoustic event detection (Chapter 7), and a chapter on Language Technologies (Chapter 8). The next chapters address the higher-level analysis of situations and human interaction. Here, we start with Chapter 9 on extracting relevant cues for human interaction analysis, such as focus of attention, body pose, and gestures. It is followed by Chapter 10 on acoustic emotion recognition. Then, Chapter 11 discusses several approaches to recognize a number of human "activities." Chapter 12 then motivates and introduces a situation model as a tool to describe a smart environment, its users, and their activities. Chapter 13 presents the "targeted audio" device, a novel audio output device with a very focused audio beam that can be used to direct audio output to individual users in scenes without disturbing others. Chapter 14 discusses the extraction and use of auditory cues, such as prosody, in order to provide for a natural flow of human-computer interaction. The chapter also discusses how multimodal output can be improved by using prosodically aware talking heads and speech synthesis. Finally, in Chapter 15, we will motivate and introduce the perceptual technology evaluations that have been conducted in the CHIL project and that have, among other things, also led to the creation of the CLEAR evaluation workshop, a new international workshop on multimodal technologies for the perception of humans. In this chapter, we will also describe the various multimodal data sets that have been produced in CHIL and made available to the research community.

3

Person Tracking

Keni Bernardin[1], Rainer Stiefelhagen[1], Aristodemos Pnevmatikakis[2], Oswald Lanz[3], Alessio Brutti[3], Josep R. Casas[4], Gerasimos Potamianos[5]

[1] Universität Karlsruhe (TH), Interactive Systems Labs, Fakultät für Informatik, Karlsruhe, Germany
[2] Athens Information Technology, Peania, Attiki, Greece
[3] Foundation Bruno Kessler, irst, Trento, Italy
[4] Universitat Politècnica de Catalunya, Barcelona, Spain
[5] IBM T.J. Watson Research Center, Yorktown Heights, NY, USA

One of the most basic building blocks for the understanding of human actions and interactions is the accurate detection and tracking of persons in a scene. In constrained scenarios involving at most one subject, or in situations where persons can be confined to a controlled monitoring space or required to wear markers, sensors, or microphones, these tasks can be solved with relative ease. However, when accurate localization and tracking have to be performed in an unobtrusive or discreet fashion, using only distantly placed microphones and cameras, in a variety of natural and uncontrolled scenarios, the challenges posed are much greater. The problems faced by video analysis are those of poor or uneven illumination, low resolution, clutter or occlusion, unclean backgrounds, and multiple moving and uncooperative users that are not always easily distinguishable. The problems faced by acoustic techniques, which rely on arrays of distant microphones to detect and pinpoint the source of human speech, are those of ambient noise, cross-talk, or rapid speaker turns, reverberations, unpredictable or hardly identifiable nonhuman noises, silent persons, etc. This chapter presents the efforts undertaken by the CHIL Consortium in the visual, acoustic, and audiovisual domains for the simultaneous and unobtrusive tracking of several users in close to real-life interaction scenarios. As contributions to this challenging and varied field are too numerous to be presented in detail, the discussion focuses on the main problems encountered and lessons learned, on higher-level design strategies and trends that proved successful. In addition, it tries to highlight individual approaches of particular interest that show the advantages of combining multiple far-field sensory sources. A noteworthy point is that almost all investigated techniques are designed to allow for fully automatic online operation, making them usable in a wide range of applications requiring quick system reaction or feedback.

Fig. 3.1. Example screenshot of a person tracking system running on CHIL data (image taken from [9]).

3.1 Goals and Challenges

During its 3.5-year period, the CHIL project carried out a considerable amount of research on detection and tracking techniques powerful enough to accomplish its ambitious goal: the detailed, unobtrusive, and discreet observation of humans engaging in natural interactions. Within the list of technological challenges addressed by the project in the "Who and Where" field, person tracking occupied a quite important position because of (1) the various subtasks it attempted, including acoustic, visual, and multimodal tracking, and (2) the high number of partners involved in friendly competition (so-called co-opetition) to create the most reliable and performant system.

Research focused mostly on the tracking of persons inside smart rooms equipped with a number of audiovisual sensors. The goal of tracking was to determine, for all points in time, the scene coordinates of room occupants with respect to a given room coordinate frame. This is in contrast to much of the visual tracking research, where only image coordinates are estimated, and to most of the acoustic or multimodal tracking research, where only relative azimuths are determined. In CHIL, the requirement, through all modalities, was to track the *xy*-coordinates of a person on the ground plane, with the aim of reaching precisions down to a few centimeters. The restriction to the plane is due in part to the fact that the goal is to track entire bodies, not, for example, heads or torsos, and in part to the ease of annotation when creating the reference ground truth. Although the estimation of a person's height was not deemed a high priority in technology evaluations, almost all developed acoustic trackers and most of the more advanced visual trackers did, in fact, perform this estimation in live systems. Whereas at the start of the project, only the tracking of a single room occupant – the main speaker in a seminar – was attempted, by the second year, attention was shifted to the simultaneous tracking of all room occupants on the visual side, and the consecutive tracking of alternating speakers on the acoustic side (see [2] for more details on tracking tasks).

The CHIL room sensors used in tracking include a minimum of four fixed cameras installed in the room corners, with highly overlapping fields of view, one wide-angle camera fixed under the ceiling overlooking the entire room, at least three T-shaped four-channel microphone arrays, and one Mark III 64-channel microphone array on the room's walls (see also Chapter 15). While the availability of a high number of sensors may be seen as an advantage, offering a great deal of redundancy in the captured information that could be exploited by algorithms, it must also be seen as a challenge, requiring one to solve problems such as data synchronization, transfer, and distributed processing, spatiotemporal fusion, modality fusion, etc. One should note that the minimal required sensor setup offers only rough guidelines on the amount and placement of sensors, leaving open such parameters as the type of cameras or microphones used, camera resolution and frame rate, color constancy across views, etc., while developed trackers are expected to operate on data from all CHIL rooms alike. From the audio point of view, it is also worth mentioning that CHIL represents one of the first attempts to perform and systematically evaluate acoustic tracking with a distributed microphone network (DMN), while most of the research in this field is focused on linear or compact microphone arrays. A DMN scenario is characterized by completely different problem specifics than that of a compact array. For example, the near-field assumption does not hold between microphone pairs from different clusters. Moreover, in contrast to what is often implicitly done when exploiting linear arrays, in the CHIL scenario, no assumption can be made on the relative position or distance of the speaker to the various sensors. In this sense, the CHIL room sensor setup itself created new and interesting research problems that required the development of original tracking and data fusion techniques.

Another factor making the CHIL tracking tasks particularly challenging is the nature of the application scenario. As mentioned above, algorithms have to automatically adapt to data coming from up to five CHIL smart rooms with very different characteristics, such as room dimensions, illumination, chromatic and acoustic signature, average person-sensor distances, camera coverage, furnishing, reverberation properties, sources of noise or occlusion, etc. The scenario is that of real seminars and meetings with sometimes large numbers of occupants free to sit around tables or on rows of chairs, stand or move around, occasionally enter or leave, laugh, interrupt or occlude each other, etc., making it hard to adopt any assumptions but the most general ones about their behavior (see Chapter 15 for a description of the CHIL seminar and meeting corpus). This prevents the application of conventional techniques such as visual tracking based on foreground blobs or generic color models, track initialization based on creation or deletion zones, temporal smoothing of sound source estimations, etc., and requires elaborate methods for combined (person or speech) detection and tracking, model adaptation, data association, feature and sensor selection, and so forth. The developed tracking systems have been extensively tested in a series of increasingly challenging evaluations using the CHIL seminar and meeting database [21, 20]. These large-scale, systematic evaluations, using realistic, multimodal data and allowing one to objectively compare systems and measure progress, also constitute a noteworthy contribution to the research field.

Fig. 3.2. Example screenshot of a person tracking system running on CHIL data.

3.2 Difficulties and Lessons Learned

In retrospect, one of the most important problems that needed to be solved in the person tracking field was the definition of a set of metrics for the tracking of multiple objects that could be used in offline evaluations. No consensus on standardized metrics existed in the research field. For the first set of CHIL tracking evaluations, only simple metrics for single object tracking, based on average spatial error, were used for the visual subtask. These metrics were also different from those used for the acoustic subtask, which made performance comparisons difficult. To remedy this, a set of concise and intuitive metrics that would be general enough to be used in all subtasks – visual, acoustic, and multimodal – was defined at the end of 2005. These metrics, usable for single- or multiple-object tracking alike, judge a tracker's performance based on its ability to precisely estimate an object's position (multiple-object tracking precision, $MOTP$) and its ability to correctly estimate the number of objects and keep correct trajectories in time (multiple-object tracking accuracy, $MOTA$) (see [3] for details). They were subsequently used for all evaluations, notably the CLEAR [21, 20] evaluations, for a broad range of tasks including acoustic, visual, and multimodal 3D person tracking, 2D visual person tracking, face detection and tracking, and vehicle tracking.

A related problem was that of creating accurate ground-truth labels, for person positions and speech activity. Whereas 3D position information could be easily obtained from the synchronized camera views by marking head centroids in each view and triangulating, the annotation of active speakers was somewhat more problematic. This information was not discernible in the video streams and had to be extracted from transcriptions that were made separately on the audio channels. The evaluation criteria required us to identify speech segments, including portions of silence, of overlapping speech, or containing active noise sources, which is why transcriptions had to be made on far-field microphone channels. In the first set of evaluations, only single person tracking was attempted and the audio and video tasks were completely separated. The advent of multimodal and multiperson tracking tasks in the 2006 evaluations made it necessary to merge audio and video annotations and tools, which until then had been developed independently. The process of defining valid annotation guidelines went on until the final year of the project. On the whole, it can be said that the transcription of audiovisual data for tracking no doubt constitutes one of the main lessons learned by the consortium.

Another smaller but no less important achievement was the definition of common standards for the calibration of room cameras and the distribution of camera and microphone calibration and position information.

Finally, the synchronization of all sensory streams was a big challenge, both for online operation in live demonstrations and for the offline processing of recorded data. This included the synchronization of all cameras, the synchronization of all microphone arrays, table-top, or close-talking microphones (these were not used in tracking), of the Mark III array with respect to the microphone clusters, and finally of all microphones with respect to the cameras. The synchronization of all T-shaped microphone clusters could easily be achieved to the sample level (at 44 KHz) by capturing all channels using a single 24-channel RME Hammerfall sound card, a solution adopted unanimously by all partners. The requirements for synchronization of video frames from different camera views or of cameras with respect to the Mark III and T-shaped microphone arrays were less stringent, so that simple solutions, such as NTP synchronization of the capturing machines' clocks, were sufficient. More precise and practical solutions worth mentioning were, however, implemented by some project partners. One example is the use of gen-locked cameras synchronized to the frame level, coupled with a dedicated board, delivering a common synchronization signal to all cameras and microphone clusters. Such a solution was applied with success, for example, by UPC, both for online operation and for flawless recording over extended lengths of time.

3.3 Results and Highlights

Throughout the duration of the project, steady progress was made, going from single-modality systems with manual or implicit initialization, using simple features, sometimes implying several manually concatenated offline processing steps and tracking at most one person, to fully automatic, self-initializing, real-time-capable systems, using a combination of features, fusing several audio and visual sensor streams and capable of tracking multiple targets. Aside from the tracking tasks, which grew more and more complex, the evaluation data, which were initially recorded only at the UKA-ISL site, also became increasingly difficult and varied, with the completion of four more recording smart rooms, the inclusion of more challenging interaction scenarios, the elimination of simplifying assumptions such as main speakers, areas of interest in the room, or manually segmented audio data that exclude silence, noise, or cross-talk. Nevertheless, the performance of systems steadily increased over the years. Figure 3.3 shows the progress made in audio, video, and multimodal tracking for the last official CLEAR evaluations [21, 20].

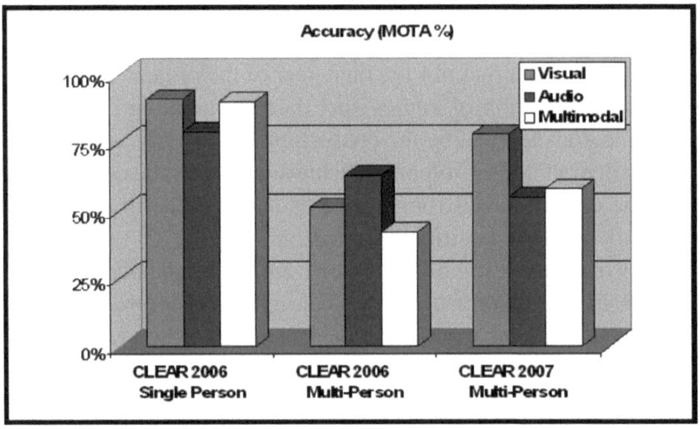

Fig. 3.3. Best system performances for the CLEAR 2006 and 2007 3D person tracking evaluations. The MOTA score measures a tracker's ability to correctly estimate the number of objects and their rough trajectories.

On the visual side, from the viewpoint of algorithm design, two distinct approaches have been followed by the various 3D tracking systems built within the Consortium:

1. A model-based approach: A 3D model of the tracked object is maintained by rendering it onto the camera views, searching for supporting evidence per view, and, based on that, updating its parameters. Such systems were used in CHIL, for example, in 2005 by UKA-ISL for particle filter-based single-person tracking [17] and in 2006 by FBK-irst and UPC for particle filter- and voxel-based multi-person tracking [14, 1]. For 2007, all approaches were further refined [4, 15, 9].

2. A data-driven approach: 2D trackers operate independently on the separate camera views; then the 2D tracks belonging to a same target are collected into a 3D one. Such systems were used throughout the course of the project by various

partners. Notable examples are systems used in 2006 by IBM [25, 23] and in 2007 by AIT [12].

The two approaches are depicted in Fig. 3.4. An important advantage of the model-based approach is that rendering can be implemented in a way that mimics the real image formation process, including effects like perspective distortion and scaling, lens distortion, etc. In the context of multibody tracking, this is particularly advantageous, since occlusions can be handled at the rendering level. This way, the update is done by looking for supporting evidence only in the image parts where the different models are visible; thus, occlusions are handled in a systematic manner. Systems that do so are [11, 18], and the FBK-irst tracker [14]. The real novelty of the FBK-irst approach is that it can do the update in a probabilistic framework (particle filtering) more efficiently than previous approaches: The computational complexity grows at most quadratically with the number of targets, rather than exponentially. This makes the system run robustly in real time with 7 to 10 occluding targets. The disadvantage, of course, is that the person models have to be initialized and occasionally updated, which in some situations may be tricky (e.g., several people entering simultaneously into the monitored room).

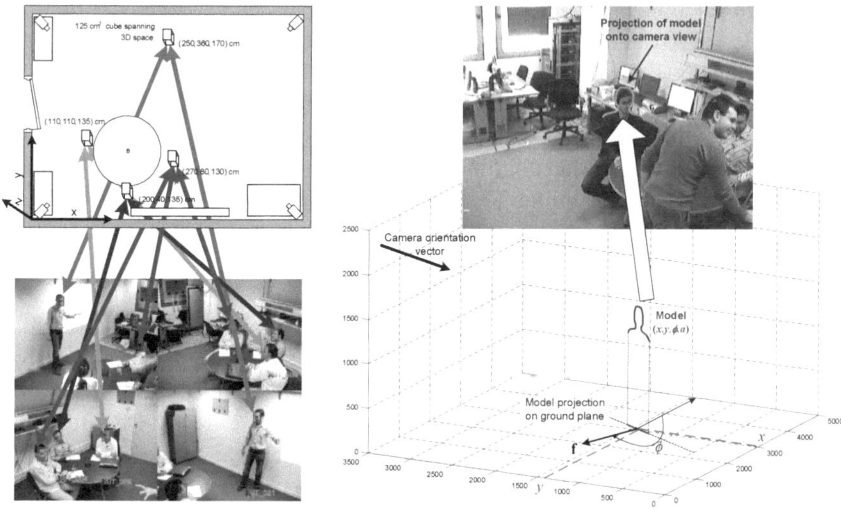

(a) (b)

Fig. 3.4. (a)Data-driven and (b) model-based approaches toward 3D tracking.

The handling of occlusions and the association of (possibly split or merged) tracks are the main drawbacks of the data-driven approach. There is not enough information in the independent camera views to efficiently address them. The work-around in this case is to work with faces and/or heads instead of bodies. In this case, the initialization problems of the direct approach are diminished, as initialization is handled by face detectors. Also, 2D face tracking can be complemented by detection to increase precision.

In terms of performance, the model-based approach generally provides for better accuracy (MOTA) but less precision (MOTP) than the data-driven one. In terms of speed, the data-driven approach is relatively slow due to the face detectors, which constitute a bottleneck in the system.

On the acoustic side, most approaches built on the computation of the generalized cross-correlation (GCC-PHAT) between microphone pairs of an array. Other methods relied on calculating the mutual information between the microphones [22] in order to integrate information about the reverberant characteristics of the environment. Tracking approaches can be roughly categorized as follows:

1. Approaches that rely on the computation of a more or less coarse global coherence field (GCF, or SRP-PHAT), on which the tracking of correlation peaks is performed. Such an approach was used, for example, by UPC [1], in combination with Kalman filtering, or by FBK-irst [7], using simple temporal filtering of the peaks.
2. Particle filter approaches, which stipulate a number of candidate person positions at every point in time and measure the agreement of the observed acoustic signals (their correlation value) to the position hypothesis. Such a system was used, for example, by UKA-ISL [16] in the 2006 evaluations for visual, acoustic, and multimodal tracking alike.
3. Approaches that feed computed time delays of arrival (TDOAs) between microphone pairs directly as observations to a Kalman or other probabilistic tracker. This approach was repeatedly followed, for example, by UKA-ISL, using an iterated extended Kalman filter [13] or a joint probabilistic data association filter (JPDAF) [10]. The advantage of the JPDAF approach is that it is designed to keep track of a number of sound sources, including noise sources, by maintaining a Kalman filter for each of them. The association of observations, in the form of peaks in correlation between microphone pairs, to tracks and the update of track positions and uncertainties is made jointly for all tracks using a probabilistic framework. This allows one to better handle rapid speaker switches or occasional sources of noise, as these do not disturb the track of the main target. Another point of merit of the UKA-ISL tracker was the incorporation of automatic speech activity detection (SAD) into the tracking framework itself. In this case, the activity of a target is detected by analyzing its state error covariance matrix. Whereas in earlier CHIL systems, SAD was often performed and evaluated separately, toward the end of the project, more and more approaches featured built-in speech detection techniques [10, 19].

It should be mentioned at this point that a direct correlation between speech source localization performance and automatic far-field speech recognition performance could be shown [24]. In these experiments, which were based on CHIL seminar data, the output of an audiovisual, particle filter-based speaker tracking system was passed to a beamforming module. The beamformer filtered the data captured by the Mark III linear array before passing it on for speech recognition. In the experiments, increasing the localization accuracy led to a measurable decrease in the word error rate (WER).

In the field of multimodal tracking, finally, efforts started in the second half of 2005. While most initial systems performed audio and video tracking separately, and combined tracker outputs in a postprocessing step, a few select approaches incorporated the multimodal streams at the feature level. These were notably particle filter-based trackers [16, 4, 6], as these allow for the probabilistic, flexible, and easy integration of a multitude of features across sensors and modalities. The underlying idea is that early fusion of the data provides more accurate results, as it eliminates the effects of wrongful hard decisions made by monomodal trackers. An important point that became clear during the last two years of the project is that multimodal fusion does not necessarily lead to higher accuracies, contrary to what may be expected in general. It is, in fact, highly dependent on the task and data at hand. For the 2006 evaluations, the multimodal tracking task was split into two conditions. The first required all participants of a meeting to be tracked audiovisually. The second required to track only the active speakers in predetermined segments of clean speech audiovisually. The result was that accuracies for the first condition were no better than using visual tracking alone, as the inclusion of acoustic features could only help improve accuracies for one participant out of many for select points in time. The achievable slight improvements made no significant difference when computing global averages. Accuracies for the second condition, in turn, were no better than using acoustic tracking alone, as the limitation to timeframes of clean speech produced an imbalance in favor of the acoustic modality, which nearly eliminated the advantages of using visual features. Due to this, the multimodal task was redesigned for the 2007 evaluations, with the inclusion of silence and noise segments in the tracking data, and the requirement to continuously track the last known speaker. Systems were thus obliged to select the right target person using acoustic features, track audiovisually, and bridge periods of silence using visual cues. With the so-defined task and data, the advantages of fusion become more clearly visible, as could be shown, for example, in [4].

On a final note, it is important to mention that many of the CHIL tracking systems were designed with realistic online operation in mind. In addition to being continuously tested in offline evaluations, the developed techniques were also integrated with great success in a number of real-time demonstrations, not only limited to the inside of CHIL smart rooms. These include the simultaneous determination of a speaker's position and orientation using only microphone clusters [8], the robust visual tracking of multiple persons through occlusions (the "SmarTrack" system [14]; see Fig. 3.5), which was demonstrated outside CHIL rooms, notably at the IST conference in Helsinki in 2006, the simultaneous audiovisual tracking and identification of multiple persons [5], which required the online integration of several of tracking and ID components, and the various Memory Jog services and their demonstrations (see Chapter 18, where accurate situation models of the CHIL room were constructed based on tracker output).

Fig. 3.5. Example screenshot of the "SmarTrack" person tracking system in live operation.

References

1. A. Abad, C. Canton-Ferrer, C. Segura, J. L. Landabaso, D. Macho, J. R. Casas, J. Hernando, M. Pardas, and C. Nadeu. UPC audio, video and multimodal person tracking systems in the CLEAR evaluation campaign. In *Multimodal Technologies for Perception of Humans, Proceedings of the First International CLEAR Evaluation Workshop*, LNCS 4122, Southampton, UK, 2007.
2. K. Bernardin. CLEAR 2007 evaluation plan – 3D person tracking task, 2007. `http://www.clear-evaluation.org/`.
3. K. Bernardin, A. Elbs, and R. Stiefelhagen. Multiple object tracking performance metrics and evaluation in a smart room environment. In *The Sixth IEEE International Workshop on Visual Surveillance (in conjunction with ECCV)*, Graz, Austria, May 2006.
4. K. Bernardin, T. Gehrig, and R. Stiefelhagen. Multi-level particle filter fusion of features and cues for audio-visual person tracking. In *Multimodal Technologies for Perception of Humans, Proceedings of the International Evaluation Workshops CLEAR 2007 and RT 2007*, LNCS 4625, pages 70–81, Baltimore, MD, May 8-11 2007.
5. K. Bernardin and R. Stiefelhagen. Audio-visual multi-person tracking and identification for smart environments. In *ACM Multimedia 2007*, Augsburg, Germany, Sept. 2007.
6. R. Brunelli, A. Brutti, P. Chippendale, O. Lanz, M. Omologo, P. Svaizer, and F. Tobia. A generative approach to audio-visual person tracking. In *Multimodal Technologies for Perception of Humans, Proceedings of the First International CLEAR Evaluation Workshop*, LNCS 4122, pages 55–68, Southampton, UK, 2007.
7. A. Brutti. A person tracking system for CHIL meetings. In *Multimodal Technologies for Perception of Humans, Proceedings of the International Evaluation Workshops CLEAR 2007 and RT 2007*, LNCS 4625, Baltimore, MD, May 8-11 2007.
8. A. Brutti, M. Omologo, and P. Svaizer. Speaker localization based on oriented global coherence field. *Interspeech*, 2006.
9. C. Canton-Ferrer, J. Salvador, J. Casas, and M. Pardas. Multi-person tracking strategies based on voxel analysis. In *Multimodal Technologies for Perception of Humans, Proceed-*

ings of the International Evaluation Workshops CLEAR 2007 and RT 2007, LNCS 4625, pages 91–103, Baltimore, MD, May 8-11 2007. Springer.
10. T. Gehrig and J. McDonough. Tracking multiple speakers with probabilistic data association filters. In *Multimodal Technologies for Perception of Humans, Proceedings of the First International CLEAR Evaluation Workshop*, LNCS 4122, Southampton, UK, 2007.
11. M. Isard and J. MacCormick. Bramble: A Bayesian multiple-blob tracker. *Proceedings of the International Conference Computer Vision*, 2003.
12. N. Katsarakis, F. Talantzis, A. Pnevmatikakis, and L. Polymenakos. The AIT 3D audio / visual person tracker for CLEAR 2007. In *Multimodal Technologies for Perception of Humans, Proceedings of the International Evaluation Workshops CLEAR 2007 and RT 2007*, LNCS 4625, pages 35–46, Baltimore, MD, May 8-11 2007.
13. U. Klee, T. Gehrig, and J. McDonough. Kalman filters for time delay of arrival-based source localization. *Journal of Advanced Signal Processing, Special Issue on Multi-Channel Speech Processing*, 2006.
14. O. Lanz. Approximate Bayesian multibody tracking. *IEEE Transactions on Pattern Analysis and Machine Intelligence*, 28(9):1436–1449, Sept. 2006.
15. O. Lanz, P. Chippendale, and R. Brunelli. An appearance-based particle filter for visual tracking in smart rooms. In *Multimodal Technologies for Perception of Humans, Proceedings of the International Evaluation Workshops CLEAR 2007 and RT 2007*, LNCS 4625, pages 57–69, Baltimore, MD, May 8-11 2007.
16. K. Nickel, T. Gehrig, H. K. Ekenel, J. McDonough, and R. Stiefelhagen. An audio-visual particle filter for speaker tracking on the CLEAR'06 evaluation dataset. In *Multimodal Technologies for Perception of Humans, Proceedings of the First International CLEAR Evaluation Workshop*, LNCS 4122, Southampton, UK, 2007.
17. K. Nickel, T. Gehrig, R. Stiefelhagen, and J. McDonough. A joint particle filter for audio-visual speaker tracking. In *Proceedings of the Seventh International Conference on Multimodal Interfaces - ICMI 2005*, pages 61–68, Oct. 2005.
18. K. Otsuka and N. Mukawa. Multiview occlusion analysis for tracking densely populated objects based on 2-D visual angles. *Proceedings of the International Conference Computer Vision and Pattern Recognition*, 2004.
19. C. Segura, A. Abad, C. Nadeu, and J. Hernando. Multispeaker localization and tracking in intelligent environments. In *Multimodal Technologies for Perception of Humans, Proceedings of the International Evaluation Workshops CLEAR 2007 and RT 2007*, LNCS 4625, pages 82–90, Baltimore, MD, May 8-11 2007.
20. R. Stiefelhagen, R. Bowers, and J. Fiscus, editors. *Multimodal Technologies for Perception of Humans, Proceedings of the International Evaluation Workshops CLEAR 2007 and RT 2007*. LNCS 4625. Springer, Baltimore, MD, May 8-11 2007.
21. R. Stiefelhagen and J. Garofolo, editors. *Multimodal Technologies for Perception of Humans, First International Evaluation Workshop on Classification of Events, Activities and Relationships, CLEAR'06*. LNCS 4122. Springer, Southampton, UK, Apr. 6-7 2006.
22. F. Talantzis, A. Constantinides, and L. Polymenakos. Estimation of direction of arrival using information theory. *IEEE Signal Processing Letters*, 12(8):561 – 564, Aug. 2005.
23. A. Tyagi, G. Potamianos, J. W. Davis, and S. M. Chu. Fusion of multiple camera views for kernel-based 3D tracking. In *Proceedings of the IEEE Workshop on Motion and Video Computing (WMVC)*, Austin, Texas, 2007.
24. M. Wölfel, K. Nickel, and J. McDonough. Microphone array driven speech recognition: influence of localization on the word error rate. *Proceedings of the Joint Workshop on Multimodal Interaction and Related Machine Learning Algorithms (MLMI)*, 2005.

25. Z. Zhang, G. Potamianos, A. W. Senior, and T. S. Huang. Joint face and head tracking inside multi-camera smart rooms. *Signal, Image and Video Processing*, pages 163–178, 2007.

4

Multimodal Person Identification

Aristodemos Pnevmatikakis[1], Hazım K. Ekenel[2], Claude Barras[3], Javier Hernando[4]

[1] Athens Information Technology, Peania, Attiki, Greece,
[2] Universität Karlsruhe (TH), Interactive Systems Labs, Fakultät für Informatik, Karlsruhe, Germany
[3] Laboratoire d'Informatique pour la Mécanique et les Sciences de l'Ingénieur (LIMSI-CNRS), Orsay, France
[4] Universitat Politècnica de Catalunya, Barcelona, Spain

Person identification is of paramount importance in security, surveillance, human-computer interfaces, and smart spaces. All these applications attempt the recognition of people based on audiovisual data. The way the systems collect these data divides them into two categories:

- Near-field systems: Both the sensor and the person to be identified focus on each other.
- Far-field systems: The sensors monitor an entire space in which the person appears, occasionally collecting useful data (face and/or speech) about that person. Also, the person pays no attention to the sensors and is possibly unaware of their existence.

Near-field person identification systems require the person's attention. The person is aware of the sensors' location, approaches the sensors and offers the system samples of his or hers face and voice. Such systems are obtrusive, but the audiovisual streams thus collected have a high signal-to-noise ratio. The images depict faces of high resolution, approximately frontal in pose and neutral in expression, while the sound is almost free of the detrimental effects of the room's acoustics: reverberation and attenuation. The typical application of such systems is access control, where these systems are expected to offer close to perfect recognition rates. A typical example of such system using near infrared face recognition can be found in [12]

Far-field person identification systems [15] employ audiovisual sensors scattered in the space the person is expected to be in. The systems do not anticipate that the person will acknowledge the existence of the sensors, nor behave in any way that will facilitate the collection of noise-free audiovisual streams. Hence, far-field data streams are corrupted with noise: The video streams contain faces viewed from arbitrary angles, distances, under arbitrary illumination, and possibly, depending on the environment of the deployment, with arbitrary expressions. Similarly, the sound streams suffer from reverberations, large attenuations, and the coexistence of background sounds. The audiovisual environment changes dramatically as the person moves around the space. As a result, the faces collected are tiny (typically of 10 pixels between the eyes; see Fig. 4.1) and with gross variations in pose, expression, and illumination. The speech samples are also attenuated, corrupted with all sorts of background noises (occasionally entirely masked by them) and reverberations. Nevertheless, far-field systems have three features that allow them to offer usable recognition rates:

- the use of multiple sensors (many cameras and microphones),
- the abundance of training data that are audiovisual streams similar to those on which the system is expected to operate,
- the possibly long periods of time that the systems can collect data on which they are going to base their identity decision.

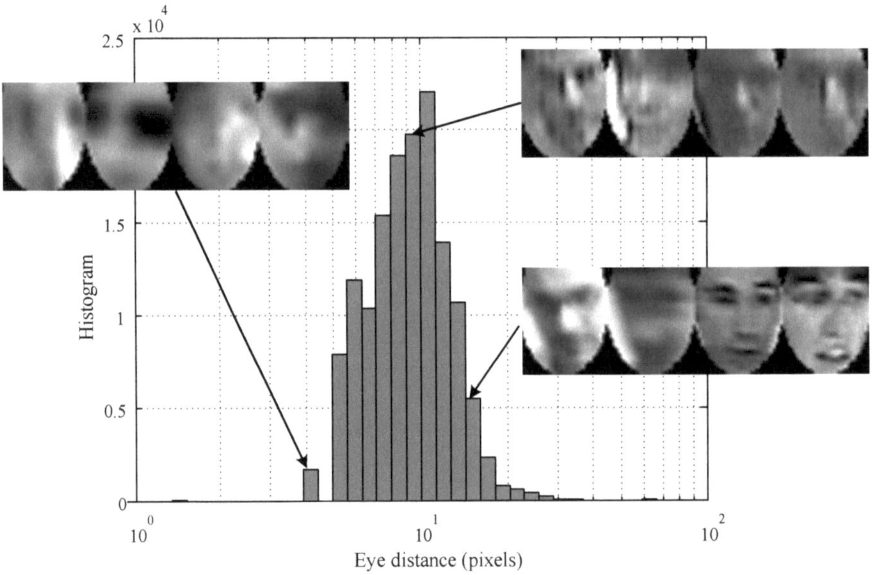

Fig. 4.1. Histogram of the distances between the eyes of the faces to be recognized in typical far-field scenarios. The faces are those of the CLEAR 2006 evaluation. Typical faces at three different eye distances are also shown.

The goal of the far-field multimodal person identification systems built within the CHIL Consortium is to equip smart spaces with a means of giving an identity to each person working in the space. Contrary to the more usual verification problems, where an identity claim is verified over hundreds of individuals, the CHIL systems perform identification of the people present from a rather limited set of possible identities. This is not a restriction in the intended application domain; the number of the people called to work in a given smartspace is indeed limited.

In order to measure progress in these two categories of video-to-video person identification systems, it is important to utilize formal evaluation protocols. This is a well-established procedure for near-field face recognition, but to the authors' knowledge, it is only through the CLEAR (2006 and 2007) person identification evaluations [18, 19] and their predecessor Run-1 evaluations [8] initiated by the CHIL project that a formal evaluation campaign focusing on far-field multimodal person identification has been launched.

The following sections outline, the different approaches of the CHIL Consortium for audio-only, video-only, and multimodal person identification systems, leading to the lessons learned and progress made in the field of far-field multimodal person identification over the course of the project.

4.1 Speaker Identification

All systems construct speaker models using Gaussian mixture models (GMM) of some short-term speech features. The features are derived from speech frames using the Mel frequency cepstral coefficients (MFCC) [11, 1, 2, 17, 16, 7, 6], the perceptually-weighted linear prediction coefficients (PLP) [17], or the frequency filtering (FF) [14, 13] approaches. Postprocessing of the features like cepstral mean normalization and feature warping are included in some systems [1, 2, 7, 6]. The models for the different speakers are trained either using the available training speech for each speaker independently, or by maximum a posteriori (MAP) adaptation of a universal background model (UBM) [1, 2]. The latter allows much larger GMMs by pooling all available speech and training a large UBM GMM.

During the recognition phase, a test segment is scored against all possible speakers and the speaker model with the highest log-likelihood is chosen.

The systems explore different options of utilizing the multiple audio streams. These fall into two categories: those that attempt to improve the quality of the signal prior to constructing or testing models, utilizing some sort of beamforming [2, 13], and those that attempt postdecision fusion of the classification outcomes derived from each standalone microphone [13].

Beamforming is not the only option for preprocessing the signals. Since the recordings are far-field and corrupted by all sorts of background noises, they can be preprocessed by a speech activity detector (SAD) to isolate speech. Also, principal components analysis (PCA) can be employed on the features prior to model construction or testing, efficiently combining different feature extraction approaches. The PCA transformation matrix can be derived globally from all speakers, or a per-speaker approach can be followed [17].

4.2 Face Identification

The face identification systems extract the faces from the video streams, aided by face labels. Only in the Run-1 evaluations were the faces automatically detected [8].

The face labels provide a face bounding box and eye coordinates every 200 ms. Some systems use only the marked faces [8], while others attempt to interpolate between the labels [17]. The labels are quite frequent in the CLEAR 2007 data set. That is not the case in the CLEAR 2006 data set, where face bounding boxes are provided every 1 sec [19]. For this data set, faces can be collected from the nonlabeled frames using a face detector in the neighborhood of the provided labels. Experiments with a boosted cascade of simple classifiers (Viola-Jones detector) [20] have shown enhanced performance over the simple linear interpolation [15].

The systems geometrically normalize the extracted faces, based either on the eyes [15, 16, 8] or on the face bounding box [15, 17, 7]. In the latter case, the faces are just scaled to some standard size. Some of the systems battle face registration errors by generating additional images, either by perturbing the assumed eye positions around the marked ones [8], or by modifying the face bounding box labels by moving the center of the bounding box by 1 pixel and changing the width or height by ± 2 pixels [7]. Geometric normalization is followed by further processing, aiming at intensity normalization. One approach is to normalize the integer values of the luminance to a mean of 128-mean and a standard deviation of 40 (luminance normalization) [17]. Other, more aggressive approaches like histogram equalization and edginess have been used in the past but have been abandoned as they degrade performance in the case of pose or expression variations [15, 8].

The processed faces obtained from the CLEAR 2007 data set lie on a highly non-linear manifold, making recognition difficult. This is shown in Fig. 4.2, where the first two PCA dimensions of the gallery faces for two different individuals are depicted. In this case, the manifold shows faces from person 1 ranging from left profile to frontal and finally to right profile, as the first PCA coefficient varies from -1,500 to +1,500. The same holds for person 2, only in this case there are not many frontal faces to occupy the range around the zero value of the first PCA coefficient. Hence, the first PCA coefficient is for pose and not person discrimination. Person 1 is discriminated from person 2 mainly by the second PCA coefficient, at the threshold value of 500, although there are many outliers from both people, since their projections fall into each Other's vicinity.

Different features are extracted by the different systems. They are all classified using a nearest-neighbor classifier, although the distance from class centers has been used by some systems in the past [15, 16, 8]. Obviously from Fig. 4.2, this is not a good choice, since the projected faces of a class are quite spread out and many times form disjoint clusters.

One approach performs block-based discrete cosine transform (DCT) to non-overlapping blocks of size 8×8 pixels [5, 4, 7, 9]. The obtained DCT coefficients are then ordered according to the zigzag scan pattern. The first coefficient is discarded for illumination normalization and the remaining first 10 coefficients in each block are selected in order to create compact local feature vectors. Furthermore, robustness against illumination variations is increased by normalizing the local feature vectors to unit norm. The global feature vector is generated by concatenating the local feature vectors.

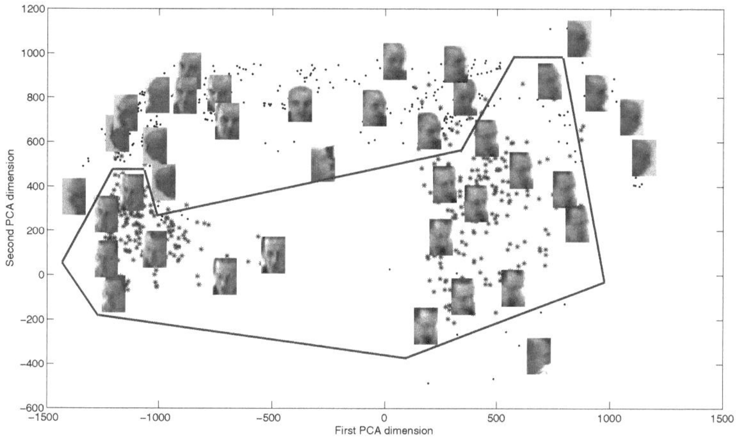

Fig. 4.2. 2D face manifold for people 1 and 2 of the CLEAR 2007 data set and representative faces centered on their respective projections. The faces of the two people lie on arcs, with the first PCA coefficient characterizing the pose and not the person. The projected faces of person 1 are depicted with black dots, while those of person 2 with blue stars. The characteristic faces belonging to person 2 are enclosed by a blue polygon.

Alternatively, PCA can be used to obtain features in a face subspace [15, 17, 16, 8]. The distance employed in the classifier is a modification of the weighted Euclidean, with the weights of each dimension depending on the neighborhood of the particular gallery point. The closest neighbors (using Euclidean distance) to the given gallery point are used to estimate the scatter matrix of gallery points in the vicinity of that gallery point [17]. The gallery point-dependent weights are the eigenvalues of that scatter matrix. Although this estimation of point-dependent weights is computationally expensive, it is performed once at system training and the weights for each of the projected gallery images are stored to be used in the recognition phase.

Linear discriminant analysis (LDA) has also been used in the past [15, 16, 8], without much success, as the face manifolds obtained under unconstrained conditions are nonseparable in a linear way (see the example of Fig. 4.2). Subclass LDA can address this problem [17]. Hierarchical clustering trees are used to automatically generate subclasses corresponding to face clusters belonging to the same class, by pruning the tree at some distance value.

Gaussian modeling of intrapersonal variations is also used to evaluate the probability that the difference of a gallery face from a probe face is indeed intrapersonal [17]. Forming all difference images is not computationally feasible; hence, a selection of the faces to be used is performed by grouping the gallery images of any person using hierarchical clustering trees. The trees are constructed using the projected gallery images onto a PCA subspace. For every person, some projected images are selected as the median of every cluster obtained by the trees. The intrapersonal differences are formed and modeled at the reduced dimension of the PCA subspace.

The fusion of the decisions obtained from all camera views within the testing period is performed using the weighted-sum rule [15, 17, 16, 7, 8, 6]. The weights are calculated using various metrics based on the distance of the best and second-best matches. Some systems also fuse decisions from different feature extraction methods [15, 17, 16], as each of these is best suited for different types of impairments in the face images.

4.3 Multimodal Person Identification

All multimodal person identification systems perform postdecision fusion. The weighted-sum rule is again utilized, with the individual modality confidences being used for the calculation of the weights.

The individual modality confidences are calculated based on the observation that the difference of the confidences between the closest and second-closest matches is generally smaller in the case of a false classification than in the case of a correct classification. A method named the *cumulative ratio of correct matches* (CRCM), which uses a nonparametric model of the distribution of the correct matches with respect to the confidence differences between the best two matches, is used. This way, the classification results with a greater confidence difference between the two best matches receive higher weights [5, 4, 7, 9]. Alternatively, the ratio between the closest and the second-closest matches [15, 17, 16], or a histogram equalization of the monomodal confidence scores [10, 3], can be utilized.

4.4 Lessons Learned

As seen in the previous sections, the unimodal person identification systems evolved in the CHIL Consortium have explored different options regarding their three parts: data extraction and preprocessing, feature extraction, and classification. Following many experiments and four evaluations (two internal to CHIL and two evaluations conducted with the CLEAR workshops 2006 and 2007, which were open to the scientific community; see also Chapter 15), many lessons have been learned.

The following can be concluded regarding audio person recognition:

- Speech extraction and preprocessing: The experiments on the use of SAD to preprocess the audio for speech extraction are not conclusive; most systems just extract features from the whole audio duration. The preprocessing of the multiple microphone channels to produce a single, hopefully cleaner, audio signal using beamforming degrades performance. Postdecision fusion of the recognition outcomes from the different channels is the way to utilize the multiple audio sources.

- Feature extraction: The FF features are very promising. The standalone PLP features, or those obtained by the combination of PLP and MFCC into a single feature vector using PCA, are better than standalone MFCC. Feature postprocessing is counterproductive in the context of the CHIL seminars, where the speaker is generally recorded in a stable acoustic context.
- Classifier: All systems employ a Bayesian classifier based on the GMMs of the features. The use of a UBM with MAP adaptation is better than estimating speaker-specific models directly. Training the UBM by pooling all the training data outperforms using other training data or a direct MLE training of the target models.

The conclusions for face recognition are as follows:

- Face extraction and normalization: It is better to extract faces from only the labeled frames, without any interpolation. Selecting just the frontal faces (using the provided labels) is detrimental to performance. There are not enough experiments with the different geometric normalization approaches to be conclusive.
- Feature extraction: The feature extraction methods can be compared only under the same face extraction and normalization methods. This makes comparison very difficult. The best system employs local appearance-based face recognition using DCT. Experiments using the same face extraction and normalization methods led to the following ranking of feature extraction methods:
  - Intrapersonal modeling (Bayesian) > Subclass LDA > PCA > LDA
  - LDA with Kernel PCA combination > Kernel PCA > LDA > PCA
- Classifier: Since CLEAR 2006,we have known that the nearest-neighbor classifier, albeit slower, greatly outperforms the distance from the class centers one. Also, there is lot to gain by exploiting the optimum distance metric for each feature extraction method. While some systems only used Manhattan or Euclidean distance metrics, experiments have shown performance gain when using modifications of the Mahalanobis distance metric for PCA-based methods, or the cosine for LDA-based methods.

Finally, regarding multimodal systems, the one with the best audio subsystem always outperforms the rest. The results show that multimodal fusion provides an improvement in the recognition rate over the unimodal systems in all train/test conditions.

Recognition performance greatly increased in the last two years of the project, where results are somewhat comparable. In CLEAR 2006, the task involved 26 people and the segments had been selected so that speech was present in them, paying no attention to the faces. In CLEAR 2007, the people to be recognized increased to 28 and the choice of the segments was more balanced across the two modalities. In both evaluations, two training conditions (15 and 30 seconds) and four testing conditions (1, 5, 10, and 20 seconds) were selected. The best performance obtained in the two evaluations is summarized in Fig. 4.3.

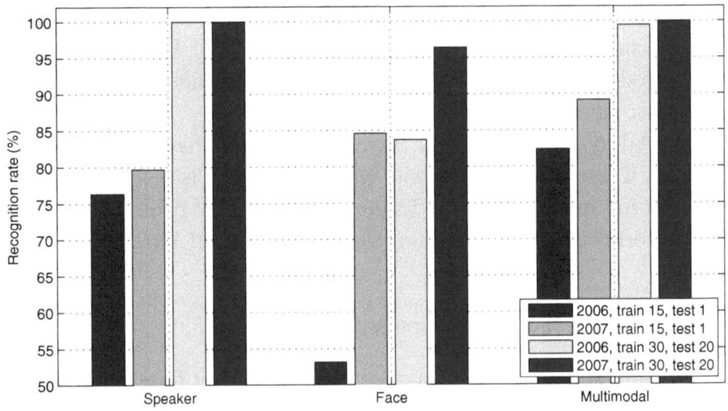

Fig. 4.3. Performance evolution of the person identification systems in the CLEAR 2006 and 2007 evaluations. Two out of the eight conditions are shown per evaluation; the shortest training and testing (15 and 1 seconds, respectively) and the longest training and testing (30 and 20 seconds, respectively).

The most impressive performance boost has been achieved in face identification. Significant improvements have also been achieved in speaker and multimodal identification for the short training and testing segments.

References

1. C. Barras, X. Zhu, J.-L. Gauvain, and L. Lamel. The CLEAR'06 LIMSI acoustic speaker identification system for CHIL seminars. In *Multimodal Technologies for Perception of Humans. First International Evaluation Workshop on Classification of Events, Activities and Relationships, CLEAR 2006*, LNCS 4122, pages 233–240, 2006.
2. C. Barras, X. Zhu, C.-C. Leung, J.-L. Gauvain, and L. Lamel. Acoustic speaker identification: The LIMSI CLEAR'07 system. In *Multimodal Technologies for Perception of Humans, Proceedings of the International Evaluation Workshops CLEAR 2007 and RT 2007*, LNCS 4625, pages 233–239, Baltimore, MD, May 8-11 2007.
3. P. Ejarque, A. Garde, J. Anguita, and J. Hernando. On the use of genuine-impostor statistical information for score fusion in multimodal biometrics. *Annals of Telecommunications, Special Issue on Multimodal Biometrics*, 62(1-2):109–129, Apr. 2007.
4. H. K. Ekenel, M. Fischer, Q. Jin, and R. Stiefelhagen. Multi-modal person identification in a smart environment. In *CVPR Biometrics Workshop 2007, IEEE Conference on Computer Vision and Pattern Recognition*, Minneapolis, Jun. 2007.
5. H. K. Ekenel, M. Fischer, and R. Stiefelhagen. Face recognition in smart rooms. In *Machine Learning for Multimodal Interaction, Fourth International Workshop, MLMI 2007*, Brno, Czech Republic, Jun. 2007.
6. H. K. Ekenel and Q. Jin. ISL person identification systems in the CLEAR evaluations. In *Multimodal Technologies for Perception of Humans. First International Evaluation*

Workshop on Classification of Events, Activities and Relationships, CLEAR 2006, LNCS 4122, pages 249–257, Southampton, UK, Apr. 6-7 2007.
7. H. K. Ekenel, Q. Jin, and M. Fischer. ISL person identification systems in the CLEAR 2007 evaluations. In *Multimodal Technologies for Perception of Humans, Proceedings of the International Evaluation Workshops CLEAR 2007 and RT 2007*, LNCS 4625, pages 256–265, Baltimore, MD, May 8-11 2007.
8. H. K. Ekenel and A. Pnevmatikakis. Video-based face recognition evaluation in the CHIL project – run 1. In *7th IEEE International Conference on Automatic Face and Gesture Recognition, FG06*, pages 85–90, 2006.
9. H. K. Ekenel and R. Stiefelhagen. Analysis of local appearance-based face recognition: Effects of feature selection and feature normalization. In *CVPR Biometrics Workshop*, New York, Jun. 2006.
10. M. Farrús, P. Ejarque, A. Temko, and J. Hernando. Histogram Equalization in SVM Multimodal Person Verification. In *ICB*, pages 819–827, 2007.
11. J.-L. Gauvain and C. Lee. Maximum a posteriori estimation for multivariate Gaussian mixture observations of Markov chains. *IEEE Transactions on Speech and Audio Processing*, 2(2):291–298, Apr. 1994.
12. S. Z. Li, L. Zhang, S. Liao, X. Zhu, R. Chu, M. Ao, and R. He. A near-infrared image based face recognition system. In *7th IEEE International Conference on Automatic Face and Gesture Recognition (FG 2006)*, pages 455–460, Southampton, UK, April 2006.
13. J. Luque and J. Hernando. Robust speaker identification for meetings: UPC CLEAR07 meeting room evaluation system. In *Multimodal Technologies for Perception of Humans, Proceedings of the International Evaluation Workshops CLEAR 2007 and RT 2007*, LNCS 4625, pages 266–275, Baltimore, MD, May 8-11 2007.
14. J. Luque, R. Morros, A. Garde, J. Anguita, M. Farrus, D. Macho, F. Marqués, C. Martínez, V. Vilaplana, and J. Hernando. Audio, video and multimodal person identification in a smart room. In *Multimodal Technologies for Perception of Humans. First International Evaluation Workshop on Classification of Events, Activities and Relationships, CLEAR 2006*, LNCS 4122, pages 258–269, Southampton, UK, Apr. 6-7 2006. Springer-Verlag.
15. A. Pnevmatikakis and L. Polymenakos. *Far-Field Multi-Camera Video-to-Video Face Recognition*. I-Tech Education and Publishing, 2007.
16. A. Stergiou, A. Pnevmatikakis, and L. Polymenakos. A decision fusion system across time and classifiers for audio-visual person identification. In *Multimodal Technologies for Perception of Humans, Proceedings of the first International CLEAR evaluation workshop, CLEAR 2006*, LNCS 4122, pages 223–232, Southampton, UK, Apr. 6-7 2006.
17. A. Stergiou, A. Pnevmatikakis, and L. Polymenakos. The AIT multimodal person identification system for CLEAR 2007. In *Multimodal Technologies for Perception of Humans, Proceedings of the International Evaluation Workshops CLEAR 2007 and RT 2007*, LNCS 4625, pages 221–232, Baltimore, MD, May 8-11 2007.
18. R. Stiefelhagen, K. Bernardin, R. Bowers, J. Garofolo, D. Mostefa, and P. Soundararajan. The CLEAR 2006 evaluation. In *Multimodal Technologies for Perception of Humans, Proceedings of the First International CLEAR Evaluation Workshop, CLEAR 2006*, LNCS 4122, pages 1–45, Southampton, UK, Apr. 6-7 2006.
19. R. Stiefelhagen, K. Bernardin, R. Bowers, R. T. Rose, M. Michel, and J. Garofolo. The CLEAR 2007 evaluation. In *Multimodal Technologies for Perception of Humans, Proceedings of the International Evaluation Workshops CLEAR 2007 and RT 2007*, LNCS 4625, pages 3–34, Baltimore, MD, May 8-11 2007.
20. P. A. Viola and M. J. Jones. Rapid object detection using a boosted cascade of simple features. In *IEEE Computer Society Conference on Computer Vision and Pattern Recognition (CVPR 2001)*, pages 511–518, Kauai, HI, Dec. 2001.

5

Estimation of Head Pose

Michael Voit[1], Nicolas Gourier[2], Cristian Canton-Ferrer[3], Oswald Lanz[4], Rainer Stiefelhagen[1], Roberto Brunelli[4]

[1] Universität Karlsruhe (TH), Interactive Systems Labs, Fakultät für Informatik, Karlsruhe, Germany
[2] INRIA, Rhône-Alpes, Saint Ismier Cedex, France
[3] Universitat Politècnica de Catalunya, Barcelona, Spain
[4] Foundation Bruno Kessler, irst, Trento, Italy

In building proactive systems for interacting with users by analyzing and recognizing scenes and settings, an important task is to deal with people's occupations: Not only do their locations or identities become important, but their looking direction and orientation are crucial cues to determine everybody's intentions and actions. The understanding of interaction partners or targeted objects is relevant in deciding whether any unobtrusive system should become aware of possible matters or engaged in conversations.

The recognition of looking directions is a subtle task of either tracking people's eye gaze or finding an approximation that allows for rather unobtrusive observations, since capturing pupils requires more or less highly detailed recordings that can be made possible only with nearby standing cameras that rather limit any user's range of movement. Such an approximation can be found in estimating people's head orientation.

Whereas eye gaze allows to perceive even the smallest changes of a respective person's looking direction, the one estimation of head orientation shows its strength, especially upon low resolution textured captures where in-depth analysis of facial features, as pupils, is not possible at all. When users are allowed to move without restrictions throughout an entire room, such a loss of detail happens quite often. Furthermore, any rotation of the head, such that only views of the back of the head or profile captures are available, makes it impossible to gather information about the user's eye gaze but still allows one to derive an albeit coarse estimation of the head's rotation and, with such, knowledge about the person's orientation. All these advantages, however, require techniques that are able to provide good generalization, especially considering the strong variance of a head's appearance, depending on the viewpoint from which the observation is made. Eventually, the dedicated head representation not only includes sharp facial features when frontal shots of heads are available but it also spans small and blurred captures as well as profile views and ambiguous shots of the back of the head, when persons are allowed to move away from the camera, which most likely is the case in intelligent environments. Compared to rather predefined single-user work spaces where a person is expected to sit in front of his or her display and a camera always delivers near-frontal observations, unrestricted movement needs to cope not only with a wider range of head rotations, and with such a stronger variance of head appearances, but also with the surrounding room's features such as the maximum distance to the camera, variance in lighting throughout the room, and occlusions.

Considering these two very different scenarios we encountered during the CHIL project, we therefore mostly distinguished between the following camera setups and expected a priori statements:

1. Single-camera settings:
   In this sensor setup, only one camera is used to classify mostly near-frontal head orientations in the range of -90° to $+90^\circ$. Usually, the person whose head orientation is to be estimated is standing or sitting right in front of that camera, thus providing face captures with rather high-resolution textures and restricted head movement.
2. Multicamera settings:
   Multicamera environments are to help overcome the limitations of single-camera settings: Observed people should have the freedom to move without boundaries – this also includes the observation of their corresponding head rotations; to cope with captures of the back of people's heads, several more cameras guarantee to capture at least one frontal shot and further profile views of the same person. This scenario not only requires state-of-the-art pose estimators, but also fusion techniques that merge multiple views or single estimates into one joint hypothesis. Head captures vary strongly in size and facial details appear mostly blurred (due to the rather high distance to the cameras) or vanish completely, as the head rotates away from a single camera's viewpoint.

In the remainder of this chapter, both sensor environments are further explored and our corresponding work in CHIL is introduced and summarized. Section 5.1 describes two techniques on the very popular topic of using only a single camera in front of a person. Section 5.2 copes with scenarios where more than one camera is available and concentrates on fusing several single-view hypotheses and joint estimation techniques. We present all approaches with their individual advantages and provide common evaluation results on INRIA's Pointing04 Database [6] for single-view head captures and on the data set of the CLEAR 2007 evaluation [11] for multiview scenarios.

5.1 Single-Camera Head Pose Estimation

Single-camera head pose estimation mostly copes with people sitting in front of a camera, showing profile or at least near-frontal face captures all the time. This leads to rather detailed captures of the user's head and face, in contrast to scenarios, where having no restrictions on people's trajectories leads to huge distances to observing cameras, thus providing only small-sized head captures where details in facial resolution are often lost or at least blurred. Most appearance-based classifiers have the ability to perform well under both circumstances since the classification is based on the whole-image representation only, no matter how detectable nostrils or lip corners are.

5.1.1 Classification with Neural Networks

A popular appearance-based approach to estimate head orientation in single views is the use of neural networks [12, 5, 10, 13]. UKA-ISL adopted this scheme under CHIL [14], where an overall accuracy of 12.3° and 12.8° could be achieved for pan and tilt estimation, respectively, on the Pointing04 database. Neural networks follow their biological counterpart and therefore mostly show their strength in generalization: After training the network on example images, this classifier has the ability to interpolate and generalize for new, unseen head images, thus allowing for almost continuous pose estimations. The network itself only receives a preprocessed head image – preprocessing usually involves the enhancement of the image's contrast to elaborate facial details – upon which the output is based. This output can either consist of a horizontal estimation only, or include further output neurons for hypothesizing the vertical orientation, too.

5.1.2 Refining Pose Estimates Using Successive Classifiers

The disadvantage of regular classifiers, even neural networks, for estimating pose is that they regularly do not allow for balancing between faster but less detailed results and deeper searches that typically deliver more accurate output. To overcome this drawback, a classification that consists of several steps, each refining the previous gathered result, is advisable. Since higher accuracy mostly goes hand in hand with higher run time, especially coarse estimates need to show a good balance between their resolution and speed. INRIA presented such a new approach in [5], where both the holistic appearance of the face as well as local details within it are combined to receive a refined classification of observed head poses. This new two-step approach performed with state-of-the-art accuracy as good as 10.1° mean error in pan and 12.6° in tilt estimation for unknown subjects.

After normalizing tracked face regions in size and slant, these normalized captures are used as a basis for a coarse estimation step by projecting them onto a *linear auto-associative memory* (LAAM), learned using the *Widrow-Hoff rule*. LAAMs are a specific case of neural networks where each input pattern is associated with each other. The Widrow-Hoff rule increases a memory's performance. At each presentation of an image, each cell of the memory modifies its weights from the others by correcting the difference between the response of the system and the desired response. A coarse head orientation estimation can then easily be made possible by searching for the prototype that best matches a current image. The advantages of using LAAMs are that they require very few parameters to be built (which allows for easy saving and reloading) but they also show robustness to partial occlusion. This allows for quite fast and easy-to-implement algorithms to gather information about a coarse pose in advance, whereas a refined classification might follow successively. Such a successive refinement can be achieved in multiple ways, by applying quite detailed classifiers to the collected face region. A trend in current research is to use wavelet families for refined estimations; however, Gaussian receptive fields proved to be less expensive than the often-used Gabor wavelets but do show interesting properties such as rotation invariance and scalability. Gaussian receptive fields motivate the construction of a model graph for each pose: Each node of the graph can be displaced locally according to its saliency in the image. These model graphs, called *salient grid graphs*, do not require a priori knowledges or heuristics about the human face and can therefore be adapted to pose estimation of other deformable objects. Whereas LAAMs only deliver a coarse estimate of the observed pose, a successive search among the coarse pose neighbors results in a final determination of the model graph that obtains the best match. The pose associated with it can then be selected as the final head orientation estimate.

5.2 Multicamera Head Pose Estimation

Single-camera head pose estimation has been well evaluated during years of research. Following the trend of focusing upon real-life scenarios and bringing computers into everyday living environments, that task changed to cope with observations that do not build on predefined restrictions for the user. The use of only one sensor to cover an entire room for following and tracking people's actions would never result in respective accuracies as achievable as with dedicated restrictions to only be presented with near-frontal shots of people's heads. A logical step is thus to equip a room with multiple cameras so that at least one observation always guarantees near-frontal views of the tracked person, no matter how he or she moves throughout the room. This introduces several new problems: Depending on the user's position in the room, his or her head size strongly varies over different camera observations: The nearer one stands to a camera, the bigger the head appears. The further away a person is moving, the smaller his head appears, whereas facial details vanish into blurriness or cannot be detected at all. Further issues arise around how to combine numerous views into one joint, final estimate. This *fusion* can be achieved by merging on either the signal level or a higher level. Fusing on the signal level allows for the overall dimension and processing overhead to be reduced by limiting the classification problem to one combined feature, whereas higher-level techniques often allow one to include (available) context information and help choose a smaller subset of advantageous camera views or at least leave the possibility open to extend the overall system with further sensors without the need to retrain underlying classifiers. This new task was first defined in the CHIL project and was later evaluated during the CLEAR 2006 and 2007 evaluations [14, 15, 1, 17, 16, 2, 3].

5.2.1 From Single-View Estimates to Joint Hypotheses

An approach that gathers several single-view classifiers into one successive combination and allows for an accuracy of up to 8.5° and 12.5° for horizontal and vertical orientation estimation, respectively, on the CLEAR 2007 data set can be found in UKA-ISL's publication [15], where neural networks were applied to every camera provided in CHIL smart rooms. All in all, hypothesizing over all interesting rotation directions, a single network was trained to classify camera-relative estimates: Due to the cameras' different locations, every view depicts highly different poses that need to be coped with. Training a classifier to camera-relative orientations, a successive transformation into the world coordinate system overcomes this discrepancy. That way, the classifier becomes invariant to location changes or any possible extension. For their advantage in generalization, neural networks show their strength both in single-view as well as multiview environments where face size differs strongly, as long as the training database includes sufficient examples of later observations. This single classifier can then be applied to every camera provided in the room for gathering as many single-view hypotheses of the same observed head as there are cameras. The fusion is kept independent from the classification itself: An intuitive approach is to build the average of all camera estimates into a merged output. Relying on a single neuron's output for each view, as suggested in Section 5.1.1, however, results in including a lot of noise and the overall estimate varies strongly over time. A far better way is to train the network not to output one continuous estimate, but, in fact, to describe a likelihood distribution over the defined range of possible head pose observations (i.e., $-180°$ to $+180°$ for the horizontal rotation). By further letting this distribution include the classifier's uncertainty, a successive merging of all views can be implemented by averaging the likelihood values of all single-view distributions in a Bayesian filter scheme. As described in [15], two such filters are used to track horizontal (pan) and vertical (tilt) head rotation separately. Following Bayes' rule, such a filter computes the likelihood of being in a given state, which corresponds to a certain rotation angle, depending on a previously observed likelihood distribution (the a priori knowledge) and a current measurement. Whereas the a priori knowledge implies some temporal smoothing by including the previous state distribution itself, the current measurement is obtained by building the average of all cameras' estimated likelihoods for every final pose state. The gathered a posteriori distribution over all states hence presents the joint hypothesis and allows us to classify a final pose estimate, given the current observations.

5.2.2 Fusing on the Signal Level

Fusion, of course, requires processing power to necessarily run multiple classifiers instead of only one classifier. It has the advantage of using one joint feature vector that is computed from all available views. Such a possible signal-level-based fusion technique can be found in combining spatial and color analysis to exploit redundancy, as shown by UPC in [4]. The technique presented there was also evaluated on the CLEAR 2007 data set and showed an overall accuracy of 20.48° mean error for pan estimation [3]. The system itself builds upon the idea of producing synthetic reconstructions of different head poses and searching through those templates with a new, currently achieved query vector. Since one of the face's very distinct features is the observable amount of skin, a first step in constructing the feature representation is to gather skin patches. The intuition behind this approach is that the combination of skin patches over all camera views allows for a reconstruction of skin distribution over all possible head poses. As described in [7], the probabilistic classification of pixels to contain skin color can be computed solely on their RGB information, where a distribution of skin color can be computed by means of offline hand-selected samples of skin pixels. The classification of all skin pixels in a head region and the backprojection from camera space into world coordinates then allow an ellipsoid reconstruction of skin distribution: an approximation of the head's shape and color information [see Fig. 5.1(c)]. For classifying such descriptors by matching with precomputed templates, a planar representation provides a saliency map that is easy to interpret and can be used as a likelihood evaluation function for the 3D ellipsoid's voxels and its derived head orientation. Depictions of such interpretations are shown in Fig. 5.2.

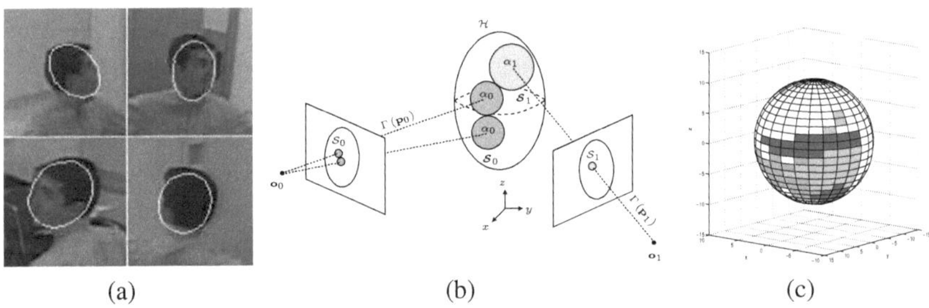

Fig. 5.1. Color and spatial information fusion process scheme. Ellipsoid describing the head and skin patches are detected on the original images in (a). In (b), skin pixels are backprojected onto the surface of the ellipsoid. The image in (c) shows the result of information fusion, obtaining a synthetic reconstruction of the face's appearance.

5.2.3 Integrated 3D Tracking and Pose Estimation

A common misconception of dedicated head pose estimating systems is the task of head alignment, that is, detecting an optimal head bounding box upon which the final estimation can be based by cropping this region of interest and interpreting it as a dedicated head capture. This detection is assumed to be coped with in external head tracking and alignment systems, which most often work independently of any further person tracking. Tightly linked modules might therefore provide both an increase in speed as well as better generalization, considering misaligned head regions, and overall an improvement in a tracker's accuracy due to possible head pose confidence feedback.

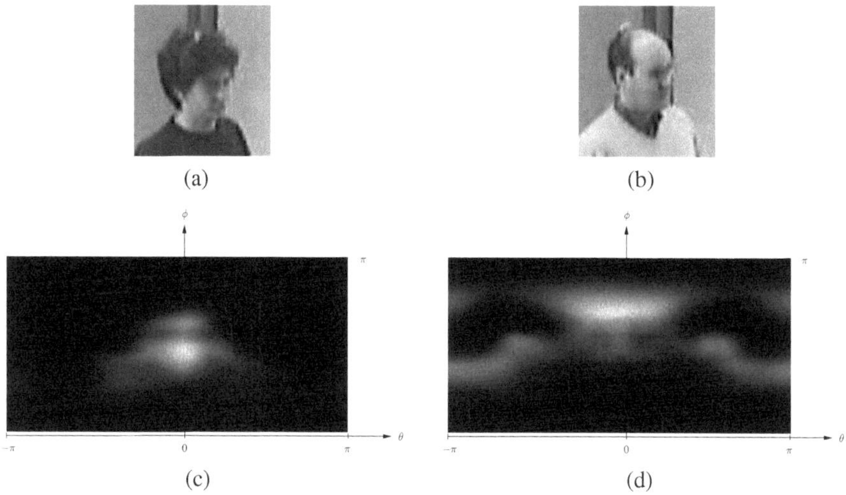

Fig. 5.2. Planar representation of the head appearance information.

One such integration was presented by FBK-irst in [8] by means of a Bayesian estimation problem that includes both 2D body position and moving velocity as well as horizontal and vertical head orientation. In every frame step, a hypothesized body position can be updated along with its corresponding velocity component according to the time elapsed between these two frames. To account for uncertainty and ambiguity, a particle filter allows one to propagate numerous hypotheses.

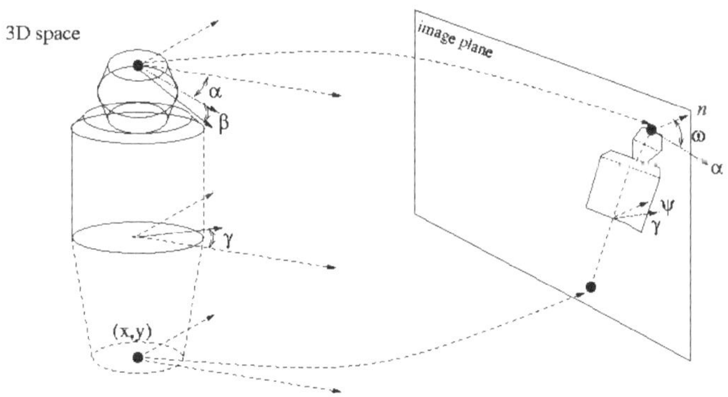

Fig. 5.3. 3D shape model parameterized by floor position (x,y), body orientation γ, and head pose (α,β) (pan and tilt angle), and an approximate, but efficient, rendering implementation that still conveys imaging artifacts such as perspective distortion and scaling. Note the offset of the head patch from the central axis; it gives a strong cue for head pose estimation. Involved in the rendering are the angular offsets ω and ψ of the body parts to the camera's optical axis n.

Low-dimensional shape and appearance models identifying image patches where the head and torso are expected to appear in each camera frame then help in computing each hypothesis' likelihood. One such adopted model is depicted in Fig. 5.3: Each hypothesis is used to construct a synthesized 3D appearance of the tracked body by assembling a set of rigid cone trunks. These trunks are positioned, scaled, and oriented according to floor location, target height, and body part orientations – a triple of 3D points, representing the centers of hips, shoulders, and top of head – that is computed from the hypothesized vector. This allows for an efficient and fast evaluation of every hypothesis. These vertices are backprojected onto every camera frame for gathering 2D segments, within which color histograms are to describe the appearance of the individual body parts. By previously collecting corresponding histograms of all body parts (which can be easily obtained upon a person's entrance into a room), potential head and upper torso patches can easily be identified within the image by comparing the corresponding histograms to their respective general counterparts. Interpolating between these templates allows for synthesizing new poses and views (Fig. 5.4). Finally, by multiplying all single-view scores, a joint multiview value can be obtained that allows one to classify for the best pose and location. Evaluated on the CLEAR 2007 data set, an overall mean error of 29.52° horizontally and 16.32° could thus be obtained [9].

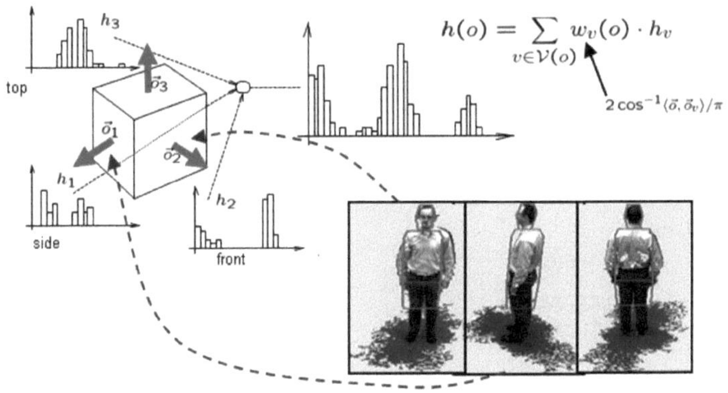

Fig. 5.4. Histogram synthesis. The cube faces represent preacquired top, front, and side views of the object as seen from a new viewpoint. A weight is assigned to each view according to its amount of visible area. A new object appearance is then generated as a weighted interpolation of these histograms.

5.3 Conclusion and Future Work

This chapter has presented an overview of CHIL's perceptive task to recognize people's head orientation under both monocular and multiview scenarios. Head pose estimation in single-view environments has already been the subject of numerous research projects. CHIL's main contribution in this field focused on extending that task toward using multiple cameras and allowing tracked persons to move and work without any limitations that eventually remained from a single camera's setup. During CLEAR evaluations, all developed systems were compared on publicly made data sets for both conditional tasks, which attracted a lot of interest for further, external participants. The results, achieving mean error rates as low as 10.1° for pan and 12.6° for tilt in single-view and 8.5° and 12.5°, respectively, for multiview environments, demonstrated our research to be competitive and state-of-the-art. Nevertheless, remaining issues such as lighting conditions, the diversity of different hair-styles when capturing people from their back, or evaluating processing speed against a possible enhancement of accuracy by increasing the number of camera views yet remain to be coped with for further robustness and increased usability. Head pose not only allows for an indication about a person's orientation, but rather makes it possible to approximate that person's eye gaze and looking direction to successively infer a target on which he or she is focusing. Our ongoing and future research in this field will thus analyze the (visual) focus of people's attention, based on head orientation, in order to continue with the unobtrusive setup of sensors and perception (see also Chapter 9). The joint combination of further modalities to estimate the visual focus of attention will also be the subject of future research and evaluation.

References

1. S. O. Ba and J.-M. Obodez. Probabilistic head pose tracking evaluation in single and multiple camera setups. In *Proceedings of the International Evaluation Workshops CLEAR 2007 and RT 2007*, Baltimore, MD, May 2007. Springer.
2. C. Canton-Ferrer, J. R. Casas, and M. Pardàs. Head pose detection based on fusion of multiple viewpoints. In *Multimodal Technologies for Perception of Humans, Proceedings of the First International Evaluation Workshop on Classification of Events, Activities and Relationships, CLEAR 2006*. Springer, Apr. 2006.
3. C. Canton-Ferrer, J. R. Casas, and M. Pardàs. Head orientation estimation using particle filtering in multiview scenarios. In *Multimodal Technologies for Perception of Humans, Proceedings of the International Evaluation Workshops CLEAR 2007 and RT 2007*, LNCS 4625, pages 317–327, Baltimore, MD, May 8-11 2007.
4. C. Canton-Ferrer, C. Segura, J. R. Casas, M. Pardàs, and J. Hernando. Audiovisual head orientation estimation with particle filters in multisensor scenarios. *EURASIP Journal on Advances in Signal Processing*, 2007.
5. N. Gourier. Machine observation of the direction of human visual focus of attention, Oct. 2006. PhD thesis.
6. N. Gourier, D. Hall, and J. L. Crowley. Estimating face orientation from robust detection of salient facial features. In *Proceedings of Pointing 2004, ICPR, International Workshop on Visual Observation of Deictic Gestures*, Cambridge, UK, 2004.
7. M. Jones and J. Rehg. Statistical color models with application to skin detection. *International Journal of Computer Vision*, 46:81–96, 2002.
8. O. Lanz and R. Brunelli. Dynamic head location and pose from video. In *IEEE Conference on Multisensor Fusion and Integration*, 2006.

9. O. Lanz and R. Brunelli. Joint Bayesian tracking of head location and pose from low-resolution video. In *Multimodal Technologies for Perception of Humans, Proceedings of the International Evaluation Workshops CLEAR 2007 and RT 2007*, LNCS 4625, pages 287–296, Baltimore, MD, May 8-11 2007.
10. R. Rae and H. J. Ritter. Recognition of human head orientation based on artificial neural networks. In *IEEE Transactions on Neural Networks*, volume 9, pages 257–265, Mar. 1998.
11. R. Stiefelhagen, K. Bernardin, R. Bowers, R. T. Rose, M. Michel, and J. Garofolo. The CLEAR 2007 evaluation. In *Multimodal Technologies for Perception of Humans, Proceedings of the International Evaluation Workshops CLEAR 2007 and RT 2007*, LNCS 4625, pages 3–34, Baltimore, MD, May 8-11 2007.
12. R. Stiefelhagen, J. Yang, and A. Waibel. Simultaneous tracking of head poses in a panoramic view. In *International Conference on Pattern Recognition - ICPR 2000*, Barcelona, Sept. 2000.
13. Y.-L. Tian, L. Brown, J. Connell, S. Pankanti, A. Hampapur, A. Senior, and R. Bolle. Absolute head pose estimation from overhead wide-angle cameras. In *IEEE International Workshop on Analysis and Modeling for Face and Gestures*, Oct. 2003.
14. M. Voit, K. Nickel, and R. Stiefelhagen. Neural network-based head pose estimation and multi-view fusion. In *Multimodal Technologies for Perception of Humans, Proceedings of the First International Evaluation Workshop on Classification of Events, Activities and Relationships, CLEAR 2006*, Southampton, UK, Apr. 6-7 2006. Springer.
15. M. Voit, K. Nickel, and R. Stiefelhagen. Head pose estimation in single- and multi-view environments - results on the CLEAR'07 benchmarks. In *Multimodal Technologies for Perception of Humans, Proceedings of the International Evaluation Workshops CLEAR 2007 and RT 2007*, LNCS 4625, pages 307–316, Baltimore, MD, May 8-11 2007. Springer.
16. S. Yan, Z. Zhang, Y. Fu, Y. Hu, J. Tu, and T. Huang. Learning a person-independent representation for precise 3D pose estimation. In *Proceedings of the International Evaluation Workshops CLEAR 2007 and RT 2007*, Baltimore, MD, May 2007. Springer.
17. Z. Zhang, Y. Hu, M. Liu, and T. Huang. Head pose estimation in seminar rooms using multi-view face detectors. In *Multimodal Technologies for Perception of Humans, Proceedings of the First International Evaluation Workshop on Classification of Events, Activities and Relationships, CLEAR 2006*. Springer, Apr. 2006.

6

Automatic Speech Recognition

Gerasimos Potamianos,[1] Lori Lamel,[2] Matthias Wölfel,[3] Jing Huang,[1] Etienne Marcheret,[1] Claude Barras,[2] Xuan Zhu,[2] John McDonough,[3] Javier Hernando,[4] Dusan Macho,[4] Climent Nadeu[4]

[1] IBM T.J. Watson Research Center, Yorktown Heights, NY, USA
[2] LIMSI-CNRS, Orsay, France
[3] Universität Karlsruhe (TH), Interactive Systems Labs, Fakultät für Informatik, Karlsruhe, Germany
[4] TALP Research Center, Universitat Politèchnica de Catalunya, Barcelona, Spain

Automatic speech recognition (ASR) is a critical component for CHIL services. For example, it provides the input to higher-level technologies, such as summarization and question answering, as discussed in Chapter 8. In the spirit of ubiquitous computing, the goal of ASR in CHIL is to achieve a high performance using far-field sensors (networks of microphone arrays and distributed far-field microphones). However, close-talking microphones are also of interest, as they are used to benchmark ASR system development by providing a best-case acoustic channel scenario to compare against.

Although ASR is a well-established technology, the CHIL scenario presents significant challenges to state-of-the-art speech recognition systems. This is due to numerous reasons, for example, the presence of speech from multiple speakers with varying accents and frequent periods of overlapping speech, a high level of spontaneity with many hesitations and disfluencies, and a variety of interfering acoustic events, such as knocks, door slams, steps, cough, laughter, and others. Note that the problem of identifying such acoustic events has also been investigated in CHIL, and it is discussed in Chapter 7. In addition, the linguistic content in CHIL scenarios, that of technical seminars, constitutes another challenge, since there exists only a relatively small amount of in-domain acoustic and language modeling data. The focus on handling far-field microphone speech exacerbates these issues, due to the low signal-to-noise ratios and room reverberation.

Of course, these challenges also affect *speech activity detection* (SAD), *speaker identification* (SID), and *speaker diarization* (SPKR) – also known as the "who spoke when" – problems. These technologies jointly address the "what", "when", and "who" of human interaction. In particular, SAD and SPKR constitute crucial components of state-of-the-art ASR systems, and as such they are briefly discussed in this chapter. In addition, they are relevant to other CHIL perceptual technologies, for example, acoustic-based speaker localization and identification, which are discussed in more detail in Chapters 3 and 4, respectively.

Progress of the ASR systems developed for the CHIL domain has been benchmarked by yearly technology evaluations. During the first two years of CHIL, the Consortium internally evaluated ASR with a first dry run held in June 2004, followed by an "official" evaluation in January 2005. Following these two internal campaigns, the CHIL sites involved in the ASR activity (IBM, LIMSI, and UKA-ISL) participated in the Rich Transcription evaluations of 2006 and 2007 – RT06s [13] and RT07 [45]. These international evaluations were sponsored by NIST and attracted a number of additional external participants.

The CHIL partner sites involved in ASR work have made steady progress in this technology in the CHIL domain. For example, in the far-field ASR task, the best system performance improved from a word error rate of over 68% in the 2004 dry run to approximately 52% in the CHIL 2005 internal evaluation, and from 51% in RT06s down to 44% in RT07. These improvements were achieved in spite of the fact that the recognition task became increasingly more challenging: Indeed, from 2005 to 2006, multiple recording sites involving speakers with a wider range of nonnative accents were introduced into the test set. Furthermore, in both 2006 and 2007, the degree of interactivity in the data was significantly increased.

This chapter documents progress in the challenging task of recognizing far-field speech in the CHIL scenarios, and it highlights the main system components and approaches followed by the three CHIL partners involved in this effort. Section 6.1 provides an overview of the ASR problem and its evaluation framework in CHIL, as well as the data resources available for system development. A brief discussion of the SAD and SPKR subsystems appears in Section 6.2. Section 6.3 describes the main components of ASR systems with highlights of specific approaches investigated in CHIL within each of these components. An example of an ASR system implementation is provided in Section 6.4, followed by a brief overview of ASR experimental results in Section 6.5. Finally, the chapter concludes with a summary, a discussion of open problems, and directions for future research (Section 6.6).

6.1 The ASR Framework in CHIL

Automatic speech recognition in CHIL constitutes a very challenging problem due to the interactive scenario and environment as well as the lack of large corpora fully matched to the CHIL specifics. In the next subsections, we briefly overview these two factors by providing more details of the ASR evaluation framework in CHIL as well as the available data resources utilized in training the ASR systems.

6.1.1 ASR Evaluation Framework

The CHIL interactive seminars were held inside smart rooms equipped with numerous acoustic and visual sensors. The former include a number of microphone arrays located on the walls (typically, at least three T-shaped four-microphone arrays and at least one linear 64-channel array) as well as a number of tabletop microphones. In addition, most meeting participants wore headset microphones to capture individual speech in the close-talking condition. This setup was installed in five CHIL partner sites and used to record data for the RT06s and RT07 speech technology evaluation campaigns. It also represents a significantly more evolved setup than the initial smart room design installed at the UKA-ISL site to provide data for the 2004 and 2005 CHIL internal evaluations. For example, in the 2004 dry-run evaluation, the far-field audio signals were captured only by a 16-channel linear array and a single tabletop microphone.

The above setup has been designed to allow data collection and ASR evaluation with main emphasis on the use of unobtrusive far-field sensors that fade into the background. These data form the basis of the *multiple distant microphone* (MDM) condition, the designated primary condition in the RT06s and RT07 ASR technology evaluations, where all tabletop microphones – typically ranging from three to five – were utilized to yield a single transcript. Additional conditions are the *single distant microphone* (SDM) one, where only one preselected tabletop microphone is used, as well as a number of conditions that involve using the linear or T-shaped arrays [45].

An interesting contrasting condition is called the *individual headset microphone* (IHM) condition, where the data recorded on the channels from the headsets worn by all lecture participants are decoded, with the purpose of recognizing the wearer's speech. This represents a close-talking ASR task, and putting aside for a moment the challenging task of robust cross-talk removal, it is designed to quantify the ASR performance degradation due to the use of far-field sensors.

Because of the reduced sensory setup in the initial smart room, the conditions were somewhat different in the 2004 and 2005 CHIL internal evaluations. In particular, only the lecturer's speech was decoded in these evaluations, for both the close-talking and far-field conditions. Furthermore, in 2004, only the 16-channel linear array was used in the far field.

6.1.2 Data

A number of seminars and interactive lectures were collected in the five state-of-the-art smart rooms of the CHIL Consortium. Nevertheless, the available amount of CHIL data remains insufficient for training ASR systems, comprising less than 10 hours of speech. This issue was, of course, even more pronounced in the early CHIL evaluations, since the data have been incrementally collected over the duration of the project. To remedy this problem, additional publicly available corpora [31] exhibiting similarities to the CHIL scenarios were utilized for system development. These are the ICSI, ISL, and NIST meeting corpora [24, 7, 14], the additional non-CHIL meeting data from prior RT evaluation runs (2004-2006), including a corpus collected by NIST in 2007, data collected by the AMI Consortium [1], and the TED corpus of lectures collected at the Eurospeech Conference in Berlin in 1993 [50, 30]. Most data sets contain close-talking (headset or lapel) and multiple far-field microphone data, with the exception of TED that contains lapel data only. In total, there are on the order of 250 hours of speech in the combined corpora.

The three CHIL sites used various parts of these sources in their acoustic model-building process, as all these corpora exhibit certain undesirable variations from the

CHIL data scenario and acoustic environment. In particular, since only a portion of the TED corpus is transcribed, the remainder was exploited by some CHIL partners (e.g., LIMSI) via unsupervised training [29]. It is also worth noting that in earlier ASR systems developed for the CHIL internal runs in 2004 and 2005, some partners also used other corpora, such as Broadcast News (LIMSI), or even proprietary data sets such as the ViaVoice and MALACH project corpora (IBM), after applying necessary model adaptation.

For language model training, transcripts from the CHIL data sets as well as the above-mentioned meeting corpora were used. In addition, LIMSI generated a set of cleaned texts from published conference proceedings; these are also very relevant to the task due to the technical nature of CHIL lectures, which were employed in some of the developed ASR systems. Furthermore, some sites used Web data, for example, data available from the EARS program [31], and possibly additional sources such as data from conversational telephone speech, e.g., the Fisher data [31].

6.2 ASR Preprocessing Steps

Two important preprocessing stages in all CHIL partners' ASR systems are the speech activity detection (SAD) and speaker diarization (SPKR) components. These locate the speech segments that are processed by the ASR systems and attempt to cluster homogeneous sections, which is crucial for efficient signal normalization and speaker adaptation techniques. As mentioned in the introduction, these components have been evaluated in separate CHIL internal and NIST-run evaluation campaigns. Within the Consortium, they have attracted significant interest among the CHIL partners in addition to the three sites involved in ASR system development.

6.2.1 Speech Activity Detection

Speech activity detection has long been an important topic as a front-end step to the ASR process, having a positive impact on ASR systems in terms of both CPU usage and ASR accuracy. This is due to the fact that the decoder is not required to operate on nonspeech segments, thus reducing the processing effort and word insertion error rate. In addition, SAD systems provide the segments used as input to the speaker diarization component (see ahead). Robust performance of SAD is also important in other technologies of interest in CHIL, such as acoustic speaker localization.

Not surprisingly, the SAD technology has been investigated by many CHIL partners. At some stage during the CHIL project, AIT, FBK-irst, IBM, INRIA, LIMSI, UKA-ISL, and UPC developed SAD systems for CHIL. For instance, all these partners participated in the 2005 CHIL internal evaluation of the technology, a campaign that followed the dry run conducted in the summer of 2004. More recently, the technology was evaluated stand alone in RT06s (AIT, FBK-irst, IBM, INRIA, LIMSI, and UPC participated), and as a component of SPKR systems in RT07 (IBM, LIMSI, and UPC took part); both campaigns were organized by NIST. At the RT06s evaluation, the best systems achieved very encouraging error rates of 4% and 8% for the conference and lecture subtasks, respectively, when calculated by the NIST diarization metric.

For SAD system development, CHIL partners followed various approaches that differed in a number of factors, for example, in feature selection [Mel frequency cepstrum coefficients (MFCCs), energy-based, combined acoustic and energy-based features, etc.], the type of classifier used [Gaussian mixture model classifiers (GMMs), support vector machines, linear discriminants, decision trees], the classes of interest (IBM initially used three broad classes), and channel combination techniques (based on signal-to-noise ratios, voting, etc.). Details can be found in a number of partner site papers, for example, [35, 38, 22].

6.2.2 Speaker Diarization

Speaker diarization, also referred to as the "who spoke when" task, is the process of partitioning an input audio stream into homogeneous segments according to speaker identity. It is useful as an ASR preprocessing step because it facilitates unsupervised adaptation of speech models to the data, which improves transcription quality. It also constitutes an interesting task per se, since structuring the audio stream into speaker turns improves the readability of automatic transcripts. A review of activities in speaker diarization can be found in [51].

Historically, SPKR systems were evaluated by NIST on Broadcast News data in English up to 2004; following that, the meeting domain became the main focus of the RT evaluations. These included CHIL lecture seminars and multisite meetings in the 2006 and 2007 evaluations [13, 45]. Similarly to ASR, a number of evaluation conditions have been defined in these campaigns (e.g., MDM and SDM conditions). Notice that in the adopted evaluation framework, the number of speakers in the recording or their voice characteristics are not known a priori; therefore, they must be determined automatically. Also, SPKR systems were evaluated as independent components. Clearly, the use of other sources of knowledge, such as output of multimodal person tracking and identification, could dramatically improve system accuracy. Without such additional information, diarization of audio data recorded by distant microphones remains a very challenging task, for the same reasons that apply to the far-field ASR problem in CHIL, as discussed in the introduction.

A number of CHIL partner sites developed SPKR systems (AIT [43], IBM [22, 20], LIMSI [59, 60], and UPC [34]). The following specific research directions were addressed in these systems:

- *Exploiting multiple distant microphone channels:* Acoustic beamforming was performed on the input channel after Wiener filtering, using a delay-and-sum technique [34, 4]. Up to 20% relative gain was obtained by using the beamformed audio, compared to using a single channel on conference data, even if the gain on lecture data was less significant [60]. Using the delays between the acoustic channels as features appears also to be a very promising direction [39].
- *Speech parameterization:* Frequency filtering, which showed good results in the CLEAR 2007 evaluation in the acoustic person identification task, was used as

an alternative to the classical MFCC features [34]. Derivative parameters were tested but did not seem of much benefit [60].
- *Speech activity detection:* SAD errors directly affect SPKR system performance. Therefore, additional effort was placed toward improving SAD models, as discussed in a previous section. When the diarization task is considered standalone, a different balance has to be chosen between missed speech and false-alarm speech than when SAD is used as an ASR preprocessing step. An explored solution was a purification of the acoustic segments using an automatic word-level alignment in order to reduce the amount of silence or noise portions, which are potentially harmful during clustering [20].
- *Segmentation and clustering:* An initial step provides an overestimated number of clusters; each cluster is modeled with a Gaussian model (typically a single Gaussian with a full covariance matrix or a GMM with diagonal covariance matrices). Clusters are further grouped following a distance (Mahalanobis, likelihood gain [20], a Bayesian information criterion (BIC) measure [34], cross log-likelihood ratio [60]) until some threshold is reached.

Results in the RT07 evaluation campaign demonstrate that the speaker diarization problem is far from being solved in the CHIL scenarios. In particular, the best SPKR system (developed by LIMSI) was benchmarked at a 26% diarization error rate for the MDM condition. This is significantly worse than the diarization error typically achieved in the Broadcast News task – about 10% [5].

6.3 Main ASR Techniques and Highlights

Although the three CHIL sites have developed their ASR systems independently of each other, all systems contain a number of standard important components, which are summarized in this section. In addition, in the spirit of collaboration through competition in technology evaluation – the so-called co-opetition paradigm that has been adopted in the CHIL project as a means to promote progress – CHIL partners have shared certain components, such as the UKA-ISL beamforming algorithm for far-field acoustic channel combination or the LIMSI proceedings text corpora for language model training.

6.3.1 Feature Extraction

Acoustic modeling requires the speech waveform to be processed in such a way that it produces a sequence of feature vectors with a relatively small dimensionality in order to overcome the statistical modeling problem associated with high-dimensional feature spaces, called the *curse of dimensionality* [6]. Feature extraction in ASR systems aims to preserve the information needed to determine the phonetic class, while being invariant to other factors including speaker differences such as accent, emotion, fundamental frequency, or speaking rate, as well as other distortions, for example, background noise, channel distortion, reverberation, or room modes. Clearly, this step is crucial to the ASR system, as any loss of useful information cannot be recovered in later processing.

Over the years, many different speech feature extraction methods have been proposed. The methods are distinguished by the extent to which they incorporate information about the human auditory processing and perception, robustness to distortions, and length of the observation window. Within the CHIL framework, different state-of-the-art feature extraction methods have been investigated. For example, ASR systems have utilized MFCCs [9] or perceptual linear prediction (PLP) [18] features. Feature extraction in CHIL partner systems often involved additional processing steps, for example, linear discriminant analysis (LDA) [17] or a maximum likelihood linear transform (MLLT) [16]. Feature normalization steps, such as variance normalization and vocal tract length normalization (VTLN) [3] were also employed by some sites.

In addition to the above, a novel feature extraction technique has been developed by UKA-ISL that is particularly robust to changes in fundamental frequency, f_0. This is important in the CHIL scenario, as public speeches have a higher variance in f_0 than do private conversations [19]. Additional advantages of the proposed approach, based on a warped-minimum variance distortionless response spectral estimation [56], are an increase in resolution in low-frequency regions relative to the traditionally used Mel filter banks, and the dissimilar modeling of spectral peaks and valleys to improve noise robustness, given that noise is present mainly in low-energy regions. To further increase the robustness to noise, a signal adaptive front end has been proposed [52], that emphasizes classification of relevant characteristics, while classification-irrelevant characteristics are alleviated according to the characteristics of the input signal; for example, vowels and fricatives have different characteristics and should therefore be treated differently. Experiments conducted by UKA-ISL have demonstrated that the proposed front ends reduce the *word error rate* (WER) on close-talking microphone data by up to 4% relative, and on distant speech by up to 6% relative, as compared to the widely used MFCC features [58].

6.3.2 Feature Enhancement

Feature enhancement manipulates speech features in order to retrieve features that are more similar to the ones observed in clean training data of the acoustic model. Thus, the mismatch between the unknown, noisy environment and the clean training data is reduced. Speech feature enhancement can be realized either as an independent preprocessing step or on the features within the front end of the ASR system. In both cases, it is not necessary to modify the decoding stage or acoustic models of the ASR system.

Most popular feature enhancement techniques for speech recognition operate in the frequency domain. Simple methods such as spectral subtraction are limited to removing stationary noise, where the spectral noise floor is estimated on noise-only regions. More advanced methods attempt to track either the clean speech or the noise for later subtraction. First approaches in this direction used Kalman filters (KFs) [25], which assume the relationship between the observations and the inner state to be linear and Gaussian, which does not hold in practice. To overcome this constraint, variants to the KF, such as the extended KF, have been proposed.

Research in CHIL has focused on the enhancement of features in the logarithmic Mel-spectra domain by tracking the noise with a particle filter (a.k.a. sequential Monte Carlo method) [44, 48]. Feedback of the ASR system into the feature enhancement process results in further improvements by establishing a coupling between the particle filter and the ASR system, which had been treated as independent components in the past [11]. Experiments conducted by UKA-ISL using a novel feature enhancement technique that is able to jointly track and remove nonstationary additive distortions and late reverberations have demonstrated that word accuracy improvements on distant recordings by more than 20% relative are possible [54], independent of whether the acoustic models of the speech recognition system are adapted.

6.3.3 Acoustic Modeling

For acoustic modeling in CHIL, *hidden Markov models* (HMMs) were exclusively used. Most sites estimated system parameters by the expectation maximization (EM) algorithm (maximum-likelihood training) [10], followed by discriminative model training using the maximum mutual information (MMI) [41] or the minimum phone error (MPE) approach [42]. A number of adaptation techniques were also employed, ranging from maximum a posteriori estimation (MAP) [15] to maximum-likelihood linear regression (MLLR) [32], feature space MLLR (fMLLR), or speaker adaptive training (SAT) [2]. The above approaches require a multipass decoding strategy, where a word hypothesis is used for unsupervised model adaptation prior to the next decoding pass. Finally, some sites developed systems with slight variations to improve final system performance through combination or cross-system adaptation.

An additional area of interest in acoustic modeling is that of pronunciation modeling. The pronunciation dictionary is the link between the acoustic and language models. All CHIL sites used phone representations in their systems, with about 40 to 50 phoneme-like units. Special phone symbols were also sometimes used to model nonspeech events such as hesitation, cough, silence, etc. Each lexical entry of the word dictionary can then be associated with one or more pronunciations to explicitly model frequent variants. It is common practice to include some acronyms, compound words, or word sequences in order to capture some of the coarticulation in spontaneous speech. These multiwords typically represent only a small part of the vocabulary. Some of the CHIL sites (e.g., LIMSI) also explored explicitly including pronunciation variants for nonnative accented speech; however, while the variants better represented the foreign accents, the overall recognition performance did not improve.

6.3.4 Language Modeling

For language modeling, different n-gram language models (LM), with $n = 3$ or 4, have been employed by the CHIL sites. These LMs were typically developed separately for various data sources and were subsequently linearly interpolated to give rise to a single model. Most often, CHIL and other meeting corpora were employed for this task. In addition, sometimes text from scientific proceedings (close to the CHIL lecture subjects) or data mined from the Web were also used. Based on these LMs, typical perplexities of the CHIL test data ranged in the order of 105 to 140. In terms of vocabulary size, CHIL sites used anywhere from 20k to 60k vocabularies, achieving out-of-vocabulary (OOV) rates in the order of 0.5 to 2.0%.

The use of a connectionist LM [47], shown to be performant when LM training data are limited, was explored at LIMSI. The basic idea is to project the word indices onto a continuous space and to use a probability estimator operating on this space. Since the resulting probability densities are continuous functions of the word representation, better generalization to unknown n-grams can be expected. A neural network LM was trained on the transcriptions of the audio data and proceedings texts and interpolated with a standard back-off LM. A significant word error reduction of 1.6% absolute was obtained when rescoring word lattices in under $0.3\times$RT.

An important part relevant to the language modeling work is the determination of the recognition vocabulary. The recognizer word list is usually determined by combining all the distinct words in the available audio transcriptions with the most frequent words in the relevant text sources. It is common practice to require a minimum number of word observations to be included in the word list. This ensures that the word occurs often enough to warrant modeling and also reduces the number of "false words" arising from typographical errors. Some text preprocessing is generally carried out to ensure conformity of the various text sources, removing undesirable data (email, addresses, mathematical formulas and symbols, figures, tables, references), formatting characters, and ill-formed lines. Acronyms, numbers, and compound words are also processed to ensure consistency and to approximate spoken language. The word list is typically selected so as to minimize the OOV rate on a set of development data. It was recently proposed to select the most probable words by linear interpolation of the unigram language models obtained from individual data sources.

6.3.5 Multiple Microphone Processing

The use of multiple microphones is an important component of far-field ASR in CHIL, as the sound pick-up quality might vary at different spatial locations and directions. An appropriate selection or combination of the different microphones can improve the recognition accuracy. The degree of success depends on the quality and variance of information provided by the microphones and the combination method used.

Speech recognition channel combination techniques can be broadly classified into signal and word-based combination methods. *Signal combination algorithms*, such as *blind source separation* and *beamforming* techniques, exploit the spatial diversity resulting from the fact that the desired and interfering signal sources are located at different points in space. This diversity can be taken advantage of by suppressing signals coming from directions other than the desired source direction. Those approaches assume that the speaker's position (time delay of arrival between different microphones) can be reliably estimated, and it might employ knowledge about the microphone positions relative to each other. Correct speaker localization is crucial for optimal recognition accuracy [57]. Due to the reduction of multiple channels into one channel, the decoding time is not significantly changed, compared to that of a single-microphone approach.

In contrast to signal combination, *word based-combination techniques*, such as ROVER [12] and confusion network combination (CNC) [36], fuse information from the recognition output of different systems that can be represented as a one-best, *n*-best, or lattice word sequence, augmented by confidence scores. Word-based approaches assume that the transcription of different microphone channels leads to different word hypotheses. Their advantage is that no spatial information of the speaker or microphones is required. However, since each microphone channel is decoded independently, these approaches are computationally expensive. A hybrid approach, where the beamformed channel is augmented by additional channels and combined with CNC, has been shown to lead to additional improvements over either of the other approaches [55].

Due to the broad variance of the different microphone channels, it may not be optimal to blindly consider all channels for combination (e.g., if a microphone is directly placed near a sound source). It may instead be preferable to measure the quality of the different microphones and select only "good" channels. A traditional measure to achieve such selection is the *signal-to-noise ratio* (SNR). More reliable measures consider the properties of the human auditory system and/or operate on the features of the recognition system. One promising approach in this direction is based on class separability [53], which shows significant improvements over SNR-based channel selection methods.

Employing multiple microphones has been shown to improve word accuracy by up to 10% absolute, which compensates for approximately one third of the reduction in WER observed when moving a single microphone from the mouth region of the speaker (close talk) into the room.

6.4 An ASR System Example

Following the overview of the main approaches used in CHIL for ASR, we proceed with a more detailed description of the ASR systems developed by one of the three CHIL partners, IBM.

The IBM ASR systems for CHIL have progressed significantly over the duration of the project. In particular, during the first two project-internal evaluations (2004 and 2005), the IBM team focused on combining in-house available ASR systems, appropriately adapted to the available CHIL data [8]. However, it soon became apparent that this approach yielded a poor performance in the CHIL task; as a result, new systems trained exclusively on meeting-like corpora were developed for the RT06s and RT07 evaluations [23, 21]. The new approach was based on developing a small number of parallel far-field ASR systems (typically three or four) with minor variations in their acoustic modeling, and combining them using ROVER (for the close-talking condition, a single system was developed). Additional work has been carried out for language modeling in order to create larger and richer LMs, suitable for the CHIL tasks. More details follow.

6.4.1 Acoustic Modeling

For acoustic modeling, first a speaker-independent (SI) model is trained, based on 40-dimensional acoustic features generated by an LDA projection of nine consecutive frames of 13-dimensional perceptual linear prediction (PLP) features, extracted at 100 Hz. The features are mean-normalized on a per-speaker basis. The SI model uses continuous-density, left-to-right HMMs with Gaussian mixture emission distributions and uniform transition probabilities. In addition, the model uses a global semi-tied covariance linear transformation [46], updated at every EM training stage. The system uses 45 phones; namely, 41 speech phones, one silence phone, and three noise phones. The final HMMs have 6k context-dependent tied states and 200k Gaussians. Since only a small part of the training data is from CHIL, MAP-adaptation of the SI model was deemed necessary to improve performance on CHIL data.

The SI features are further normalized with a voicing model (VTLN) with no variance normalization. The most likely frequency warping is estimated among 21 candidate warping factors ranging from 0.8 to 1.2. A VTLN model is subsequently trained on features in the VTLN warped space. The resulting HMMs have 10k tied states and 320k Gaussians. Following VTLN, a SAT system is trained on features in a linearly transformed feature space resulting from applying speaker-dependent fMLLR transforms to the VTLN-normalized features. Following SAT, feature space minimum phone error (fMPE) transforms are estimated [40], followed by MPE training [42] and MAP-MPE on the available amount of CHIL-only data [23, 21].

Following the above training procedure, two systems are built, one with the VTLN step present, and one with VTLN removed. Based on the latter, two additional SAT systems are built using a randomized decision-tree approach [49].

In contrast to the far field, only one system has been developed for the close-talking condition. This is identical in both RT06s and RT07 evaluations and is a 5k-state, 240k Gaussian mixture HMM system with both VTLN and variance normalization present [23].

6.4.2 Language Modeling

Five separate four-gram LMs were built. The first four were also used in the IBM RT06s system and were based on CHIL data (0.15M words), non-CHIL meetings (2.7M), scientific conference proceedings (37M), and Fisher data (3M words) [31]. A novel fifth LM used 525M words of Web data available from the EARS program [31]. For decoding, two interpolated LMs were used based on these five models. A reduced-size model was pruned to about 5M n-grams and was employed for static decoding, whereas a larger 152M n-gram model was used in conjunction with an on-the-fly dynamic graph expansion decoding. A 37k-word vocabulary was used.

6.4.3 Recognition Process

After speech segmentation and speaker clustering, a final system output was obtained in three decoding passes for each microphone: (a) an initial SI pass using MAP-adapted SI models to decode; (b) employing output from (a), warp factors using the voicing model and fMLLR transforms are estimated for each cluster using the SAT model. The VTLN features after applying the fMLLR transforms are subjected to the fMPE transform, and a new transcript is obtained by decoding, using the MAP-adapted MPE model and the fMPE features. (c) The output transcripts from step (b) are used in a cross-system fashion to estimate MLLR transforms on the MPE model. The adapted MPE model together with the large LM is used for final decoding with a dynamic graph expansion decoder.

In the far field, where multiple ASR systems have been developed, ROVER over these systems is applied for obtaining the SDM condition output. For the MDM condition, ROVER is first applied over all available tabletop microphones, followed by ROVER over the available four systems.

6.5 Experimental Results

The work described in Section 6.4 has resulted in significant progress over the duration of the CHIL project. For example, in the far-field condition, the initial IBM approach yielded 64.5% and 70.8% WER in the 2004 and 2005 CHIL-internal evaluations, respectively. In contrast, performance improved significantly in RT06s and RT07, reaching 50.1% and 44.3% WER, respectively, for the IBM MDM systems. In the close-talking task, under manual segmentation, system improvement has been less dramatic: In 2004, a 35.1% WER was achieved by the IBM system, whereas in 2005, a 36.9% WER was recorded. The new system development for the RT evaluation runs improved performance to 27.1% in 2006 and 31.7% in 2007. The latter represented a slight degradation, due to the more challenging nature of the 2007 data, and the lack of time for retraining the close-talking acoustic model in the IBM system.

In addition to IBM, ASR systems developed by LIMSI and UKA-ISL also achieved significant milestones over the duration of the CHIL project. For example, LIMSI achieved the lowest WERs in the 2005 CHIL-internal evaluation for both close-talking and far-field conditions, whether the UKA-ISL ASR system yielded a 26.7% WER in the IHM (close-talking) condition at the RT06 evaluation.

Overall, as mentioned at the beginning of this chapter, the CHIL Consortium consistently improved ASR technology over the duration of the project. This is clearly depicted in Fig. 6.1, where the lowest WER of all developed and evaluated far-field ASR systems is depicted over the four technology evaluations. This progress is especially noteworthy due to the fact that the ASR task has become increasingly more challenging over time. In particular, between 2005 and 2006, the number of recording sites increased from one to five, and the task modified to cover ASR for all seminar participants. Furthermore, between 2006 and 2007, the seminar interactivity increased significantly, with more focus placed on meeting-like, interactive seminars.

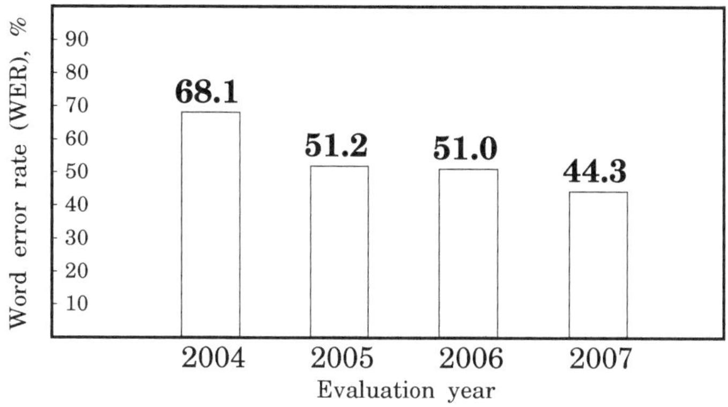

Fig. 6.1. Far-field ASR performance over the duration of the CHIL project.

6.6 Conclusions and Discussion

This chapter has presented an overview of the progress achieved in the automatic transcription of CHIL meetings over the past three and half years. Over this period, ASR technology developed by IBM, LIMSI, and UKA-ISL was evaluated four times, twice in internal consortium benchmarks, and twice in the Rich Transcription international campaigns, overseen by NIST. The latter also attracted significant interest by external parties. In these evaluations, the CHIL partner sites demonstrated significant improvements in ASR system accuracy over time and competitive performance compared to non-CHIL site systems.

Nevertheless, ASR word error rates remain high, particularly in the far-field task for the CHIL scenarios. The continued accuracy improvements indicate that further improvements are to be expected, driven by better acoustic and language modeling as well as further data availability. Future research will also address the modeling of disfluencies in spontaneous speech and pronunciation modeling for nonnative speech. In particular, better addressing the channel combination problem and concentration on advanced noise-removal techniques should benefit system performance.

Future challenges will focus on blind dereverberation, which is still a very difficult task, and the development of systems able to separate target speech from interference speech [26], the so-called cocktail party effect. This describes the ability of humans to listen to a single talker among a mixture of conversations.

It is also expected that in the future, visual speech information could be robustly extracted from the participants in CHIL interactive seminars and lectures, employing appropriately managed, active pan-tilt-zoom cameras in the CHIL smart rooms. Such information can then be fused with acoustic speech input to better address the CHIL ASR problem and its components, including speech activity detection, speaker diarization, and source separation. CHIL partners IBM and UKA-ISL have already expended a significant effort in this area and have focused on two problems of particular interest in the CHIL scenarios: the issue of visual feature extraction from nonfrontal views [33, 28] and the problem of robust audiovisual speech integration [37, 27].

References

1. AMI – Augmented Multiparty Interaction, `http://www.amiproject.org`.
2. T. Anastasakos, J. McDonough, R. Schwartz, and J. Makhoul. A compact model for speaker adaptation training. In *International Conference on Spoken Language Processing (ICSLP)*, pages 1137–1140, Philadelphia, PA, 1996.
3. A. Andreou, T. Kamm, and J. Cohen. Experiments in vocal tract normalisation. In *Proceedings of the CAIP Workshop: Frontiers in Speech Recognition II*, 1994.
4. X. Anguera, C. Wooters, and J. Hernando. Acoustic beamforming for speaker diarization of meetings. *IEEE Transactions on Audio, Speech, and Language Processing*, 15(7):2011–2022, 2007.
5. C. Barras, X. Zhu, S. Meignier, and J.-L. Gauvain. Multi-stage speaker diarization of Broadcast News. *IEEE Transactions on Audio, Speech, and Language Processing*, 14(5):1505–1512, 2006.
6. R. E. Bellman. *Adaptive Control Processes.* Princeton University Press, Princeton, NJ, 1961.
7. S. Burger, V. McLaren, and H. Yu. The ISL meeting corpus: The impact of meeting type on speech style. In *Proceedings of the International Conference on Spoken Language Processing*, Denver, CO, 2002.
8. S. M. Chu, E. Marcheret, and G. Potamianos. Automatic speech recognition and speech activity detection in the CHIL seminar room. In *Joint Workshop on Multimodal Interaction and Related Machine Learning Algorithms (MLMI)*, pages 332–343, Edinburgh, United Kingdom, 2005.
9. S. Davis and P. Mermelstein. Comparison of parametric representations of monosyllabic word recognition in continuously spoken sentences. *IEEE Transactions on Acoustics, Speech, and Signal Processing*, 28(4):357–366, 1980.
10. A. P. Dempster, M. M. Laird, and D. B. Rubin. Maximum likelihood from incomplete data via the EM algorithm. *Journal of the Royal Statistical Society Series B (Methodological)*, 39:1–38, 1977.
11. F. Faubel and M. Wölfel. Coupling particle filters with automatic speech recognition for speech feature enhancement. In *Proceedings of Interspeech*, 2006.
12. J. G. Fiscus. A post-processing system to yield reduced word error rates: Recogniser output voting error reduction (ROVER). In *Automatic Speech Recognition and Understanding Workshop (ASRU)*, pages 347–352, Santa Barbara, CA, 1997.
13. J. G. Fiscus, J. Ajot, M. Michel, and J. S. Garofolo. The Rich Transcription 2006 Spring meeting recognition evaluation. In S. Renals, S. Bengio, and J. G. Fiscus, editors, *Machine Learning for Multimodal Interaction*, LNCS 4299, pages 309–322. 2006.
14. J. S. Garofolo, C. D. Laprun, M. Michel, V. M. Stanford, and E. Tabassi. The NIST meeting room pilot corpus. In *Proceedings of the Language Resources Evaluation Conference*, Lisbon, Portugal, May 2004.
15. J.-L. Gauvain and C. Lee. Maximum a posteriori estimation for multivariate Gaussian mixture observations of Markov chains. *IEEE Transactions on Speech and Audio Processing*, 2(2):291–298, Apr. 1994.
16. R. Gopinath. Maximum likelihood modeling with Gaussian distributions for classification. In *Proceedings of the International Conference on Acoustics, Speech, and Signal Processing (ICASSP)*, pages 661–664, Seattle, WA, 1998.

17. R. Haeb-Umbach and H. Ney. Linear discriminant analysis for improved large vocabulary continuous speech recognition. In *Proceedings of the International Conference on Acoustics, Speech, and Signal Processing (ICASSP)*, volume 1, pages 13–16, 1992.
18. H. Hermansky. Perceptual linear predictive (PLP) analysis of speech. *Journal of the Acoustic Society of America*, 87(4):1738–1752, 1990.
19. R. Hincks. *Computer Support for Learners of Spoken English.* PhD thesis, KTH, Stockholm, Sweden, 2005.
20. J. Huang, E. Marcheret, and K. Visweswariah. Improving speaker diarization for CHIL lecture meetings. In *Proceedings of Interspeech*, pages 1865–1868, Antwerp, Belgium, 2007.
21. J. Huang, E. Marcheret, K. Visweswariah, V. Libal, and G. Potamianos. The IBM Rich Transcription 2007 speech-to-text systems for lecture meetings. In *Multimodal Technologies for Perception of Humans, Proceedings of the International Evaluation Workshops CLEAR 2007 and RT 2007*, LNCS 4625, pages 429–441, Baltimore, MD, May 8-11 2007.
22. J. Huang, E. Marcheret, K. Visweswariah, and G. Potamianos. The IBM RT07 evaluation systems for speaker diarization on lecture meetings. In *Multimodal Technologies for Perception of Humans, Proceedings of the International Evaluation Workshops CLEAR 2007 and RT 2007*, LNCS 4625, pages 497–508, Baltimore, MD, May 8-11 2007.
23. J. Huang, M. Westphal, S. Chen, O. Siohan, D. Povey, V. Libal, A. Soneiro, H. Schulz, T. Ross, and G. Potamianos. The IBM Rich Transcription Spring 2006 speech-to-text system for lecture meetings. In *Machine Learning for Multimodal Interaction*, pages 432–443. LNCS 4299, 2006.
24. A. Janin, D. Baron, J. Edwards, D. Ellis, D. Gelbart, N. Morgan, B. Peskin, T. Pfau, E. Shriberg, A. Stolcke, and C. Wooters. The ICSI meeting corpus. In *Proceedings of the International Conference on Acoustics, Speech, and Signal Processing (ICASSP)*, Hong Kong, 2003.
25. N. S. Kim. Feature domain compensation of nonstationary noise for robust speech recognition. *Speech Communication*, 37:231–248, 2002.
26. K. Kumatani, T. Gehrig, U. Mayer, E. Stoimenov, J. McDonough, and M. Wölfel. Adaptive beamforming with a minimum mutual information criterion. *IEEE Transactions on Audio, Speech, and Language Processing*, 15:2527–2541, 2007.
27. K. Kumatani, S. Nakamura, and R. Stiefelhagen. Asynchronous event modeling algorithm for bimodal speech recognition. *Speech Communication*, 2008. (submitted to).
28. K. Kumatani and R. Stiefelhagen. State-synchronous modeling on phone boundary for audio visual speech recognition and application to multi-view face images. In *International Conference on Acoustics, Speech, and Signal Processing (ICASSP)*, volume 4, pages 417–420, Honolulu, HI, 2007.
29. L. Lamel, J. L. Gauvain, and G. Adda. Lightly supervised and unsupervised acoustic model training. *Computer, Speech and Language*, 16(1):115–229, 2002.
30. L. F. Lamel, F. Schiel, A. Fourcin, J. Mariani, and H. Tillmann. The translanguage English database (TED). In *Proceedings of the International Conference on Spoken Language Processing (ICSLP)*, Yokohama, Japan, 1994.
31. The LDC Corpus Catalog. `http://www.ldc.upenn.edu/Catalog`.
32. C. J. Leggetter and P. C. Woodland. Maximum likelihood linear regression for speaker adaptation of continuous density hidden Markov models. *Computer Speech and Language*, 9(2):171–185, 1995.
33. P. Lucey, G. Potamianos, and S. Sridharan. A unified approach to multi-pose audio-visual ASR. In *Interspeech*, Antwerp, Belgium, 2007.

34. J. Luque, X. Anguera, A. Temko, and J. Hernando. Speaker diarization for conference room: The UPC RT07s evaluation system. In *Multimodal Technologies for Perception of Humans, Proceedings of the International Evaluation Workshops CLEAR 2007 and RT 2007*, LNCS 4625, pages 543–554, Baltimore, MD, May 8-11 2007.
35. D. Macho, C. Nadeu, and A. Temko. Robust speech activity detection in interactive smart-room environments. In *Machine Learning for Multimodal Interaction*, LNCS 4299, pages 236–247. 2006.
36. L. Mangu, E. Brill, and A. Stolcke. Finding consensus in speech recognition: Word error minimization and other applications of confusion networks. *Computer, Speech and Lanuage*, 14(4):373–400, 2000.
37. E. Marcheret, V. Libal, and G. Potamianos. Dynamic stream weight modeling for audio-visual speech recognition. In *International Conference on Acoustics, Speech, and Signal Processing (ICASSP)*, volume 4, pages 945–948, Honolulu, HI, 2007.
38. E. Marcheret, G. Potamianos, K. Visweswariah, and J. Huang. The IBM RT06s evaluation system for speech activity detection in CHIL seminars. In *Machine Learning for Multimodal Interaction*, LNCS 4299, pages 323–335. 2006.
39. J. M. Pardo, X. Anguera, and C. Wooters. Speaker diarization for multi-microphone meetings using only between-channel differences. In *Machine Learning for Multimodal Interaction*, pages 257–264. LNCS 4299, 2006.
40. D. Povey, B. Kingsbury, L. Mangu, G. Saon, H. Soltau, and G. Zweig. fMPE: Discriminatively trained features for speech recognition. In *International Conference on Acoustics, Speech, and Signal Processing (ICASSP)*, volume 1, pages 961–964, Philadelphia, PA, 2005.
41. D. Povey and P. Woodland. Improved discriminative training techniques for large vocabulary continuous speech recognition. In *International Conference on Acoustics, Speech, and Signal Processing (ICASSP)*, Salt Lake City, UT, 2001.
42. D. Povey and P. C. Woodland. Minimum phone error and I-smoothing for improved discriminative training. In *International Conference on Acoustics, Speech, and Signal Processing (ICASSP)*, pages 105–108, Orlando, FL, 2002.
43. E. Rentzeperis, A. Stergiou, C. Boukis, A. Pnevmatikakis, and L. C. Polymenakos. The 2006 Athens Information Technology speech activity detection and speaker diarization systems. In *Machine Learning for Multimodal Interaction*, pages 385–395. LNCS 4299, 2006.
44. C. P. Robert and G. Casella. *Monte Carlo Statistical Methods*. Springer, New York, 2nd edition, 2004.
45. Rich Transcription 2007 Meeting Recognition Evaluation. `http://www.nist.gov/speech/tests/rt/rt2007`.
46. G. Saon, G. Zweig, and M. Padmanabhan. Linear feature space projections for speaker adaptation. In *International Conference on Acoustics, Speech, and Signal Processing (ICASSP)*, pages 325–328, Salt Lake City, UT, 2001.
47. H. Schwenk. Efficient training of large neural networks for language modeling. In *Proceedings of the International Joint Conference on Neural Networks*, pages 3059–3062, 2004.
48. R. Singh and B. Raj. Tracking noise via dynamical systems with a continuum of states. In *International Conference on Acoustics, Speech, and Signal Processing (ICASSP)*, Hong Kong, 2003.
49. O. Siohan, B. Ramabhadran, and B. Kingsbury. Constructing ensembles of ASR systems using randomized decision trees. In *International Conference on Acoustics, Speech, and Signal Processing (ICASSP)*, volume 1, pages 197–200, Philadelphia, PA, 2005.

50. *The Translanguage English Database (TED) Transcripts (LDC catalog number LDC2002T03, ISBN 1-58563-202-3).*
51. S. E. Tranter and D. A. Reynolds. An overview of automatic speaker diarization systems. *IEEE Transactions on Audio, Speech, and Language Processing*, 14(5):1557–1565, 2004.
52. M. Wölfel. Warped-twice minimum variance distortionless response spectral estimation. In *Proceedings of the European Signal Processing Conference (EUSIPCO)*, 2006.
53. M. Wölfel. Channel selection by class separability measures for automatic transcriptions on distant microphones. In *Proceedings of Interspeech*, 2007.
54. M. Wölfel. A joint particle filter and multi-step linear prediction framework to provide enhanced speech features prior to automatic recognition. In *Joint Workshop on Hands-free Speech Communication and Microphone Arrays (HSCMA)*, Trento, Italy, 2008.
55. M. Wölfel and J. McDonough. Combining multi-source far distance speech recognition strategies: Beamforming, blind channel and confusion network combination. In *Proceedings of Interspeech*, 2005.
56. M. Wölfel and J. W. McDonough. Minimum variance distortionless response spectral estimation, review and refinements. *IEEE Signal Processing Magazine*, 22(5):117–126, 2005.
57. M. Wölfel, K. Nickel, and J. McDonough. Microphone array driven speech recognition: influence of localization on the word error rate. *Proceedings of the Joint Workshop on Multimodal Interaction and Related Machine Learning Algorithms (MLMI)*, 2005.
58. M. Wölfel, S. Stüker, and F. Kraft. The ISL RT-07 speech-to-text system. In *Multimodal Technologies for Perception of Humans, Proceedings of the International Evaluation Workshops CLEAR 2007 and RT 2007*, LNCS 4625, pages 464–474, Baltimore, MD, May 8-11 2007.
59. X. Zhu, C. Barras, L. Lamel, and J. L. Gauvain. Speaker diarization: from Broadcast News to lectures. In *Machine Learning for Multimodal Interaction*, pages 396–406. LNCS 4299, 2006.
60. X. Zhu, C. Barras, L. Lamel, and J.-L. Gauvain. Multi-stage speaker diarization for conference and lecture meetings. In *Multimodal Technologies for Perception of Humans, Proceedings of the International Evaluation Workshops CLEAR 2007 and RT 2007*, LNCS 4625, pages 533–542, Baltimore, MD, May 8-11 2007.

7

Acoustic Event Detection and Classification

Andrey Temko[1], Climent Nadeu[1], Dušan Macho[1], Robert Malkin[2], Christian Zieger[3], Maurizio Omologo[3]

[1] TALP Research Center, Universitat Politècnica de Catalunya, Barcelona, Spain
[2] interACT, Carnegie Mellon University, Pittsburgh, PA, USA
[3] FBK-irst, Foundation Bruno Kessler, Povo, Italy

The human activity that takes place in meeting rooms or classrooms is reflected in a rich variety of acoustic events (AE), produced either by the human body or by objects handled by humans, so the determination of both the identity of sounds and their position in time may help to detect and describe that human activity. Indeed, speech is usually the most informative sound, but other kinds of AEs may also carry useful information, for example, clapping or laughing inside a speech, a strong yawn in the middle of a lecture, a chair moving or a door slam when the meeting has just started. Additionally, detection and classification of sounds other than speech may be useful to enhance the robustness of speech technologies like automatic speech recognition.

Acoustic event detection and classification (AED/C) is a recent discipline that may be included in the broad area of computational auditory scene analysis [1]. It consists of processing acoustic signals and converting them into symbolic descriptions corresponding to a listener's perception of the different sound events present in the signals and their sources.

This chapter presents the evolution of the pioneering works that start from the very first CHIL dry-run evaluation in 2004 and end with the second international CLEAR evaluation campaign in 2007. During those years of the CHIL project, the task was significantly modified, going from the classification of presegmented AEs to the detection and classification in real seminar conditions where AEs may have temporal overlaps. The changes undergone by the task definition along with a short description of the developed systems and their results are presented and discussed in this chapter.

It is worth mentioning several factors that make the tasks of AED/C in meeting-room environments particularly challenging. First, detection and classification of sounds have been usually carried out so far with a limited number of classes, e.g., speech and music, but there were no previously published works about meeting-room AEs when the CHIL project was launched. Consequently, the sound classes and what is considered an AE instance had to be defined. Second, the nature of AEs is different from speech, and features describing speech's spectral structure are not necessarily suitable for AED/C, so the choice of appropriate features becomes an important issue. Third, the chosen classifiers have to face the data scarcity problem caused by the lack of large specific databases. Fourth, the low signal-to-noise ratio of the recorded AEs signals is responsible for a high error rate. It is due to both the use of far-field microphones (required by the unobtrusiveness condition) and the fact that the criterion of natural interaction produced a lot of temporal overlaps of AEs with co-occurring speech and other AEs. Fifth, the evaluation of those new AED/C tasks required the establishment of an (agreed on) protocol to develop databases and appropriate metrics. In particular, recordings had to find a balance between people's natural behavior and a large enough number of recorded AEs. And six, the evaluation data come from up to five different smart rooms with different acoustic conditions and microphone locations.

The chapter is organized as follows. We start with AEC as this task was first pursued in the CHIL project because it is easier than AED. Section 7.1 presents the work carried out on AEC within three evaluations organized by CHIL, namely, the dry-run evaluation in 2004, the CHIL international evaluation in 2005, and the CLEAR international evaluation campaign in 2006 (see also Chapter 15). The definition of a set of meeting-room AEs, the task, and metrics are given in this section. The developed AEC systems are briefly described and their results are presented and discussed. While AEC is performed on previously segmented AEs, AED is a more difficult task because the identity of sounds and their position in time have to be obtained. In Section 7.2, we overview the two international evaluation campaigns where the AED task was presented, namely, CLEAR 2006 [5] and CLEAR 2007 [4]. The AED task definition, metrics, and evaluation setup are described. A summary of the developed systems and their results are proposed. Section 7.3 overviews AED demonstrations developed by CHIL project partners. Conclusions and remaining challenges are presented in Section 7.4.

7.1 Acoustic Event Classification

On the way toward the detection of AEs, it was first decided to perform classification of AEs. Let us note that classification deals with events that have been already extracted or, alternatively, the temporal positions of acoustic events in an audio stream are assumed to be known. The AEC task has been evaluated in the dry-run evaluation in 2004, the first CHIL evaluation in 2005, and the CLEAR 2006 evaluation campaign. Specifically, Section 7.1.1 presents the classes and DBs used in the evaluations. Evaluation protocol, participants, and metrics are given in Section 7.1.2. The various approaches, their results, and a discussion are presented in Section 7.1.3.

7.1.1 Acoustic Event Classes and Databases

Table 7.1 shows the acoustic event classes considered in the three evaluations. In the very first dry-run evaluations, seminar data recorded at University of Karlsruhe (UKA-ISL), which were also used for other CHIL evaluations, were transcribed according to 36 sound classes. The instances were transcribed with tight temporal bounds, allowing isolated-sound tests to be performed. In that corpus, 25 classes were found to be suitable for use in the evaluation based on the number of instances for both training and testing. That was an ad hoc set, in the sense that we did not know a priori what kinds of sounds would be present in the database, since the recordings were designed without planning the occurrence of acoustic events, and we relied on the transcriber's judgment to select the whole sound set. This resulted in a large number of class overlaps (e.g., "bang" and "bump"), but it allowed us to determine which classes were actually present and their relative number of instances. In this way, 2805 segments were taken in total and divided into training (1425 segments) and evaluation sets (1380) at random.

Due to the low performance shown in the dry-run evaluations and in order to limit the number of classes to those that were both easy to identify for transcribers and semantically relevant to the CHIL task, the number of classes for the CHIL evaluations in 2005 was decreased from 25 to 15. A few semantically similar classes were combined; e.g., the three classes "breath", "inspiration", and "expiration" were grouped into one class "breath", as can be seen in Table 7.1. Additionally, several new classes were added, like the music/phone ring class or the generic mouth noise class. However, as we wished to retain a label for every sound in the corpus, whether it was easily identifiable or not, we split the label set into two types: semantic labels and acoustic labels. Semantic labels corresponded to specific named sounds with CHIL relevance, e.g., door slamming, speech, etc. Acoustic labels corresponded to unnamed sounds that were either unidentifiable or had no CHIL relevance. Based on the availability of samples per class, 15 semantic classes were finally chosen as shown in Table 7.1. Table 7.2 shows the acoustic labels used in the CHIL evaluations in 2005. The acoustic labels used names describing both tonal and rhythmic features of a sound. For acoustic evaluation, all semantically labeled classes were mapped to the corresponding acoustic classes, as shown in Table 7.2. From seminars recorded at UKA-ISL, 7092 segments were extracted; 3039 were used for training (1946 "speech", 333 "breath", 333 "silence", 232 "unknown", etc.), 1104 for development (749 "speech", 183 "unknown", 139 "silence", 9 "cough", etc.), and 2949 for evaluation (2084 "speech", 348 "silence", 274 "unknown", 123 "breath", etc.).

<table>
<tr><th>Classes / Year</th><th>Breath</th><th>Expiration</th><th>Inspiration</th><th>Conversation</th><th>Speech</th><th>Cough</th><th>Throat</th><th>Applause</th><th>Bang</th><th>Bump</th><th>Click</th><th>Chair moving</th><th>Crump</th><th>Door slam</th><th>Door knock</th><th>Keyboard</th><th>Key jingle</th><th>Laughter</th><th>Metal</th><th>Microphone</th><th>Mouth noise</th><th>Music/phone ring</th><th>Movements</th><th>Paper work</th><th>Pen writing</th><th>Pop</th><th>Shh...</th><th>Silence</th><th>Smack</th><th>Snap</th><th>Sniff</th><th>Spoon-cup jingle</th><th>Squeak</th><th>Steps</th><th>Tap</th><th>Unknown</th><th>Whirring</th><th>Total</th></tr>
<tr><td>2004</td><td>√</td><td>√</td><td>√</td><td>√</td><td>√</td><td>√</td><td>√</td><td></td><td>√</td><td>√</td><td>√</td><td></td><td>√</td><td>√</td><td></td><td>√</td><td></td><td></td><td>√</td><td>√</td><td></td><td></td><td>√</td><td></td><td>√</td><td>√</td><td>√</td><td>√</td><td>√</td><td>√</td><td></td><td></td><td>√</td><td>√</td><td>√</td><td></td><td></td><td>25</td></tr>
<tr><td>2005</td><td colspan="3">√</td><td colspan="2">√</td><td colspan="2">√</td><td></td><td></td><td></td><td>√</td><td>√</td><td></td><td>√</td><td></td><td>√</td><td></td><td></td><td></td><td></td><td>√</td><td>√</td><td></td><td>√</td><td></td><td></td><td></td><td>√</td><td></td><td></td><td>√</td><td></td><td></td><td>√</td><td></td><td>√</td><td>√</td><td>15</td></tr>
<tr><td>2006</td><td colspan="3"></td><td colspan="2"></td><td colspan="2">√</td><td>√</td><td></td><td></td><td></td><td>√</td><td></td><td>√</td><td>√</td><td>√</td><td>√</td><td>√</td><td></td><td></td><td></td><td>√</td><td></td><td>√</td><td></td><td></td><td></td><td></td><td></td><td></td><td></td><td>√</td><td></td><td>√</td><td></td><td></td><td></td><td>15</td></tr>
</table>

Table 7.1. Evaluated acoustic event classes for evaluations that took place in years 2004, 2005, and 2006.

Semantic Label	Acoustic Label
laughter, speech	continuous tone
breath, sniff, paper, whirring, cough	continuous sound without tone
click, chair, door slam, mouth noise	single noncontinuous sound
steps	regular repetitive noncontinuous sound
keyboard typing	irregular repetitive noncontinuous sound
–	generic noise, all other kinds of sounds

Table 7.2. Semantic-to-acoustic mapping.

Due to the low number of instances for most semantic classes, it was not possible to obtain statistically significant results [2]. Besides, they were biased by prevailing classes like "speech" and "silence". For the CLEAR international evaluation campaign in 2006, we redefined the set of AE classes, including new ones that were judged as relevant for the meeting-room scenario, and reducing the number to 12 in order to make them more easily identifiable by transcribers. Besides, to get rid of the bias to "speech" and "silence" classes, the latter were discarded from evaluation, as Table 7.1 shows. To increase the number of instances for the chosen 12 classes, two partners, UPC and FBK-irst, recorded databases of isolated acoustic events [7]. From them each of 12 classes had about 100 instances, from which around two thirds were taken for training and the rest for testing.

7.1.2 Evaluation Protocol, Participants, and Metrics

In all evaluations, audio was collected with a combination of microphones. In the dry-run evaluations in 2004 and the CHIL evaluation in 2005, a channel of a wall-mounted 64-microphone array (Mark III array) was used, while in the CLEAR 2006 evaluation, any of around 90 far-field microphones (from the microphone array, several T-shaped clusters, and also tabletop microphones) was allowed to be used. Participants were permitted to use for system training only the data distributed for this evaluation and identified as training data. No other data were allowed for training purposes. Only an isolated sound test was performed. The sounds that were overlapped with other sounds were discarded. The Technical University of Catalonia (UPC) and Carnegie Mellon University (CMU) research groups participated in all three evaluations on the AEC. The Bruno Kessler Foundation (FBK-irst) group participated in the CLEAR 2006 evaluation. In the dry-run evaluations, accuracy and recall were used as metrics. The former is defined as the number of correctly classified acoustic events divided by the total number of acoustic events. Recall is defined as a mean of the individual class recalls calculated as the number of correctly classified events of a particular class divided by the number of events of that class. In the CHIL evaluations, the used error rate was defined as 100 minus accuracy (%). In the CLEAR 2006, a metric that counts the number of insertion, substitution, and deletion errors was introduced for the AED task; however, for the AEC, where only substitution errors are possible, the same error rate defined for the CHIL evaluations in 2005 was used.

7.1.3 Systems, Results, and Discussion

The systems developed at CMU were based on HMM/GMM classifiers. Several HMM topologies were tried and the submitted system was obtained with a topology given by BIC for evaluations in 2004/2005 and by k-variable k-means algorithm for the CLEAR 2006 [2]. PCA was applied to a set of features composed of MFCC in 2004 and MFCC plus a few perceptual features in the 2005/2006 evaluations. For CLEAR 2006 evaluations, after training a complete set of HMMs, site-specific feature space adaptation matrices were further trained, resulting in two systems [7].

At UPC, after comparing the preliminary results obtained with GMM and SVM classifiers, the latter were finally chosen. A set of features composed of several perceptual features and frequency-filtered bank energies was exploited. Several statistical parameters were calculated from the frame-level features over all frames in an acoustic event. The obtained set of segment-level features was fed to the SVM classifier [7].

At FBK-irst, three state HMM models with MFCC features were used. All of the HMMs had a left-to-right topology and used output probability densities represented by means of 32 Gaussian components with diagonal covariance matrices. Two different sets of models were created to fit, respectively, the FBK-irst and UPC rooms; each set was trained with data from only one of the two isolated AE databases [7].

The results of the AEC evaluation are shown in Table 7.3. The low system performance obtained in 2004 was attributable to the relatively large number of classes. Additionally, the large number of classes made it rather difficult to correctly transcribe the seminars, resulting in a number of outliers. The scores from the accuracy measure are much higher than those from recall, showing that the results are biased by prevailing classes like "speech" and "silence".

<table>
<tr><th>Evaluations</th><th colspan="2">Database</th><th>Metrics(%)</th><th>Baseline</th><th colspan="2">CMU</th><th>UPC</th><th colspan="2">FBK-irst</th></tr>
<tr><td rowspan="2">2004</td><td colspan="2" rowspan="2">Seminars 2004</td><td>Recall</td><td>4</td><td colspan="2">26.4</td><td>24.15</td><td colspan="2">–</td></tr>
<tr><td>Accuracy</td><td>47.8</td><td colspan="2">61.6</td><td>55.1</td><td colspan="2">–</td></tr>
<tr><td rowspan="3">2005</td><td colspan="2">Seminars 2004</td><td>Accuracy</td><td>47.8</td><td colspan="2">63.2</td><td>62.9</td><td colspan="2">–</td></tr>
<tr><td rowspan="2">Seminars 2005</td><td>Semantic</td><td rowspan="2">Error rate</td><td>29.4</td><td colspan="2">27.2</td><td>23.5</td><td colspan="2">–</td></tr>
<tr><td>Acoustic</td><td>29.3</td><td colspan="2">26.7</td><td>27.8</td><td colspan="2">–</td></tr>
<tr><td rowspan="2">2006</td><td rowspan="2">Databases of isolated AEs</td><td>UPC</td><td rowspan="2">Error rate</td><td>–</td><td>7.5</td><td>–</td><td>4.1</td><td>12.3</td><td>–</td></tr>
<tr><td>FBK</td><td>–</td><td>–</td><td>5.8</td><td>5.8</td><td>–</td><td>6.2</td></tr>
</table>

Table 7.3. Results of the AEC evaluations

The AEC 2005 evaluation unfortunately still suffered from problems of corpora. Transcription mistakes were still quite frequent; hence, the labels were not as reliable as they could have been. Further, the corpus was extremely unbalanced; a large majority of the segments were made of speech. Many interesting classes had only a few examples in the training or test sets, and therefore they were not well-modeled by our systems. For the 2004/2005 evaluations, Table 7.3 shows additionally the baseline scores obtained as if a system just chose the most frequent class: "speech" in the 2004 evaluations; "speech" and "continuous tone" for semantic and acoustic sets, respectively, in the 2005 evaluations. Besides, Table 7.3 shows that systems submitted in 2005 improved the results obtained with the systems submitted in 2004 for the evaluation corpus used in 2004. For the 2006 evaluation, the error rate was smaller since the database collection had been designed specifically for acoustic events, and the number of instances per class was high enough and homogeneous among classes. The SVM-based system from UPC obtained the best error rate despite the fact that it was not database-specific.

7.2 Acoustic Event Detection

The AED task was evaluated in two international evaluation campaigns: CLEAR 2006 [5] and CLEAR 2007 [4]. In CLEAR 2006, we first focused on the easier task of detection of isolated AEs and made a preliminary attempt to perform AED in spontaneous conditions. In CLEAR 2007, only the task of spontaneous AED was tackled. Specifically, Section 7.2.1 presents the classes and DBs used in the evaluations. The evaluation setup is given in Section 7.2.2. The approaches, results, and discussion are presented in Section 7.2.3.

7.2.1 Acoustic Event Classes and Databases

In the AED 2006 and 2007 evaluations, the set of acoustic events used in the AEC 2006 evaluations was kept, which appears in Table 7.1. It is worth noting that, apart from the 12 evaluated acoustic classes, there are also three other classes that have not been evaluated though the present: "speech", "unknown", and "silence".

For CLEAR 2006 and CLEAR 2007, CHIL partners turned to recordings of scripted interactive seminars. Each seminar usually consists of a presentation of 10 to 30 minutes to a group of three to five attendees in a meeting room. During and after the presentation, there are questions from the attendees with answers from the presenter. There is also activity in terms of people entering/leaving the room, opening and closing the door, standing up and going to the screen, discussion among the attendees, coffee breaks, etc. Each meeting can be conditionally decomposed into acoustic scenes: "beginning", "meeting", "coffee break", "question/answers", and "end". The recorded interactive seminars contained a satisfactory number of acoustic events, so it was possible to perform AED tests that are statistically meaningful. The AEs are often overlapped with speech and/or other AEs.

In CLEAR 2006, two main series of experiments were performed: AED in the isolated condition and AED in the real environment condition. For the task of isolated AED, the databases of isolated acoustic events were split into training and testing parts in the same way as it has been done for the AEC task. For the task of AED in real environments, five 20-minute seminars recorded at UPC were chosen as the richest ones in terms of the number of instances per class, from which one whole seminar was used for training along with all databases of isolated acoustic events, and four 5-minute extracts from the remaining four seminars were used for testing.

In CLEAR 2007, the seminars recorded at UKA-ISL, Foundation Bruno Kessler (FBK-irst), Athens Institute of Technology (AIT), and UPC were found suitable for evaluations, forming a set of 20 interactive seminars. Four interactive seminars (one from each site) were assigned for system development. Along with the seminar recordings, the whole databases of isolated AEs recorded at UPC and FBK-irst were used for development. In total, development seminar data consisted of 7495 seconds, where 16% of total time is AEs, 13% is silence, and 81% is "speech" and "unknown" classes.

From the remaining seminars, 20 5-minute segments were extracted for testing. In total, the test data consisted of 6001 seconds, where 36% is AE time, 11% is "silence", and 78% is "speech" and "unknown". Noticeably, about 64% of the time, the AEs are overlapped with "speech", and 3% of the times they overlap with other AEs.

7.2.2 Evaluation Protocol, Participants, and Metrics

The primary evaluation task in CLEAR 2006 was defined as AED evaluated on both the isolated databases and the seminars.

In order to have systems comparable across sites, a set of evaluation conditions was defined for CLEAR 2007: The evaluated system must be applied to the whole CLEAR 2007 test database; only primary systems are submitted to compete; the evaluated systems must use only audio signals, though they can use any number of microphones.

There were three participants in CLEAR 2006: CMU, UPC, and FBK-irst. In CLEAR 2007, however, eight sites signed up to participate; six sites submitted the results, while two withdrew their applications. The participating partners were AIT, Institute for Infocomm Research (IIR), FBK-irst, Tampere University of Technology (TUT), University of Illinois at Urbana-Champaign, (UIUC), and UPC. As mentioned above, the acoustic events that happen in a real environment may have temporal overlaps. In order to be able to properly score the output of the systems, appropriate metrics were developed by UPC and agreed upon by the other CHIL partners involved in the evaluations (FBK-irst, CMU, and ELDA).

For CLEAR 2006, the metric protocol consisted of two steps: (1) projecting all levels of overlapping events into a single-level reference transcription, and (2) comparing a hypothesized transcription with the single-level reference transcription. It is defined as the acoustic event error rate:

$$\mathrm{AEER} = (D + I + S)/N * 100,$$

where N is the number of events to detect, D are deletions, I insertions, and S substitutions.

However, this metric had some disadvantages. The main assumption was that one reference acoustic event can cause only one error. In some cases, the metric was ambiguous, e.g., when one part of an acoustic event is detected correctly, another part of the same event causes a substitution error, and the rest is deleted, so the final score of the metric is affected by the last decision made about the acoustic event. For this purpose, for CLEAR 2007, this metric was decomposed into two other metrics: an F-score measure of detection accuracy (AED-ACC), and an error rate measure that focuses more on the accuracy of the endpoints of each detected AE (AED-ER).

The aim of the AED-ACC metric is to score the detection of all instances of what is considered as a relevant AE. With this metric, it is not important to reach a good temporal coincidence of the reference and system output timestamps of the AEs, but rather to detect their instances. It is oriented to applications like real-time services for smart rooms, audio-based surveillance, etc. The AED-ACC is defined as the F-score (the harmonic mean between precision and recall):

$$\text{AED-ACC} = \frac{(1+\beta^2) * \text{Precision} * \text{Recall}}{\beta^2 * \text{Precision} + \text{Recall}},$$

where

$$\text{Precision} = \frac{\text{number of correct system output AEs}}{\text{number of all system output AEs}},$$

$$\text{Recall} = \frac{\text{number of correctly detected reference AEs}}{\text{number of all reference AEs}},$$

and β is a weighting factor that balances precision and recall. In this evaluation, the factor β was set to 1. A *system output (reference) AE* is considered *correct or correctly produced (correctly detected)* either (1) if there exists at least one reference (system output) AE whose temporal center is situated between the timestamps of the system output (reference) AE and the labels of the system output AE and the reference AE are the same, or (2) if the temporal center of the system output (reference) AE lies between the timestamps of at least one reference (system output) AE and the labels of the system output AE and the reference AE are the same.

For some applications, it is necessary to have a good temporal resolution of the detected AEs. The aim of this metric is to score AED as a task of general audio segmentation. Possible applications can be content-based audio indexing/retrieval, meeting-stage detection, etc. In order to define AED-ER, the NIST metric for speaker diarization was adapted to the task of AED. The audio data are divided into adjacent segments whose borders coincide with the points whether either a reference AE or a system output AE starts or stops, so that, along a given segment, the number of reference AEs and the number of system output AEs do not change. The AED-ER score is computed as the fraction of time, including regions of overlaps, in which a system output AE is not attributed correctly to a reference AE, in the following way:

$$(\text{AED - ER}) = \frac{\sum_{allseg.} dur(seg) * (max(N_{REF}, N_{SYS} - N_{correct}(seg)))}{\sum_{allseg.} dur(seg) * N_{REF}(seg)}$$

where, for each segment seg, dur(seg) is a duration of seg, N_{REF} (seg) is the number of reference AEs in seg, N_{SYS} (seg) is the number of system output AEs in seg, and $N_{correct}$ (seg) is the number of reference AEs in seg that correspond to system output AEs in seg. Notice that an overlapping region may contribute several errors.

7.2.3 Systems, Results, and Discussion

In CLEAR 2006, the systems used for AED were essentially those used for AEC. Two main directions were taken: (1) first performing segmentation and then classification (in the UPC and CMU systems), and (2) merging the segmentation and classification in one step as usually performed by the Viterbi search in the current state-of-the-art ASR systems (in the FBK-irst system). Specifically, the CMU system includes an HMM-based events/others segmentation before feeding the HMM event classifier. At UPC, silence/nonsilence segmentation was done on a sliding window of 1s, and the nonsilence portions were then fed to the SVM classifier with a subsequent smoothing and postprocessing. At FBK-irst, however, a standard ASR system was used. All systems submitted to CLEAR 2006 used a single-microphone channel, although CMU and UPC used a Mark III channel, while FBK-irst used a channel taken from a T-shaped microphone cluster.

In CLEAR 2007, the UPC system used in CLEAR 2006 was modified by eliminating the segmentation step, performing signal normalization, and using voting-based fusion of the decisions taken from four microphones. The FBK-irst systems remained almost the same from CLEAR 2006, only changing the number of Gaussians from 32 to 128. Site adaptation was also used in the FBK-irst system in CLEAR 2007. The system submitted by AIT used a hierarchical approach for event classification based on an HMM classifier with different HMM topologies per class and also building a specific system for each site. Signals taken from five microphones were averaged in order to cope with the ambient noise. IIR used a system based on HMM/GMM classifiers. For the AED task, a multichannel system based on HMM was implemented at TUT. Observation probabilities were calculated separately for one channel from each T-shaped microphone array. After that, all the observation probabilities were combined and the optimal path through all models was decoded. An HMM-based system with lattice rescoring using features selected by AdaBoost was implemented at UIUC. Only data from one microphone in the development seminar data were used in the system.

As a summary, five participants (FBK-irst, IIR, AIT, TUT, and UIUC) exploited HMM-based systems, and UPC used an SVM-based system. Three systems were multimicrophone (AIT / TUT / UPC), and three were single-microphone (FBK-irst/IIR/UIUC).

Table 7.4 shows the results obtained for the AED task in the CLEAR 2006 and CLEAR 2007 evaluations. As can be seen from the table, the lowest detection error rates for CLEAR 2006 [7] were obtained by the FBK-irst systems, which did not use the segmentation step. Notice that both the CMU and UPC systems achieved better results than the FBK-irst systems in the classification task (Table 7.3). Although a number of reasons might explain the differences across the systems, we conjectured that the initial segmentation step included in both the UPC and CMU systems, but not in the FBK-irst systems, was the main cause of those systems' lower overall detection performance. Further investigation would be needed in that direction. Besides, as can be seen in Table 7.4, in the 2006 evaluations, the error rates increased significantly for the UPC seminar database. One possible reason for such a poor performance was the difficulty in detecting low-energy acoustic classes that overlap with speech, such as "chair moving", "steps", "keyboard typing", and "paper work". Actually, these classes cover the majority of the events in the UPC seminars and probably were the cause of the poor results obtained in the seminar task. Using multiple microphones might be helpful in this case.

<table>
<tr><th>Evals</th><th colspan="2">Database</th><th>Metrics (%)</th><th colspan="2">CMU</th><th>UPC</th><th colspan="4">FBK-irst</th><th colspan="4">AIT</th><th>IIR</th><th>TUT</th><th>UIUC</th></tr>
<tr><td rowspan="4">2006</td><td rowspan="3">Databases of IAE seminars</td><td>UPC</td><td rowspan="4">AEER</td><td>45.2</td><td>–</td><td>58.9</td><td>23.6</td><td>–</td><td colspan="2">–</td><td colspan="4">–</td><td>–</td><td>–</td><td>–</td></tr>
<tr><td>FBK</td><td>–</td><td>52.5</td><td>64.6</td><td>–</td><td>33.7</td><td colspan="2">–</td><td colspan="4">–</td><td>–</td><td>–</td><td>–</td></tr>
<tr><td>UPC</td><td>–</td><td>177.3</td><td>97.1</td><td>–</td><td>–</td><td colspan="2">99.3</td><td colspan="4">–</td><td>–</td><td>–</td><td>–</td></tr>
<tr><td colspan="2">Overall</td><td colspan="2">80.5</td><td>69.6</td><td colspan="4">46.8</td><td colspan="4">–</td><td>–</td><td>–</td><td>–</td></tr>
<tr><td rowspan="10">2007</td><td rowspan="4">Seminars</td><td>AIT</td><td rowspan="5">AED-ACC</td><td colspan="2">–</td><td>18.6</td><td>16.8</td><td>–</td><td>–</td><td>–</td><td>2</td><td>–</td><td>–</td><td>–</td><td>19.4</td><td>15.7</td><td>27.8</td></tr>
<tr><td>FBK</td><td colspan="2">–</td><td>25.1</td><td>–</td><td>30</td><td>–</td><td>–</td><td>–</td><td>3.9</td><td>–</td><td>–</td><td>22</td><td>20</td><td>37.3</td></tr>
<tr><td>UKA</td><td colspan="2">–</td><td>23.3</td><td>–</td><td>–</td><td>11.8</td><td>–</td><td>–</td><td>–</td><td>8.5</td><td>–</td><td>16.7</td><td>4</td><td>35.4</td></tr>
<tr><td>UPC</td><td colspan="2">–</td><td>23.9</td><td>–</td><td>–</td><td>–</td><td>29</td><td>–</td><td>–</td><td>–</td><td>3.5</td><td>30.8</td><td>19.4</td><td>42</td></tr>
<tr><td colspan="2">Overall</td><td colspan="2">–</td><td>23</td><td colspan="4">23.4*</td><td colspan="4">4.4*</td><td>22.9</td><td>14.7</td><td>33.6</td></tr>
<tr><td rowspan="4">Seminars</td><td>AIT</td><td rowspan="5">AED-ER</td><td colspan="2">–</td><td>128</td><td>103</td><td>–</td><td>–</td><td>–</td><td>184</td><td>–</td><td>–</td><td>–</td><td>238</td><td>105</td><td>99</td></tr>
<tr><td>FBK</td><td colspan="2">–</td><td>157</td><td>–</td><td>87</td><td>–</td><td>–</td><td>–</td><td>155</td><td>–</td><td>–</td><td>161</td><td>92</td><td>99</td></tr>
<tr><td>UKA</td><td colspan="2">–</td><td>147</td><td>–</td><td>–</td><td>157</td><td>–</td><td>–</td><td>–</td><td>212</td><td>–</td><td>142</td><td>299</td><td>99</td></tr>
<tr><td>UPC</td><td colspan="2">–</td><td>120</td><td>–</td><td>–</td><td>–</td><td>103</td><td>–</td><td>–</td><td>–</td><td>246</td><td>145</td><td>105</td><td>100</td></tr>
<tr><td colspan="2">Overall</td><td colspan="2">–</td><td>137</td><td colspan="4">109*</td><td colspan="4">203*</td><td>170</td><td>139</td><td>99</td></tr>
</table>

Table 7.4. Results of the AED evaluations. * means a different system is used for each site.

In CLEAR 2007 evaluation results [4], even the best scores (those from UIUC's) are very low: only 36.3% of accuracy, and almost 100% of error rate, due to temporal overlaps. The system with the best scores shows a relatively low error in the "steps" class, which accounts for 40% of the total number of AEs. On average, more than 71% of all error time occurs in overlapped segments. If they were not scored, the error rate of most submitted primary systems would be around 30-40%. These results indicate that there is still much room for system improvement in meeting-room spontaneous AED. The choice of the system structure may be important. For instance, none of the AED systems presented to the evaluations was built as a set of isolated detectors. This approach can be addressed in the future [6]. An improvement in performance can be expected from the use of several modalities. Some sounds can be detected by video technologies; e.g., "steps" can be assumed when detecting from the video an object moving around the smart room, etc.

7.3 Demonstrations of Acoustic Event Detection

7.3.1 FBK-irst Acoustic Event Detection and Classification Demo

This section describes the acoustic event detection and classification demo developed by FBK-irst under the CHIL project. First, the implemented system detects the acoustic event through a detector, that is based on spectral variation functions. Then it derives its position using a localizer based on the global coherence field (GCF). Finally, it identifies the type, choosing among all possible CHIL events through an HMM-based classifier. The video of the demo has been recorded[1].

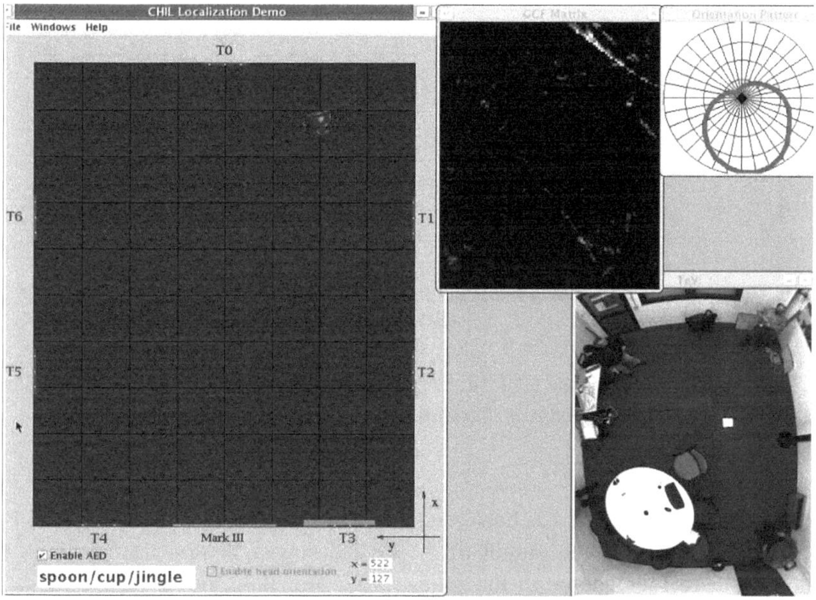

Fig. 7.1. Snapshot of the video showing the acoustic event detection and classification demo developed by FBK-irst.

A snapshot of the video appears in Fig. 7.1. The window on the left shows the position of the detected and classified event in the room map. The result of the classification process is represented through an icon placed in its estimated location in the room map and a text box on the bottom. The events in the room are shot by a camera whose output is reported in the window at the lower right corner. At the right upper corner are two windows showing the GCF and the oriented GCF used by the sound source localizer to estimate the location and directivity pattern of the acoustic event.

[1] The FBK-irst AED demo is available at http://www.youtube.com/watch?v=5_ZgrGL3CnU.

7.3.2 TALP-UPC Acoustic Event Detection and Classification Demo

The AED SVM-based system, written in the C++ programming language, is part of the smartAudio++ software package developed at UPC, which includes other audio technology components (such as speech activity detection and speaker identification) for the purpose of real-time online activity detection and observation in the smart room environment. That AED system implemented in the smart room has been used in demos about technology services developed in CHIL. Also, a specific GUI-based demo has been built that shows the detected isolated events and the positions of their sources in the room. The positions are obtained from the acoustic source localization (ASL) system developed at the TALP center [3].

(a) (b)

Fig. 7.2. The two screens of the GUI: (a) real-time video, and (b) the graphical representation of the AED and ASL functionalities ("keyboard typing" is being produced).

The video of the demo has been recorded while the demo was running in UPC's smart room[2]. It shows the output of the GUI during a session lasting about two minutes, where four people in the room speak, are silent, or make one of the 12 meeting-room sounds defined in CHIL and a few others. There are two screens in the GUI output, as shown in Fig. 7.2. One corresponds to the real video captured from one of the cameras installed in UPC's smart room, and the other is a graphical representation of the output of the two technologies. The two functionalities are simply juxtaposed in the GUI, so, e.g., it may happen that the AED output is correct but the output of acoustic source localization is not, thus showing the right event in a wrong place. The AED technology includes an "unknown" output, symbolized with "?". There are two different situations where the "unknown" label may appear. First and most frequently, it appears when the AED algorithm does not have enough confidence to assign a detected nonsilent event to one of the above-mentioned 12 classes. Second, the "unknown" label is produced when an out-of-list (never-seen-before) AE is detected.

[2] The UPC-TALP AED demo is available at http://www.youtube.com/watch?v=UBSxBd_HYeI.

7.4 Conclusions and Remaining Challenges

The work presented in this chapter has focused on several pioneering evaluation tasks concerning the detection and classification of acoustic events that may happen in a lecture/meeting-room environment. In this context, two different tasks have been evaluated: acoustic event classification and acoustic event detection. Two kinds of databases have been used: two databases of isolated acoustic events, and a database of interactive seminars containing a significant number of acoustic events of interest.

The evolution of task definition, the metrics, and the set of acoustic events have been presented and discussed. The detection and classification systems that submitted results to the evaluation campaigns have been succinctly described, and their results have been reported. In the classification task, the error rates shown in the last performed AEC evaluation (2006) were low enough to convince us to turn to the AED task, where the results obtained with acoustic events in isolated conditions have been satisfactory. However, the performances shown by the AED systems in meeting-room spontaneous conditions indicate that there is still much room for improvement.

References

1. G. B. D. Wang. *Computational Auditory Scene Analysis: Principles, Algorithms and Applications*. Wiley-IEEE Press, 2006.
2. R. Malkin, D. Macho, A. Temko, and C. Nadeu. First evaluation of acoustic event classification systems in the CHIL project. In *Joint Workshop on Hands-Free Speech Communication and Microphone Array, HSCMA'05*, March 2005.
3. C. Segura, A. Abad, C. Nadeu, and J. Hernando. Multispeaker localization and tracking in intelligent environments. In *Multimodal Technologies for Perception of Humans, Proceedings of the International Evaluation Workshops CLEAR 2007 and RT 2007*, LNCS 4625, pages 82–90, Baltimore, MD, May 8-11 2007.
4. R. Stiefelhagen, R. Bowers, and J. Fiscus, editors. *Multimodal Technologies for Perception of Humans, Proceedings of the International Evaluation Workshops CLEAR 2007 and RT 2007*. LNCS 4625. Springer, Baltimore, MD, May 8-11 2007.
5. R. Stiefelhagen and J. Garofolo, editors. *Multimodal Technologies for Perception of Humans, First International Evaluation Workshop on Classification of Events, Activities and Relationships, CLEAR'06*. LNCS 4122. Springer, Southampton, UK, Apr. 6-7 2006.
6. A. Temko. *Acoustic Event Detection and Classification*. PhD thesis, Universitat Politècnica de Catalunya, Barcelona, 2007.
7. A. Temko, R. Malkin, C. Zieger, D. Macho, C. Nadeu, and M. Omologo. Evaluation of acoustic event detection and classification systems. In *Multimodal Technologies for Perception of Humans. First International Evaluation Workshop on Classification of Events, Activities and Relationships CLEAR 2006*, LNCS 4122, pages 311–322. Springer-Verlag, Southampton, UK, Apr. 6-7 2006.

8

Language Technologies: Question Answering in Speech Transcripts

Jordi Turmo[1], Mihai Surdeanu[1], Olivier Galibert[2], Sophie Rosset[2]

[1] TALP Research Center, Universitat Politècnica de Catalunya, Barcelona, Spain
[2] LIMSI-CNRS, Orsay, France

The question answering (QA) task consists of providing short, relevant answers to natural language questions. Most QA research has focused on extracting information from text sources, providing the shortest relevant text in response to a question. For example, the correct answer to the question, "How many groups participate in the CHIL project?" is "15", whereas the response to "Who are the partners in CHIL?" is a list of them. This simple example illustrates the two main advantages of QA over current search engines: First, the input is a natural-language question rather a keyword query; and second, the answer provides the desired information content and not simply a potentially large set of documents or URLs that the user must plow through.

One of the aims of the CHIL project was to provide information about what has been said during interactive seminars. Since the information must be located in speech data, the QA systems had to be able to deal with transcripts (manual or automatic) of spontaneous speech. This is a departure from much of the QA research carried out by natural-language groups, who have typically developed techniques for written texts that are assumed to have a correct syntactic and semantic structure. The structure of spoken language is different from that of written language, and some of the anchor points used in processing such as punctuation must be inferred and are therefore error-prone. Other spoken-language phenomena include disfluencies, repetitions, restarts, and corrections. If automatic processing is used to create the speech transcripts, an additional challenge is dealing with the recognition errors. The response can be a short string, as in text-based QA, or an audio segment containing the response.

This chapter summarizes the CHIL efforts devoted to QA for spoken language carried out at UPC and at CNRS-LIMSI. Research at UPC was directed at adapting a QA system developed for written texts to manually and automatically create speech transcripts, whereas at LIMSI an interactive oral QA system developed for the French language was adapted to the English language. CHIL organized the pilot track on question answering in speech transcripts (QAST), as part of CLEF 2007, in order to evaluate and compare QA technology on both manually and automatically produced transcripts of spontaneous speech.

8.1 Question Answering

Two main paradigms are used to search for information: document retrieval and precise information retrieval. In the first approach, documents matching a user query are returned. The match is often based on some keywords that were extracted from a query, and the underlying assumption is that the documents best matching the topic of the query provide a data pool from which the user might find information that suits their need. This need can be very specific (e.g., "Who is presiding the Senate?"), or it can be topic-oriented (e.g., "I'd like information about the Senate"). The user is left to filter through the returned documents to find the desired information, which is quite appropriate for general topic-oriented questions, and less well-adapted to more specific queries.

The second approach to search, which is better suited to the specific queries, is embodied by so-called question answering systems, which return the most probable answer given a specific question (e.g., the answer to "Who won the 2005 Tour de France?" is "Lance Armstrong".).

In the QA and information retrieval domains, progress has been assessed via evaluation campaigns [1, 10, 18, 13, 5]. In the question answering evaluations, the systems handle independent questions and should provide one answer to each question, extracted from textual data, for both open or restricted domains.

Recently, there has been growing interest in extracting information from multimedia data such as meetings, lectures, etc. Spoken data are different from textual data in various ways. The grammatical structure of spontaneous speech is quite different from written discourse and includes various types of disfluencies. The lecture and interactive meeting data of interest to CHIL are particularly difficult due to run-on sentences and interruptions.

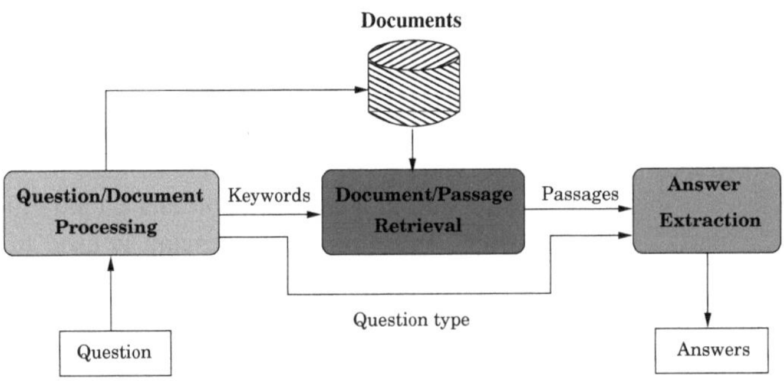

Fig. 8.1. Standard system architecture.

Figure 8.1 shows an example of a standard architecture for a QA system. Typical QA systems (mostly textual) are composed of question processing, document processing, document or passage retrieval, and answer extraction components [17, 9].

The answer extraction component is quite complex and often involves natural-language analysis, pattern matching, and sometimes even logical inference [8]. Most of the existing natural-language tools are not designed to handle spoken phenomena.

8.2 Question Answering: From Written to Spoken Language

This section describes the QA approaches implemented by UPC. Both systems follow the general architecture presented in Fig. 8.1. In the second system, the passage retrieval and answer extraction modules have been modified to better handle automatic transcriptions.

8.2.1 QA System for Manual Transcriptions (UPC1)

The architecture of the first UPC system is comprised of four modules [14]:

1. Document processing. The document collection is preprocessed with linguistic tools to tokenize and mark part-of-speech (POS) tags, lemmas, and named entities (NE) in the text. This information is used to build an index consisting of two parts. The first part, built on the lemmatized text, is built using the word lemmas and the recognized named entities. This text is indexed and used by the passage retrieval module to find relevant passages. The second part, built on the original text, is built using only the recognized named entities. This text is returned by the passage retrieval module when a query succeeds on the indexed lemmatized text.
2. Question processing. The main goal of this subsystem is to detect the expected answer type. The question type is determined among a set of 53 types by using a maximum entropy multiclass classifier. In addition, this subsystem selects the non-stop words from the question as keywords that are used by the passage retrieval subsystem.
3. Passage retrieval. The goal of this subsystem is to retrieve a set of passages from the document collection, which are relevant to the keywords selected from the question that is a two-step procedure. First, a document retrieval application (Lucene; http://jakarta.apache.org/lucene) is executed in order to filter out those documents in the collection that are not relevant to the question. Second, an iterative procedure based on query relaxation is performed to retrieve enough relevant passages. To address this issue, the algorithm adjusts the number of keywords used for passage retrieval and their proximity until the quality of the recovered information is satisfactory. The proposed algorithm makes limited use of syntax – only POS tagging is essential – which makes it very robust for speech transcriptions.
4. Answer extraction. The goal of this latter subsystem is to identify the exact answer to the question in the passages retrieved in the previous step. First, answer candidates are identified from the set of named entities occurring within the passages and being the same type as the answer type detected by the question processing step. Then these candidates are ranked using a scoring function based on a set of heuristics. The first candidate can be considered the most plausible answer.

As stated above, answer candidates are named entities occurring in the relevant passages. In order to recognize and classify the transcribed named entities, two different NERC models are combined, together with a rule set: a model learned from the CoNLL data (written text corpora) and a model learned from the QAST development data (see Section 8.4). A model learned from written texts is included in the combination since in previous experiments it was shown to improve the results [14]. In order to learn the QAST model, UPC used the same learning strategy that was used to train the CoNLL model: the same feature set and the same perceptron-based learning algorithm [14]. Previously, the named entities occurring in the QAST development corpus were manually annotated with their types. The learned models and the rule set were combined by applying them in the following order: (1) the rule set; (2) the QAST model; (3) the CoNLL model. In case of ambiguity in the decisions of the individual steps, the first decision is taken. The reasons for this specific order are, on the one hand, that rules tend to achieve better precision than learned models and, on the other hand, that the QAST model is specific for the type of documents being dealt with (be i.e.; manual transcripts of CHIL seminars).

8.2.2 QA System for Imperfect ASR Output (UPC2)

State-of-the-art ASR technology produces transcripts of spontaneous speech events with some incorrectly recognized words. This fact may have a negative effect on the performance of the QA system developed for manual transcripts when faced with the automatic ones. There are two important potential problems due to ASR errors. First, the occurrences of keywords in the question (found by the question processing step) are used both for detecting relevant documents and relevant passages, and for extracting the answer. Second, proper nouns are the most frequent kind of named entities found in documents in answer to factual questions (e.g., persons, organizations, locations, systems, methods, etc). Any ASR errors on proper nouns and keywords will have an impact on QA performance, where other errors may not.

In order to deal with QA in automatic transcripts of spontaneous speech, UPC has developed a preliminary QA system by changing both the passage retrieval and the answer extraction modules. The main differences between these new subsystems and those used to deal with manual transcripts are the strategy to search keywords and the NERC approach.

The UPC hypothesis for the searching of keywords in automatic transcripts is that an approximate match between the automatic transcripts and phonetically transcribed keywords can perform better than classical IR techniques for written text. Under such an assumption, each automatic transcript is converted to a phone sequence, as well as each keyword obtained by the question processing subsystem. Then a new IR engine named PHAST [2], inspired by the BLAST algorithm from genomics, is applied. Briefly, given the collection of phonetically transcribed documents D_i and the set of phonetically transcribed keywords k_i, PHAST is an iterative procedure. In the first step, when an approximate match is detected between subsequences of k_i and D_i, the algorithm finds the best extension of the match. If the extension is relevant enough, it is considered to be a match, and the relevance of the document is updated. Figure 8.2 shows an example of two possible approximate matches between the phonetic transcriptions of *first* and *linguistics*. The first match is not considered relevant enough to be taken as a match, whereas the second is. In the second step, each document in the collection is ranked based on its relevance computed in step 1, and the best ones are returned. The results of the first step of this algorithm are also taken into account when applying passage retrieval and answer extraction.

Global alignment
f - - - - ɪ s t - - -
l ɪ ŋ g w ɪ s t ɪ k s

Semi-local alignment
- - - f - ɪ s t - - -
l ɪ ŋ g w - ɪ s t ɪ - k s

Fig. 8.2. How global and semilocal affect the alignment of the phonetic transcription of the words "fist" and "linguistics".

Regarding the NERC problem when dealing with automatic transcripts, different models were trained for the two domains included in the QAST evaluation described in Section 8.4. For such tasks, the same perceptron-based algorithm as described above has been used. However, the set of features has been enriched with phonetic attributes related to the words and their prefixes and suffixes. Our hypothesis is that, in the worst case, an NE, $w_1, \ldots, w_k$, can be wrongly transcribed into a sequence of words, $v_1, \ldots, v_s$ $(l > s)$, so that the w_1 is the suffix of v_1 and the w_k is the prefix of v_s. In order to deal with these phonetic attributes, it is mandatory to model the similarities between phone sequences as features. For such a goal, clustering techniques were used wherein similar phone sequences were clustered together. The clusters corresponding to the best clustering were taken as features.

8.3 Fast Question Answering

This section presents the architecture of the two QA systems developed at LIMSI partially in the context of the CHIL project. The French version of these QA systems is part of an interactive QA platform called Ritel [16] that specifically addressed speed issues. The two systems share a common *noncontextual analysis* module and differ in the actual *question answering* back end.

8.3.1 Question and Document Processing

The same complete and multilevel analysis is carried out on both queries and documents. One motivation for this is that an incorrect analysis due to the lack of context or limitations of hand-coded rules is likely to happen with both data types, so using the same strategy for document and utterance analysis helps to reduce its negative impact. To do so, the query and the documents (which may come from different modalities – text, manual transcripts, automatic transcripts) are transformed into a common representation. This normalization process converts *raw* texts to a form where words and numbers are unambiguously delimited, punctuation is separated from words, and the text is split into sentence-like segments. Case and punctuation are reconstructed using a fully cased, punctuated four-gram language model [3] applied to a word graph covering all the possible variants (all possible punctuations permitted between words, all possible word cases). The language model was estimated on the House of Commons Daily Debates, the final edition of the European Parliament Proceedings, and various newspaper archives. The final result, with uppercase only on proper nouns and words clearly separated by white spaces, is passed to the analysis component.

The general objective of this analysis is to find the bits of information that may be of use for search and extraction, called *pertinent information chunks*. These can be of different categories: named entities, linguistic entities (e.g., verbs, prepositions), or specific entities (e.g., scores). All words that do not fall into such chunks are automatically grouped into chunks via a longest-match strategy. The full analysis comprises some 50 steps and takes roughly 4 ms on a typical user or document sentence. The analysis identifies 96 different types of entities. Figure 8.3 shows an example of the analysis on a query (top) and a transcription (bottom).

```
<_Qorg> which organization </_Qorg> <_action> provided </_action>
<_det> a </_det> <_NN> significant amount </_NN>
<_prep> of </_prep> <_NN> training data </_NN> <_punct> ? </_punct>
```

```
<_pro> it </_pro> <_verb> 's </_verb> <_adv> just </_adv>
<_prep_comp> sort of </_prep_comp> <_det> a </_det>
<_NN> very pale </_NN> <_color> blue </_color> <_conj> and </_conj>
<_det> a </_det> <_adj> light-up </_adj> <_color> yellow </_color>
<_punct> . </_punct>
```

Fig. 8.3. Example annotation of a query: *"Which organization provided a significant amount of training data?"* (top) and of a transcription: "It it's just sort of a very pale blue and a light-up yellow" (bottom).

8.3.2 QA System Based on Handcrafted Extraction Rules (LIMSI1)

The *question answering* system handles search in documents of any types (news articles, Web documents, transcribed broadcast news, etc.). For speed reasons, the documents are all available locally and preprocessed: They are first normalized, and then analyzed with the analysis module. The (type, values) pairs are then managed by a specialized indexer for quick search and retrieval. This bag-of-typed-words system [16] works in three steps:

1. Document query list creation: The entities found in the question are used to select a document query and an ordered list of back-off queries from a predefined handcrafted set.
2. Snippet retrieval: Each query, according to its rank, is submitted to the indexation server, stopping as soon as document snippets (sentence or small groups of consecutive sentences) are returned.
3. Answer extraction and selection: The previously extracted answer type is used to select matching entities in the returned snippets. The candidate answers are clustered according to their frequencies, with the most frequent answer winning, and where the distribution of the counts provides an idea of the system's confidence in the answer.

8.3.3 QA System Based on Automatically Built Search Descriptors (LIMSI2)

The LIMSI2 system was designed to solve two main problems with the first system: (1) The list of back-off queries will never cover all of the combinations of entities that may be found in the questions and (2) the answer selection uses only occurrence frequencies.

The first step of system 2 is to build a research descriptor (data descriptor record, DDR) that contains the important elements of the question, and the possible answer types with associated weights. Some elements are marked as *critical*, which makes them mandatory in future steps, while others are *secondary*. The element extraction and weighting is based on an empirical classification of the element types in importance levels. Answer types are predicted through rules based on combinations of elements of the question. Figure 8.4 shows a sample DDR.

```
question: in which company does Bart work as a project manager ?
ddr:
{ w=1, critical, pers, Bart},
{ w=1, critical, NN, project manager },
{ w=1, secondary, action, works },
answer_type = {
  { w=1.0, type=orgof },
  { w=1.0, type=organisation },
  { w=0.3, type=loc },
  { w=0.1, type=acronym },
  { w=0.1, type=np },
}
```

Fig. 8.4. Example DDR constructed for a question: Each element contains a weight `w`, its importance for future steps, and the pair (type,value). Each possible answer type contains a weight `w` and the type of the answer.

Documents are selected using this DDR. Each element of the document is scored with the geometric mean of the number of occurrences of all the DDR elements that appear in it, and sorted by score, keeping the *n*-best. Snippets are extracted from the document using fixed-size windows and scored using the geometrical mean of the number of occurrences of all the DDR elements that appear in the snippet, smoothed by the document score. In each snippet, all the elements whose type is one of the predicted possible answer types are candidate answers. A score $S(A)$ is associated to each candidate answer A:

$$S(A) = \frac{[w(A)\sum_E \max_{e=E} \frac{w(E)}{(1+d(e,A))^\alpha}]^{1-\gamma} \times S_{snip}^\gamma}{C_d(A)^\beta C_s(A)^\delta}$$

where $d(e,A)$ is the distance to each element e of the snippet, instantiating a search element E of the DDR; C_s and C_d are the number of occurrences of A in the extracted snippets and in the entire document collection, respectively; S_{snip} is the extracted snippet score; $w(A)$ is the weight of the answer type; $w(E)$ the weight of the element E in the DDR; and α, β, γ, and δ are tuning parameters estimated by systematic trials on the development data ($\alpha, \beta, \gamma \in [0,1]$ and $\delta \in [-1,1]$).

An intuitive explanation of the formula is that each element of the DDR increases the score of the candidate ($\sum_E$) proportionally to its weight ($w(E)$) and inversely proportionally to its distance of the candidate($d(e,A)$). If multiple instances of the element are found in the snippet, only the best one is kept ($\max_{e=E}$). The score is then smoothed with the snippet score (S_{snip}) and compensated in part with the candidate frequency in all the documents (C_d) and in the snippets (C_s). The scores for identical (type,value) pairs are added together and give the final scoring for all possible candidate answers.

8.4 The QAST 2007 Evaluation

The are no existing evaluation frameworks that convert factual QA systems for speech transcripts. In order to evaluate the QA approaches developed in CHIL and to compare them with those developed by external researchers, UPC, LIMSI, and ELDA organized an evaluation framework named QAST (question answering in speech transcripts; `http://www.lsi.upc.edu/~qast`) as a pilot track in the Cross Language Evaluation Forum 2007 (CLEF 2007). The specific tasks, metrics, and results are described in the following sections.

8.4.1 Tasks

The design of the QAST tasks takes into account two different viewpoints. First, automatic transcripts of speech data contain word errors that can lead to wrongly answered questions or unanswered questions. In order to measure the loss of the QA systems due to ASR technology, a comparative evaluation was introduced for both manual and automatic transcripts. Second, the type of speech data could be from a single speaker (a monologue) or from multiple speakers in an interactive situation. With the aim of comparing the performance of QA systems for both situations, two scenarios were introduced: lectures and meetings in English. From the combination of these two viewpoints, QAST covered the following four tasks:

T1: Question answering in manual transcripts of lectures,
T2: Question answering in automatic transcripts of lectures,
T3: Question answering in manual transcripts of meetings,
T4: Question answering in automatic transcripts of meetings.

The data used consisted of two different resources. For the lecture scenario, the CHIL corpus was used, with a total of 25 hours (around 1 hour per lecture) both manually transcribed and automatically transcribed by the LIMSI ASR system [11]. A set of lattices and confidences for each lecture was also provided for those QA systems that chose to explore alternate information than can be found in the 1-best ASR output. For the meeting scenario, the AMI corpus was used, with a set of 100 hours (168 meetings from the whole set of 170), both manually transcribed and automatically transcribed using an ASR system from the University of Edinburgh [7]. The domain of the meetings was *design of television remote controls*.

These resources were divided into disjoint development and evaluation sets with associated questions written in English. The development set consisted of 10 lectures with 50 questions and 50 meetings with 50 questions. The evaluation set included 15 lectures with 100 questions and 118 meetings with 100 questions. Since QAST 2007 dealt with factual questions, the possible types of answers were *person*, *location*, *organization*, *language*, *system/method*, *measure*, *time*, *color*, *shape*, and *material*.

8.4.2 Metrics

Two metrics were used to evaluate the performance of the systems in QAST: the *mean reciprocal rank* (MRR), which measures how well the correct answer is ranked on average in the list of five possible answers, and the *accuracy* (ACC), which is the fraction of the correct answers ranked in the first position in the list of five possible answers. Both metrics depend on what is considered a correct answer. A correct answer was defined in QAST as the token sequence comprised of the smallest number of tokens required to contain the correct answer in the audio stream. For instance, consider the following manual transcript of an audio recording, and its corresponding automatic transcript:

> *uhm this is joint work between the University of Karlsruhe and Carnegie Mellon, so also here in these files you find uh my colleagues and uh Tanja Schultz.*

> *{breath} {fw} and this is, joint work between University of Karlsruhe and coming around so {fw} all sessions, once you find {fw} like only stringent custom film canals communicates on on {fw} tongue initials.*

For the question, " Which organization has worked with the University of Karlsruhe on the meeting transcription system?" the answer found in the manual transcription is "Carnegie Mellon", whereas in the automatic transcription, it is "coming around".

8.4.3 Results

Five groups participated in the QAST evaluation: The Center for Language Technology (CLT, Australia), the German Research Center for Artificial Intelligence (DKFI), the Tokyo Institute of Technology (Tokyo, Japan), and the CHIL partners LIMSI (France) and UPC (Spain). Each group was permitted to submit two different runs per task. Table 8.1 summarizes the accuracy and mean reciprocal rank (MRR) obtained on the four tasks.

Range of Results	T1	T2	T3	T4
Best acc.	0.51	0.36	0.25	0.21
Best MRR	0.53	0.37	0.31	0.22
Worst acc.	0.05	0.02	0.16	0.06
Worst MRR	0.09	0.05	0.22	0.10

Table 8.1. Best and worst accuracy and MRR on the four tasks: T1, T2, T3, and T4. The results do not all come from the same system but summarize the best system results for the various conditions.

The best results for the lecture scenario were achieved by the UPC1 system, with accuracy values of 0.51 and 0.36 in tasks T1 and T2, respectively. Although both of the UPC systems achieved better results than the other ones in the tasks they participated in, only UPC1 (explicitly designed to deal with manual transcripts) was significantly better than the others in both tasks. UPC1 performed better than UPC2 in task T2 (the one in which UPC2 was tested) mainly because the use of phonetic similarity functions in the selection of answer candidates introduced too much noise. The LIMSI2 system was the second-best approach in T1 and the third one in T2 (LIMSI1 achieved the same results in this task), with accuracy values of 0.39 and 0.21, respectively. The remaining systems performed significantly worse, with maximum accuracy values of 0.16 and 0.09 in tasks T1 and T2, respectively. An important drawback of the LIMSI2 system is the loss in accuracy, which can be shown, for instance, by the difference between the MRR value (0.45) and the accuracy value (0.39) obtained when dealing with the manual transcripts. This means that the system was not able to accurately rank the extracted right answers in the first place. With the exception of the UPC1 system, the systems showed similar behavior in both tasks.

The results for the scenario of meetings show that this was a significantly harder problem. For task T3, the best result in accuracy was 0.25, achieved by both the LIMSI1 and LIMSI2 systems. However, after submitting their results, the UPC group located a bug in the script to generate the output format required by QAST. After correcting this bug, a new run was evaluated unofficially for task T3 and the UPC1 system obtained the same accuracy as the LIMSI systems. Considering the official results, the LIMSI2 system's MRR was nine points higher than UPC1's, which reinforces the hypothesis that LIMSI2 requires a more effective ranking function. In contrast, the lowest MRR value was obtained by the UPC1 system, which was mainly due to the loss in the NERC and the candidate selection steps.

For task T4, the best result was achieved by the UPC1 system (0.21 in accuracy); however, there were no significant differences between the UPC1 and LIMSI systems, which had accuracy values of 0.18 (LIMSI1) and 0.17 (LIMSI2). However, the UPC2 system was significantly worse than UPC1 (0.13 in accuracy), reinforcing the hypothesis that noise is introduced when using phonetic similarity functions.

8.5 Conclusions and Discussion

This chapter has presented an overview of the progress achieved in question answering in spoken data. A brief description of the work carried out at UPC and LIMSI was presented. An evaluation, question answering on speech transcripts (QAST), was organized by UPC, LIMSI, and ELDA as part of the CLEF evaluation and the results of the 2007 evaluation have been presented.

QAST 2007 demonstrated that the technology developed at UPC and LIMSI is more appropriate for dealing with question answering in speech transcripts than that of the external participants. The difference in the accuracy of systems applied to the manual and automatic transcripts, that is, between T1 and T2 (from 0.22 to 0.16 in average) and T3 and T4 (from 0.21 to 0.13), drops by over 36% when applied to automatic transcriptions. These observations (and others) led to several changes in the 2008 evaluation campaign [12]. One contrast is using multiple recognizer hypotheses with different word error rates (WER) with the objective of assessing the dependency of QA performance on WER. Another extension is to evaluate different subtasks (information retrieval for QA and answer extraction), with an objective of studying which part of a QA system is the most sensitive to word error rate. The evaluation has also been extended to include two additional languages (French and Spanish) and data types (Broadcast News and European Parliament data). This is a big change in the type of spoken data in terms of both content and speaking style. The Broadcast News and European Parliament discourses are less spontaneous than the lecture and meeting speech, as they are typically prepared in advance and are closer in structure to written texts. While meetings and lectures are representative of *spontaneous speech*, Broadcast News and European Parliament sessions are usually referred to as *prepared speech*. Although they typically have few interruptions and turn-taking problems when compared to meeting data, many of the characteristics of spoken language are present (hesitations, breath noises, speech errors, false starts, mispronunciations, and corrections) are still present. Two reasons for including the additional types of data were to be closer to the type of textual data used to assess written QA, and to benefit from the availability of multiple speech recognizers that have been developed for these languages and tasks in the context of European or national projects [6, 4, 15].

References

1. C. Ayache, B. Grau, and A. Vilnat. Evaluation of question-answering systems : The French EQueR-EVALDA Evaluation Campaign. In *LREC'06*, Genoa, Italy, 2006.
2. P. Comas and J. Turmo. Spoken document retrieval using phonetic similarity. Technical report, LSI-UPC, 2007.
3. D. Déchelotte, H. Schwenk, G. Adda, and J.-L. Gauvain. Improved machine translation of speech-to-text outputs. In *Proceedings of Interspeech'07*, Antwerp. Belgium, Aug. 2007.
4. S. Galliano, E. Geoffrois, G. Gravier, J. Bonastre, D. Mostefa, and K. Choukri. Corpus description of the ESTER Evaluation Campaign for the Rich Transcription of French Broadcast News. In *Proceedings of LREC'06*, Genoa, 2006.
5. D. Giampiccolo, P. Forner, A. Peñas, C. Ayache, D. Cristea, V. Jijkoun, P. Osenova, P. Rocha, B. Sacaleanu, and R. Sutcliffe. Overview of the CLEF 2007 Multilingual Question Answering Track. In *Working Notes for the CLEF 2007 Workshop*, Budapest, Hungary, Sept. 2007.
6. G. Gravier, J. Bonastre, S. Galliano, E. Geoffrois, K. McTait, , and K. Choukri. The ESTER Evaluation Campaign of Rich Transcription of French Broadcast News. In *Proceedings of LREC'04*, Lisbon, 2004.
7. T. Hain, L. Burget, J. Dines, G. Garau, M. Karafiat, M. Lincoln, J. Vepa, and V. Wan. The AMI system for the transcription of meetings. In *Proceedings of ICASSP'07*, Hawaii, 2007.
8. S. Harabagiu and A. Hickl. Methods for using textual entailment in open-domain question-answering. In *COLING'06*, Sydney, Australia, July 2006.
9. S. Harabagiu and D. Moldovan. *The Oxford Handbook of Computational Linguistics*, chapter Question-Answering. Oxford University Press, 2003.
10. N. Kando. Overview of the Sixth NTCIR Workshop. In *Proceedings of the 6th NTCIR Workshop Meeting*, Tokyo, Japan, 2006.
11. L. Lamel, G. Adda, and J.-L. G. E. Bilinski. Transcribing lectures and seminars. In *Proceedings of InterSpeech'05*, Sept. 2005.
12. L. Lamel, S. Rosset, C. Ayache, D. Mostefa, J. Turmo, and P. Comas. Question answering on speech transcriptions: the QAST evaluation in CLEF. In *LREC'08*, Marrakech, Morocco, 2008.
13. G. M. D. Nunzio, N. Ferro, T. Mandl, and C. Peters. CLEF 2007: Ad hoc track overview. In *Working Notes for the CLEF 2007 Workshop*, Budapest, Hungary, Sept. 2007.
14. M. Surdeanu, D. Dominguez-Sal, and P. Comas. Performance analysis of a factoid question answering system for spontaneous speech transcriptions. In *Proceedings of Interspeech'07*, Sept. 2006.
15. TC-Star. `http://www.tc-star.org`, 2004-2008.
16. B. van Schooten, S. Rosset, O. Galibert, A. Max, R. op den Akker, and G. Illouz. Handling speech input in the Ritel QA dialogue system. In *Proceedings of Interspeech'07*, Antwerp. Belgium, Aug. 2007.
17. E. M. Voorhees and L. P. Buckland, editors. *The Fifteenth Text Retrieval Conference Proceedings (TREC 2006)*. 2006.
18. E. M. Voorhees and L. P. Buckland. The Sixteenth Text Retrieval Conference Proceedings (TREC 2007). In Voorhees and Buckland, editor, *NIST Special Publication 500-274*, 2007.

9

Extracting Interaction Cues: Focus of Attention, Body Pose, and Gestures

Oswald Lanz[1], Roberto Brunelli[1], Paul Chippendale[1], Michael Voit[2], Rainer Stiefelhagen[2]

[1] Foundation Bruno Kessler, irst, Trento, Italy
[2] Universität Karlsruhe, Interactive Systems Labs, Fakultät für Informatik, Karlsruhe, Germany

Studies in social psychology [7] have experimentally validated the common feeling that nonverbal behavior, including, but not limited to, gaze and facial expressions, is extremely significant in human interactions. Proxemics [4] describes the social aspects of distance between interacting individuals. This distance is an indicator of the interactions that occur and provides information valuable to understanding human relationships. Another important perceptual variable in the analysis of group behavior is given by the *focus of attention*, which indicates the object or person one is *attending to*. It often happens that people turn toward their focus of attention, thereby physically expressing their attention by means of posture, gaze, and/or head orientation. Even if physically observable attention does not necessarily coincide with psychological attention, it is highly correlated with it. Tracking people's location supports proxemics studies, while reliable computation of head orientation helps in determining people's focus of attention, a key variable in the analysis of group behavior.

Another form of nonverbal communication develops along body language, including explicit gestures, such as raising a hand to catch attention, or nodding the head to manifest agreement, and body postures. Their fine-grained nature and the large variety with which different individuals execute them pose major challenges to the development of technologies for their automatized detection. Also, in dynamic gestures, it is often difficult to discern when exactly a gesture begins and ends. However, they play a crucial role in the study of human interaction, and may even be exploited as a natural way to catch the attention of the system just before a user request, much in the same way one would do with a human helper.

This chapter discusses approaches for analyzing nonverbal communication in a meeting scenario. These approaches include a neural network-based approach to estimate focus of attention [6] (which correctly determined who is looking at whom around 70% of the time on a CHIL meeting recording), a particle filter-based head and body pose tracker to estimate focus of attention in highly dynamic scenes [2], and algorithms for detecting gestures at various spatiotemporal scales [2].

9.1 From Head Pose to Focus of Attention

In static scenes, it is possible to determine an observer's focus of attention through a statistical analysis of previously collected head pose observations, without requiring explicit knowledge on their spatial location. This approach is supported by the fact that if both the observer and the object of interest do not change their location in time, the head poses observed when pointing the attention toward such an object are also persistent over time. Foci of attention, which are defined as frequently watched targets, show up as peaks in the collected head pose data and can therefore be detected using statistical tools such as data clustering.

To capture the statistically relevant spots, we describe data with a Gaussian mixture model (GMM). Each component of the GMM aims at identifying an attention target and models the distribution of the observer's pose over time when watching it. Such a distribution is well described by a Gaussian if one considers that (1) head pose observations are noisy (they are estimated from the video recordings using the methods presented in Chapter 5) and (2) head orientation and gaze direction are only roughly aligned in the human subject. By varying the number of components in the model, it is possible to control the resolution at which the visual field is partitioned. Such a number can be either supplied (e.g., the number of participants in a meeting, plus one for the whiteboard) or learned from the data.

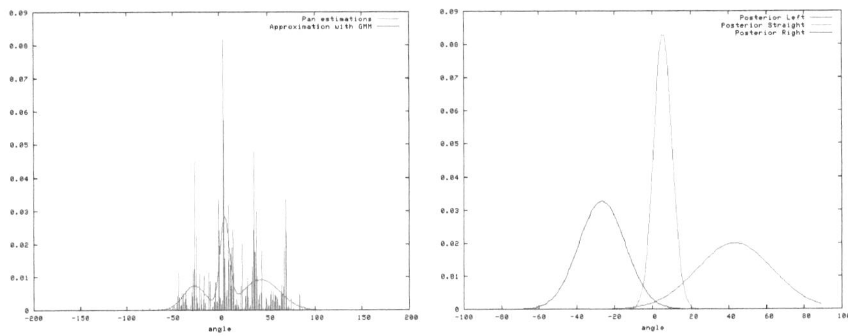

Fig. 9.1. Estimated horizontal head orientation and learned GMM for one participant extracted from the recordings of a CHIL meeting (20 min).

The best-fitting mixture model can be extracted from a collection of observed head pose data by applying the expectation-maximization (EM) algorithm. Figure 9.1 shows the distribution of estimated horizontal head orientations for one participant during a CHIL meeting recording, and the corresponding EM-learned Gaussian mixture model. The high-probability regions of the obtained distribution identify three compact and nonoverlapping classes, with each class accurately centered at the visual location of one of the other meeting participants. The observer's visual attention can then be determined at every time instant by assigning the specific class label

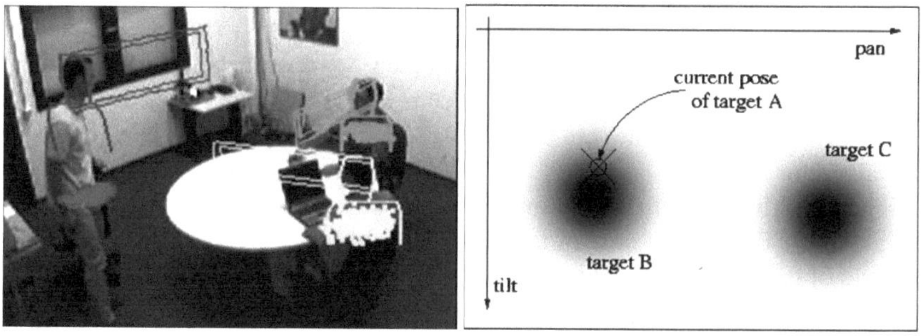

(a) (b)

Fig. 9.2. (a) Real-time location and pose tracker output, and (b) attention GMM built on-the-fly for a meeting attendee: Target A (the standing person) is drawing attention to target B (sitting in the upper right corner of the image).

the measured head pose falls within, or an "unknown" label if none of the classes applies.

9.2 Determining Focus of Attention in Dynamic Environments

The above-described approach cannot be applied to dynamic environments, where attention clusters move with the targets and the observer and are therefore mixed in the pose statistics. To overcome this problem, location information provided by the person tracker can be exploited to extract a number of shorter, nonoverlapping segments within which all the participants remain roughly at the same location. Under the assumption that only tracked targets move, each observer's focus may be found using the approach described in the previous section.

In some situations of interest, however, observers and/or targets may move frequently. Extracted segments may therefore be too short to collect a sufficient number of observations to train a specific GMM classifier for that time interval. At the same time, static segments may represent only a fraction of the whole recording; thus, information relevant to the study of actors' behavior is likely to be overlooked.

Most of the information required to detect people's focus of attention in a meeting, however, can be derived from the output of a multiperson head location and pose tracker, such as the particle filter presented in Section 5.2.3. Figure 9.2(a) shows the real-time output of the tracker during a meeting: It provides the spatial location of the participants, together with their head pose expressed as pan and tilt angles relative to a global reference frame. A bivariate attention GMM can then be built on-the-fly for each observer from these data, as follows. At a given time instant, a 3D ray is computed that connects a target with the observer's current location. The corresponding GMM component is centered at the pan and tilt angles computed from this ray. It has a diagonal covariance matrix with predefined, equal variance σ^2 in both directions. As discussed, the variability in the head pose data comes from two independent sources: the inaccuracy of the head pose estimator, say it has variance σ_P^2, and the alignment error between head pose and gaze, say with variance σ_G^2. The resulting error covariance is then computed as the sum of the two individual contributions $\sigma^2 = \sigma_P^2 + \sigma_G^2$. Typical values are $\sigma_P = 25°$ and $\sigma_G = 15°$; thus, a value of about 29° is obtained for σ.

Figure 9.2(b) shows the attention GMM computed from the pose tracker output in Fig. 9.2(a). The current pose of target A (denoted by a cross) maps into the attention field of target B, thereby inferring *"A attends to B."*. Other potential foci such as a whiteboard or a computer screen can be allocated by the user through the addition of a GMM component per object at their projected 3D positions. An important advantage of this approach is that it delivers online information about people's visual attention, information that may be required to support human-computer interaction.

9.3 Tracking Body Pose

Another cue that is relevant to the study of human interaction is body posture. To obtain such information automatically, the perceptual capabilities of the visual multiperson body and head pose tracker presented in Chapter 3 and Section 5.2.3 can be extended to provide a more detailed description of the targets' body configurations. To this purpose, the state space is extended to include a parameterization of body posture, thereby assembling the target's 2D position and, optionally, its velocity on the floor, horizontal torso orientation, torso inclination, head pan and tilt, plus a binary label for sitting and standing.

Both the dynamical model and likelihood model of the filter need to be redesigned to account for the larger state space. For the temporal propagation, the autoregressive model presented in Section 5.2.3 is applied to obtain the predicted body location and head pose. Torso orientation and inclination are derived from previous values by adding Gaussian noise with a variance proportional to the time elapsed. The probability of the transition from sitting to standing pose changes with the target's velocity module $|x_v|$ (in m/s) according to

$$\mathrm{Pr}_{\mathrm{jump}} = 0.2 - \min(0.18, 0.4 \cdot |x_v|).$$

The pose appearance likelihood extends the original 3D shape and rendering model (see Section 5.2.3), with the "backbone" now constructed using four 3D anchor points: the center of feet, hip, and shoulder, and the top of the head. These points are computed for a specific target pose as in Section 5.2.3, this time taking into account body orientation and inclination when determining the 3D center of the shoulder. An exception is made for a sitting pose: The height of the center of the hip is sampled uniformly from a window (20 cm large) centered at the typical, predefined, chair height. The backbone is assembled from the segments joining the projections of the anchor points onto the image plane, and the body profile is drawn around it as in Section 5.2.3. Figure 9.3 shows the described shape-rendering procedure. The appearance likelihood is computed in the usual manner, by matching

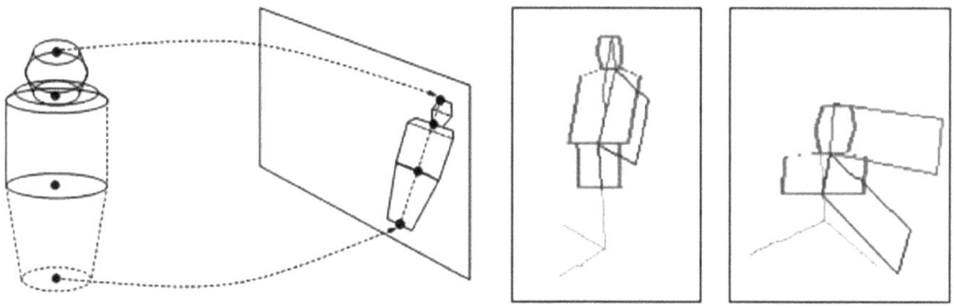

Fig. 9.3. 3D shape model of a standing person and its rendered silhouette, conveying perspective distortion and scaling effects. To the right: rendering of a standing pose with an inclined upper body and misaligned head orientation, and a sitting pose.

extracted body-part color histograms with the preacquired model histograms using Bhattacharrya-coefficient-based distance. For a sitting pose, the contributions of the leg trunks are ignored.

9.4 Pointing Gesture and Hand-Raising Detection

To locate a person's hands in 3D and subsequently detect hand gestures of interest, the output of the visual multibody tracker is used to identify candidate skin blobs extracted from the input image using a statistical skin model filter [1, 8][1]. Candidate blobs must lie within a predefined distance from the image projection of shoulder joints, which are derived from backbone anchor points and torso orientation and inclination based on the pose tracker output. The centroid of each appropriately sized blob is reprojected into 3D space in the form of a ray originating from the camera's origin. This is performed for all skin regions pertaining to the same target in each camera. The minimum distance between rays from other cameras is computed using an SVD algorithm. Provided that the closest intersection distance is small (say <30 cm), and that the hypothesized 3D position of the hand lies inside the spheres, a hypothetical 3D hand position is created. All permutations of skin regions are tested in order to find the best hypotheses, subsequently labeled as the left and right hands. Left hands are assumed to be the leftmost with respect to the torso's orientation, and likewise rightmost for the right. The same process is repeated for all targets within the room.

To decide whether a gesture relates to pointing or raising, the spatial stability of the 3D hand is examined. To compensate for noise, an acceptable variance of 30 cm in diameter (roughly equal to two hand widths) is empirically selected. When the hand centroid does not move more than this amount during a fixed time interval, and its mean Euclidean distance from the shoulder joint measures less than 75% of an

[1] The skin model is created in YCbCr color space through the prior observation and manual segmentation of hundreds of skin regions from our camera video streams.

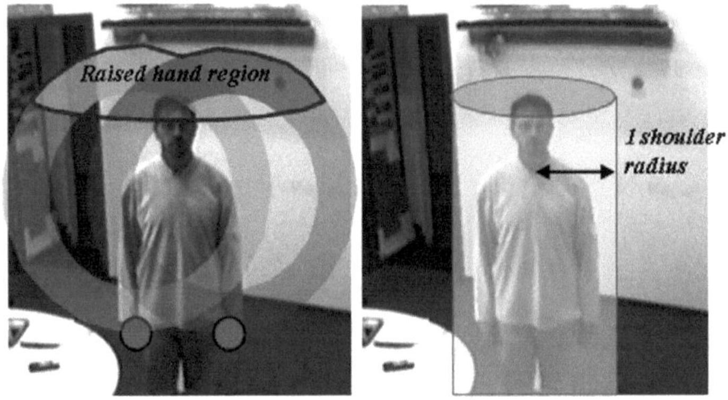

Fig. 9.4. (a) Pointing detection; (b) fidgeting volume.

arm's length, and it does not lie inside the 3D cylinder seen in Fig. 9.4(b), a pointing gesture is logged. If the height of the hand above the floor plane is greater than that of the head, then the gesture is deemed to be a raised hand; if not, an estimate of the pointing gesture vector is calculated. The direction of a pointing gesture is derived by joining the head centroid of the person with the hand centroid and then projecting a vector away from the body.

9.5 Detection of Fine-Scale Gestures

Fidgeting, which is defined as "a condition of restlessness as manifested by nervous movements", can reveal important clues about an individual's emotional state and activity [5]. Using visual means, fidgeting signatures can be detected by employing techniques such as optical flow or motion history images (MHIs) [3].

To pinpoint fidgeting, the search efforts are concentrated to the same imaginary cylindrical volume seen in Fig. 9.4(b). The estimated 3D positions of the head and hands are projected onto the images, and temporally unstable skin pixels are sought to construct an MHI representing a temporal record of skin-repetitive motion. The more often a pixel within the hand or head regions of interest changes from being skin to nonskin, the brighter the corresponding MHI pixel becomes. As well as observing periodic motion, in order to detect such actions as finger fidgeting, the spatial persistence of the hand is also taken into account. The way in which a quantitative measure of fidgeting is defined and normalized from the MHI is detailed in [1]. In essence, it is evaluated as being a ratio of fluctuating pixels to stable pixels in a given skin blob.

To detect whether a head is nodding or shaking, it is sufficient to correlate head-fidgeting events with the head orientation information supplied by the pose tracker. Although the fidgeting detector can tell us that a repetitive head movement is taking place, it is not always straightforward to determine simply from the shape of the fidgeting event whether the head is panning or tilting. To provide the missing information, the pan and tilt values are taken from the head tracker during the previous few frames and their trend used to make a joint decision.

References

1. P. Chippendale. Towards automatic body language annotation. In *7th IEEE International Conference on Automatic Face and Gesture Recognition, FG06*, pages 487–492, Southampton, UK, Apr. 2006.
2. P. Chippendale and O. Lanz. Optimised meeting recording and annotation using real-time video analysis. In *5th Joint Workshop on Machine Learning and Multimodal Interaction, MLMI08*, Utrecht, The Netherlands, Sept. 2008.
3. J. W. Davis. Hierarchical motion history images for recognizing human motion. In *IEEE Workshop on Detection and Recognition of Events in Video*, page 39, 2001.
4. E. T. Hall. *The Hidden Dimension: Man's Use of Space in Public and Private.* Bodley Head, London, 1969.
5. B. L. M. Zancanaro and F. Pianesi. Automatic detection of group functional roles in face to face interactions. In *Proceedings of the International Conference on Multimodal Interfaces, ICMI06*, pages 28–34, 2006.
6. M.Voit and R.Stiefelhagen. Tracking head pose and focus of attention with multiple far-field cameras. In *International Conference on Multimodal Interfaces - ICMI 2006*, Banff, Canada, Nov. 2006.
7. K. Parker. Speaking turns in small group interaction: A context-sensitive event sequence model. *Journal of Personality and Social Psychology*, 54(6), 1988.
8. S. Phung, A. Bouzerdoum, and D. Chai. A novel skin colour model in ycbcr colour space and its application to human face detection. In *International Conference on Image Processing 2002*, volume 1, pages 289–292, Sept. 2002.

10

Emotion Recognition

Daniel Neiberg[1], Kjell Elenius[1], Susanne Burger[2]

[1] Kungl Tekniska Högskolan, Centre for Speech Technology, Stockholm, Sweden
[2] Carnegie Mellon University, InterACT, School of Computer Science, Pittsburgh, PA, USA

Studies of expressive speech have shown that discrete emotions such as anger, fear, joy, and sadness can be accurately communicated, also cross-culturally, and that each emotion is associated with reasonably specific acoustic characteristics [8]. However, most previous research has been conducted on acted emotions. These certainly have something in common with naturally occurring emotions but may also be more intense and prototypical than authentic, everyday expressions [6, 13]. Authentic emotions are, on the other hand, often a combination of different affective states and occur rather infrequently in everyday life. They are, moreover, often restricted to only a few emotions, such as joy or frustration [3, 5, 7]. In the CHIL project, we are interested in acoustic emotion recognition in two of the given scenarios: the Socially Supportive Workspaces and the Connector agent. In the first one, we want to monitor people attending a lecture to give the speaker feedback on the attentive states of the audience; are they positive and laughing, ignorant and bored, or negative and irritated? In the Connector scenario, in which somebody tries to reach a person on the phone via the Connector agent, it is of interest to know whether the caller is starting to show some frustration. Since there was no relevant CHIL material recorded and available for our research, we decided to use two other corpora that we considered to be rather close to what we needed. They are recorded in similar circumstances and environments as the CHIL scenarios. For the Socially Supportive Workspaces, we used a corpus of small-group interaction collected at CMU and known as the ISL Meeting Corpus [2]. For the Connector agent, we used a database from the Swedish telephone service company Voice Provider that contains recordings of people interacting with automatic voice response centers [12]. The emotional context of the ISL Corpus is human-human interaction and the emotions conveyed are mainly positive, often associated with laughter. The emotional context of the Voice Provider Corpus is human-machine interaction. The emotions are rather rare and mostly negative due to frustration with the performance of the automatic voice response system. Thus, both databases contain authentic and not acted emotions in settings similar to the CHIL scenarios we want to explore. However, their diverse contexts make the task for the emotion recognizer quite different. In the following, we will first describe work within the Socially Supportive Workspaces scenario, and then we will discuss the work with the Connector agent scenario.

10.1 Emotion Recognition for the Socially Supportive Workspaces Scenario

In Section 10.1.1, the ISL Meeting Corpus is presented. In Section 10.1.2, a number of emotion recognition systems are described, while Section 10.1.3 presents the results of our experiments.

10.1.1 The ISL Meeting Corpus

The ISL Meeting Corpus [2] consists of 18 meetings, conducted in English, with an average number of 5.1 participants per meeting and an average duration of 35 minutes. The audio is of multichannel close-talk microphone 16 bit with 16-kHz quality, recorded with lapel microphones. The corpus contains a large number of participants, including nonnative speakers, and care was taken to minimize the presence of the same speakers in the training and test portions of the data. Annotation of emotion was performed by identifying a closed set of *emotionally relevant behaviors* [9]; three annotators were asked to apply the resulting scheme to the entire corpus at the utterance level. Subsequently, each annotator was asked to manually classify every utterance as expressing one of NEGATIVE, NEUTRAL, and POSITIVE speaker valence, potentially relying on their previous emotionally relevant behavior assignment. Interlabeler agreement on three-way valence, between the two annotators who agreed the most, was $\kappa = 0.68$. A single valence assignment per utterance was produced using majority voting; the experiments presented involve inference of the correct emotional valence class for those 12,758 utterances for which a majority label existed (a majority did not exist for 0.8% of the utterances in the complete development and evaluation data). For our purposes, this corpus was split into a development set and an evaluation set, as shown in Table 10.1. Each set contained roughly the same distribution of emotion classes. Finally, for tuning of parameters, the development set was split into two subsets that were used for twofold cross-validation.

10.1.2 Emotion Recognizers for the ISL Corpus

This section describes a set of state-of-the-art emotion recognition systems. The first five systems were designed in order to evaluate the performance of different feature sets separately, using no knowledge of the prior distribution of classes. This facilitates comparison with other work. In a real application, it is advisable to combine systems to improve performance. It would be reasonable to add a cost function to weight the cost of different errors. However, in our case no cost function is given, why we decided to use the maximum accuracy as our design criterion for the combined systems. This will guarantee an equal or better performance than an all-neutral

Development Set	Utterances	Percentage
NEUTRAL	6544	80.9
NEGATIVE	293	3.6
POSITIVE	1255	15.5
Total	8092	

Evaluation Set	Utterances	Percentage
NEUTRAL	3671	78.7
NEGATIVE	151	3.2
POSITIVE	844	18.1
Total	4666	

Table 10.1. The distribution of emotion categories for utterances as decided by majority voting from listeners for each set in the ISL Meeting Corpus.

classifier. It will also be conservative in the sense that a non-neutral emotion is biased to be classified as neutral rather than a neutral one being classified as non-neutral. A typical application using these detectors should switch from its default state and take an appropriate action when a non-neutral emotion is detected. However, if a non-neutral emotion is falsely detected, this response is probably more inconvenient than a neutral response to a non-neutral emotion. This is the motivation for our choice of a conservative detector. The combined systems are constructed using multiple least mean square (LMS) regression, where the log-likelihoods from each individual classifier serve as input and training labels as output. Details regarding these systems are found in [12].

System 1: GMM/MFCC. thirty-nine standard MFCCs, with 24 Mel filters in the 300-8000-Hz region calculated every 10 ms with a 25-ms window. We also used RASTA processing. The features were modeled by a GMM with 512 Gaussians for each emotion.

System 2: GMM/MFCC-LOW. A Mel-filter bank in the 20-300-Hz region was used in order to model prosody. These features were computed in the same way as the MFCCs in System 1 except that a 64-ms window was computed every 25 ms. These features were modeled by GMMs with 512 Gaussians.

System 3: GMM/Pitch. The algorithm for pitch tracking uses the average magnitude difference function (AMDF). Pitch was extracted on a logarithmic scale and the utterance mean was subtracted. Also, delta features were added. These features are modeled with GMMs using 64 Gaussians.

System 4: Trigrams of words and human noises. In this case, we used average log-likelihoods of trigrams, computed from manual orthographic transcriptions. Only human noises and words from human transcriptions were included.

System 5: Naive laugh detector. A naive laugh detector was implemented as follows: If the utterance contained laughter, as specified in the manual orthographic transcription, then the utterance was labeled as positive; otherwise, it was neutral.

Combined systems. The output from several systems was combined using multiple linear regression. The transform matrix was estimated on the training data of the development set. Three combined systems were compiled. The first used the output from all three acoustic classifiers, the second also used the trigrams, and the third used the naive laugh classifier instead of the trigrams.

10.1.3 ISL Emotion Recognition Results

Systems 1-5 used no information of the prior distribution of the emotion classes. These systems were therefore evaluated using average recall, defined as the equally weighted mean of the recall rate for all classes. These results are shown in Fig. 10.1. The results of the combined systems optimized for maximum accuracy given the priors of the emotion classes are shown in Table 10.2, along with the naive laughter detector for comparison. These results were presented at the spring 2006 CHIL Emotion Recognition Evaluation. See Section 10.3.1 for a discussion of the results.

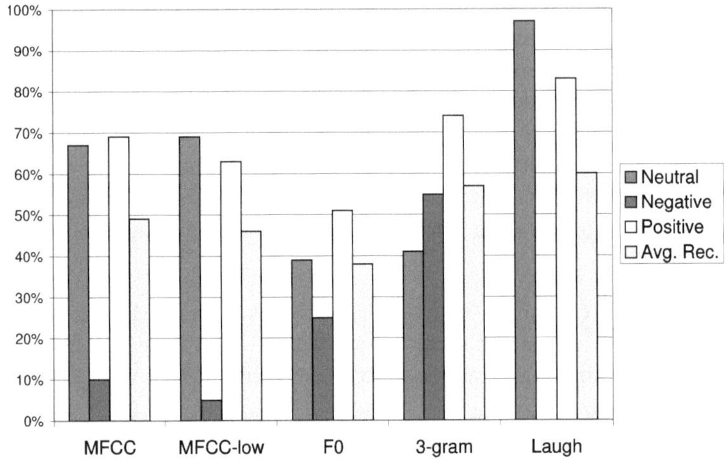

Fig. 10.1. Emotion recall rates and average recall for systems 1-5 for the ISL Meeting Corpus.

System	Accuracy
All-neutral	79%
Acoustic	82%
Acoustic + lexical trigrams	85%
Acoustic + naive laughter classifier	91%
Naive laughter classifier	91%

Table 10.2. Accuracy given the prior distribution of emotions, for combined systems and for a naive laugher detector for the ISL Meeting Corpus. Accuracy for an all-neutral classifier is included for comparison.

10.2 Emotion Recognition for the Connector Agent Scenario

As found in the literature [1, 4], emotion in spoken dialogs is not limited to human-human interaction but is also found in human-computer interaction. An important aspect in human-computer interaction is the naturalness of the interface. By adopting natural human-human interaction as a gold standard in spoken dialog systems, maximum efficiency and user satisfaction is assumed to be achieved. If the user shows frustration or anger, it is likely due to communication problems; it may be due to a recognition error, a dialog system inflexibility, or the user's inability to use the service. In these cases, an emotion recognizer may help the Connector agent generate a proper response.

In Section 10.2.1, we describe the Voice Provider Corpus that we used for our work on the Connector agent. Two emotion recognizers for this corpus are presented in Section 10.2.2, and the recognition results are found in Section 10.2.3.

10.2.1 The Voice Provider Corpus

The speech material in this study is telephone speech recorded at 8 kHz by the Swedish company Voice Provider, which runs more than 50 different voice-controlled telephone services, such as timetable and pricing information, travel booking, and postal services. The large majority of utterances are neutral (nonexpressive), but some percent show frustration, often after misrecognitions by the speech recognizer. Parts of this corpus has been used in other studies [10, 12].

Annotation

The utterances are labeled by an experienced, senior voice researcher into neutral, emphasized, or negative (frustrated) speech. When labeling a speaker's dialog, it is at times obvious that the speaker is emphasizing, or hyperarticulating, an utterance rather than expressing frustration. This is, however, not obvious without taking the dialog's context into account. Other emotions than frustration are very rare and therefore not meaningful to handle. Since the utterances are recorded in real-life telephone services, the emotions expressed must be regarded as natural. This means that labeling them is not straightforward, since it is impossible to be sure which emotion (if any) the speaker was expressing. However, our results below indicate that the annotator's decisions were consistent with some expressive content across all utterances. Also, a subset of the material was labeled by five different speech researchers and the pairwise interlabeler kappa was 0.75 - 0.80.

Description of the Corpus

The corpus consists of 20,807 dialogs with 61,078 utterances, giving an average number of 2.9 utterances per dialog. The median is two utterances and the maximum is 206 utterances. The total proportion of neutral utterances is 96.1%, while the

Development Set	Utterances	Percentage
NEUTRAL	38,229	95.8
EMPHATIC	685	1.7
NEGATIVE	977	2.5
Total	39891	
Evaluation Set	Utterances	Percentage
NEUTRAL	20,445	96.5
EMPHATIC	370	1.7
NEGATIVE	372	1.8
Total	21,187	
Total	61,078	

Table 10.3. The distribution of emotion categories for utterances in each set of the Voice Provider Corpus.

number of negative and emphatic utterances is 2.2% and 1.7%, respectively. Thus, emotional utterances are rare in our data, which may be seen as an indication that the recorded services work reasonably well. We have chosen to classify a dialog as follows: NEUTRAL if all turns are neutral, NEGATIVE if one or more turns are negative, and otherwise EMPHATIC. The proportion of neutral, negative, and emphatic dialogs as a function of the number of utterances is shown in Fig. 10.2. The negative and emphatic dialogs account for a growing proportion of utterances up to dialogs of three utterances, after which they decrease. This is most probably because the callers become frustrated and decide to hang up. Note that the curves for NEGATIVE and EMPHATIC are very similar. This indicates that when there is a problem, people will either speak louder or get frustrated. Of the negative dialogs, 5% are always negative and 12% start with a negative utterance. Longer connected sequences of negative utterances are rare; 77% of the negative dialogs have one negative utterance, 51% end with one negative, 11% end with two negatives and 4% end with three negative utterances followed by hang-up, 35% become negative and calm down to neutral, 30% are preceded by an emphatic turn somewhere in the dialog, and 13% have an emphatic utterance immediately before a negative utterance. Although there is a tendency for users to hang up after getting angry, their behavior spans a wide range of patterns. The utterances of the corpus was split into two sets for development and evaluation (Table 10.3). The speakers' identities were not known and the splits were made between dialogs.

10.2.2 Emotion Recognizers for the Voice Provider Corpus

This section describes two emotion recognition systems. More details are found in [11].

System 6: LDA. In this system, we calculate features frequently used for emotion recognition. The features can be categorized into the following broad classes: pitch, intensity, formants, voice source, and duration. For the first two categories, we used minimum, maximum, mean, range, quantiles, slope, framewise delta,

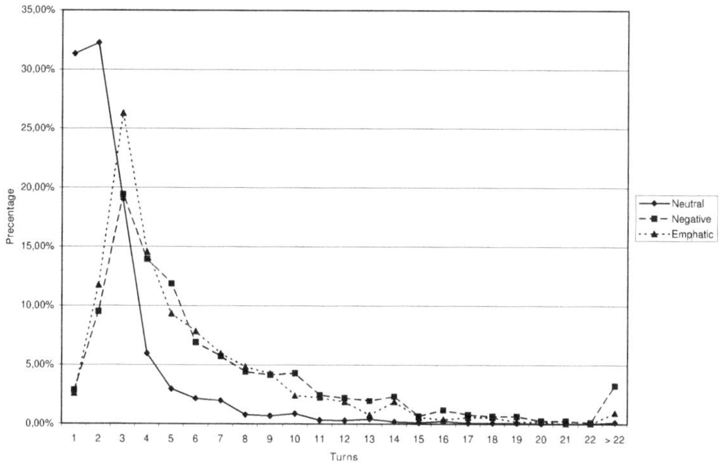

Fig. 10.2. Relative proportion of neutral, negative, and emphatic dialogs as a function of number of utterances per speaker. Since neutral utterances constitute 96.1% of all utterances, the distributions are normalized as described in Section 10.2.1.

framewise delta-delta, standard deviation, the relative position of the minimum and the maximum, and the percentage of frames with rise/level/fall. For pitch, the quantiles, the standard deviation and the range of pitch were extracted after first subtracting the average slope of pitch over the segments. The formant features used are the mean, standard deviation, and median bandwidth for formants 1 to 4. The duration features are silence, mean duration, mean syllable duration, and proportion of voiced frames. For the voice source features, the following spectral approximations are used:

- Open quotient: F0 amplitude - 2nd F0 harmonic amplitude,
- Glottal opening: F0 amplitude - F1 amplitude,
- Amplitude of voicing: F0 amplitude,
- Rate of closure: F0 amplitude - F3 amplitude,
- Skewness: F0 amplitude - F2 amplitude,
- Completeness of closure: F1 mean bandwidth,
- Logarithmic approximative normalized amplitude quotient: F3 amplitude + 20*log (F0 mean).

In addition, jitter and shimmer were added to the voice source measures. Since the LDA model is computationally inexpensive, a brute-force forward selection scheme to find the most discriminative features was adopted. Essentially, each feature was tested and the one that maximized average recall was kept in an incremental way. Each part of the development set was used as training and testing material in a crosswise fashion, resulting in two ranked lists of features. Finally, a joint list was created, where the rank of each feature was calculated as the weighted mean rank from the two lists. The feature order in the joint list was then used to add features one by one in crosswise tests on the development set. As the features were added incrementally, the number of necessary features was determined as the number where the maximum average recall was achieved. Finally, a LDA model was trained on the full development set using these features and tested on the evaluation set.

System 7: GMM/MFCC+MFCC-LOW We use a 24 Mel-filter bank in the 300-3400 Hz range for capturing spectral shape and formants. In addition, a filter bank in the 20-600 Hz region is used in order to model prosody. The latter MFCCs are computed in the same way as the MFCCs in the higher frequency region except that a 48 Mel-filter bank is used. This is followed by a cosine transformation and truncation to 12 dimensions. Each GMM has 512 Gaussians, and the log-likelihoods of the GMMs for each MFCC set were combined using linear addition.

10.2.3 Voice Provider Emotion Recognition Results

For the Voice Provider Corpus, we chose to do the experiments without taking the prior distribution of emotional categories into account. Experiments using two-fold cross-validation on the development set showed that it was difficult to distinguish between the emphatic and the neutral class, and therefore these two classes were merged. For System 6, the maximum average recall on the development set was achieved with 19 features, listed in Table 10.4. The confusion matrices for System 6 and System 7 are shown in Table 10.5 and Table 10.6. Results for different feature sets in System 6 are shown in Table 10.7, and results for each type of filter bank in System 6 are shown in Table 10.8. These results are discussed in Section 10.3.2.

10.3 Discussion

In this section, the results of our various experiments on emotion recognition in two CHIL scenarios are discussed. First, we address emotion recognition for human-human interaction in the context of Socially Supportive Workspaces. Then we address emotion recognition for human-computer interaction in the context of the Connector agent scenario.

10.3.1 The Socially Supportive Workspaces Scenario

In this human-human interaction domain, automatic emotion recognition experiments were conducted using the ISL Meeting Corpus. Experiments and observations have led to the following tentative conclusions:

1. While 16.5% of participant contributions exhibit positive emotional valence, only a small minority (3.4%) exhibits negative emotional valence; the remaining 80.1% of participant contributions are considered neutral by a labeler majority.

1	mean delta-delta F0
2	mean delta intensity
3	mean delta F0
4	logANAQ
5	mean delta-delta intensity
6	fraction F0 rise
7	syllable mean duration
8	relative intensity max. position
9	silence mean duration
10	F1 std.
11	F4 bandwidth median
12	fraction voiced
13	fraction intensity rise
14	F3 mean
15	F1 mean
16	intensity range
17	F2 mean
18	intensity quantile 5
19	F2 bandwidth median

Table 10.4. Ranked list of the best discriminative features, estimated on the Voice Provider Corpus development set.

System 6: LDA		
	Neutral	Negative
Neutral	0.73	0.27
Negative	0.20	0.80

Table 10.5. Confusion matrix for system 6, evaluated on the Voice Provider Corpus.

System 7: GMM/MFCC+MFCC-LOW		
	Neutral	Negative
Neutral	0.87	0.13
Negative	0.22	0.78

Table 10.6. Confusion matrix for system 7, evaluated on the Voice Provider Corpus.

2. MFCC-LOW features, based on a 20-300-Hz filter bank, together with their first- and second-order differences, led to an average recall classification rate of 46% in the context of a GMM log-likelihood ratio classifier — similar to that of standard MFCC features; both feature sets outperformed utterance-normalized pitch and delta pitch. These experiments suggest that there exists as much emotion information in the 20-300-Hz region as in the 300-8000-Hz region.
3. Lexical features (words, word fragments, and nonverbal vocalizations) from manual transcription, in the context of a trigram log-likelihood ratio classifier, led to an average recall of 57%. It was shown that the recall of negative valence

Feature Set	Average Recall	Neutral Recall	Negative Recall
Pitch	0.68	0.70	0.67
Intensity	0.72	0.66	0.77
Voice source	0.74	0.70	0.78
Duration	0.69	0.73	0.65
19 Best	0.76	0.73	0.80

Table 10.7. Detailed results for system 6, evaluation set.

Filter Bank	Average Recall	Neutral Recall	Negative Recall
MFCC	0.80	0.85	0.75
MFCC-LOW	0.83	0.85	0.81
Combined	0.83	0.87	0.78

Table 10.8. Detailed results for system 7, evaluation set.

was 55% and that the rates for the other emotions were also higher than random (33%).

4. Combination of the MFCC, MFCC-LOW, and pitch GMM likelihoods led to an accuracy of 82%, which is 3% higher than choosing the majority class. Note that the recall for the positive class for this combination was 30%. Adding lexical trigrams further improved accuracy, to 85%. However, a combination of the acoustic classifiers and the naive laugh detector did not perform better than a naive laughter detector by itself.
5. Reliance on the presence of transcribed laughter alone led to the highest classification accuracy observed, at 91%.

10.3.2 The Connector Agent Scenario

In the Connector agent experiments, we used the Voice Provider Corpus. While naturally occurring emotions are usually spare, the large amount of utterances ensured enough emotional material for our study. In this human-computer interaction domain, we tested an LDA-based system and a system based on GMMs using two different spectral inputs. The results in Section 10.2.3 led to the following observations:

1. For the LDA-based classifier, the voice set (voice source and formants) performed best, followed in order by the intensity, the duration, and the pitch sets. The order may be seen as an indication of how sensitive, the respective feature sets are to anger; the more sensitive the better the classification. However, the best 19 features were better than each of the four broad class feature sets.
2. Frame-based delta and delta-delta measurements of pitch and intensity were found among the highest-ranked features in the LDA-based system.
3. MFCC-LOW features performed as well as standard MFCC features; the first yielded an average recall of 83% and the second an average recall of 80%. These recall rates were achived using GMMs.

10.4 Conclusion

This chapter has explored the applicability of emotion recognition within the broader CHIL context. We have argued that successful recognition of emotional state is predicated upon the anticipated range of emotional expression and that the latter is a strong function of the intended application scenario. Research under CHIL has therefore been guided by the potential for technology deployment in specific CHIL services; we have chosen to focus on naturally occurring human-human interaction, as observed in meetings and (interactive) seminars, and on naturally occurring human-computer interaction, as observed in telephony services. Based on these studies, we believe that it is possible to recognize emotions with reasonable accuracy in these contexts.

References

1. A. Batliner, K. Fischer, R. Hubera, J. Spilkera, and E. Nöth. How to find trouble in communication. *Speech Communication*, 40:117–143, 2003.
2. S. Burger, V. McLaren, and H. Yu. The ISL meeting corpus: The impact of meeting type on speech style. In *Proceedings of the International Conference on Spoken Language Processing*, Denver, CO, 2002.
3. N. Campbell. Getting to the heart of the matter: Speech as the expression of affect; rather than just text or language. *Language Resources and Evaluation*, 39:109–118, 2005.
4. M. L. Chul and S. Narayanan. Toward detecting emotions in spoken dialogs. *IEEE, Transactions on Speech and Audio Processing*, 13(2):293–303, Mar. 2005.
5. R. Cowie and R. R. Cornelius. Describing the emotional states that are expressed in speech. *Speech Communication*, 40:5–32, 2003.
6. J. R. Davitz. *The Communication of Emotional Meaning*. McGraw-Hill, New York, 1964.
7. L. Devillers, L. Vidrascu, and L. Lamel. Challenges in real-life emotion annotation and machine learning based detection. *Neural Networks*, 18:407–422, 2005.
8. P. N. Juslin and P. Laukka. Communication of emotions in vocal expression and music performance: Different channels, same code? *Psychological Bulletin*, 129:770 814, 2003.
9. K. Laskowski and S. Burger. Annotation and Analysis of Emotionally Relevant Behavior in the ISL Meeting Corpus. In *Proceedings of LREC*. Genoa, Italy, 2006.
10. P. Laukka, K. Elenius, M. Fredriksson, T. Furumark, and D. Neiberg. Expression in spontaneous and experimentally induced affective speech: Acoustic correlates of anxiety, irritation and resignation. In *Workshop on Corpora for Research on Emotion and Affect*, Marrakesh, Morocco, 2008.
11. D. Neiberg and K. Elenius. Automatic recognition of anger in spontaneous speech. In *Proceedings of Interspeech 2008*, Brisbane, Australia, (in press).
12. D. Neiberg, K. Elenius, and K. Laskowski. Emotion Recognition in Spontaneous Speech Using GMMs. In *Proceedings ICSLP-2006*, pages 809–812. Pittsburgh, 2006.
13. K. R. Scherer. Vocal affect expression: A review and a model for future research. *Psychological Bulletin*, 99:143–165, 1986.

11

Activity Classification

Kai Nickel[1], Montse Pardàs[2], Rainer Stiefelhagen[1], Cristian Canton[2], José Luis Landabaso[2], Josep R. Casas[2]

[1] Universität Karlsruhe (TH), Interactive Systems Labs, Fakultät für Informatik, Karlsruhe, Germany
[2] Universitat Politècnica de Catalunya, Barcelona, Spain

When a person enters a room, he or she immediately develops a mental concept about "what is going on" in the room; for example, people may be working in the room, people may be engaged in a conversation, or the room may be empty. The CHIL services depend on just the same kind of semantic description, which is termed *activity* in the following. The "Connector" or the "Memory Jog", for example, could provide support that is appropriate for the given context if it knew about the current activity at the user's place. This kind of higher-level understanding of human interaction processes could then be used, e.g., for rating the user's current availability in a certain situation.

The recognition of activities depends on many factors such as the location and number of people, speech activity, and the location and state of certain objects. The perceptual technologies like person tracking, identification, or acoustic event detection provide important information upon which the higher-level analysis of the activity can be based. Due to the complexity of the scene, there are, however, potentially relevant phenomena such as a door being half-opened, which – due to their high number and variability – cannot be addressed by manually designed detectors at large. Therefore, activity recognition may need to directly analyze the observation in order to find out what is relevant and what is not to detect a certain activity.

Activity recognition may be facilitated by the detection of *events*, which are semantic descriptions for actions that have a short duration and/or are limited to a small area of the room, such as "a person enters/leaves the room". In this case, activities are being recognized by their characteristic sequence of events.

This chapter describes three systems that have been implemented for (1) the recognition of events inside the CHIL room based on person tracking and a probabilistic syntactic approach, (2) person activity classification using body gestures, and (3) event recognition and room-level tracking in multiple office rooms based on low-level audiovisual features.

11.1 Visual Activities Recognition in a Smart-Room Environment Using a Probabilistic Syntactic Approach

In a smart-room environment, it is necessary to determine the type of interactions among the people in the room in order to perform context-dependent actions. Moreover, some specific situations need to be identified by the system. For this aim, a system based on [6], which analyzes activities using stochastic parsing, has been developed by adapting it to the specific requirements of smart-room environments. The fundamental idea of [6] is to divide the recognition problem into two levels. The lower-level detections are performed using standard independent probabilistic event detectors to propose candidate detections of low-level features. The outputs of these detectors provide the input stream for a stochastic, context-free, grammar-parsing mechanism. The system can be divided into three modules: the tracking system, the events generator, and the parser. The tracking system [7] takes as input the multi-camera video sequence, reconstructs the 3D objects in the room using a foreground detection for each camera, and performs the tracking of the various detected objects. Thus, for each frame in the video sequence, we require N views from the calibrated cameras. Foreground regions are obtained for each camera using an algorithm based on Stauffer and Grimson's background learning and subtraction technique [12]. A Shape from Silhouette procedure is used next in order to generate a discrete occupancy representation of the 3D space (voxels) to decide whether a voxel is in the foreground or background by checking the spatial consistency of the N segmented silhouettes. Afterwards, a connectivity filter is introduced in order to remove isolated voxels and the remaining multiple RoIs are labeled in accordance with the results of a tracking procedure, as described in Chapter 3. The events generator and the parser are described in the following.

11.1.1 Events Generation

The main objective of the events generator is to provide the chain of events that the parser will take as input. The inputs to the events generator are the ones delivered by the tracking system: object identifier, number of frame, position (x, y, z), velocity, and volume of each object. Moreover, the output of the multicamera 3D person and object tracker is enriched by (1) an algorithm that is able to distinguish between an object and a person – assuming an average range of physical properties of adult humans – and (2) an algorithm that analyzes human body posture (standing, sitting, etc.) with a standard model of the human body that is aligned to the 3D regions of interest earlier classified as a person. This information is used together with some configuration information about the room (table, chair, and whiteboard position, dimensions of the room) to produce the events detection using simple grammars. The list of events detected by this module is the following: {"Person enters in the room", "Person exits from the room", "A person is lost in the room", "A person is found in the room", "A person sits down", "A person moves inside the room", "A person stops", "A person stands up", "A person is detected in the whiteboard area", "A new object is detected in the room", "An object disappears from the room", "The volume of a person increases", "The volume of a person decreases", "A group of people is detected", "A person is divided in two"}.

11.1.2 Video Activity Recognition

The video activity recognition is performed by the parser. Its function is as follows: Given a chain of events and a stochastic, context-free grammar, find the chain derivation with the maximum probability, if it exists. The parser we have used is based on the CYK algorithm [14]. It performs an ascendant analysis, considering subtrees from the leaves up to the root. The activities the system currently recognizes are the following:

- meeting,
- presentation,
- conversation between two people,
- leave an object ,
- take an object.

The video sequence is first analyzed with the tracking and event generators system. The generated chain of events is input to the parser, which analyzes the chain using the grammar corresponding to these classes. In order to generate more appropriate grammars, we have implemented a training system to generate the grammars for the different classes we want to learn. The Inside-Outside algorithm [8] is used to estimate the probability of the production rules of the stochastic, context-free grammars using the video sequences created for training. The production rules have been manually designed.

11.1.3 Experiments

To test the system, we have used 50 recordings where the five defined activities occur. The recordings have been done in a smart room with four fixed cameras in the corners plus a zenithal camera. An example of four frames corresponding to a "presentation" recording is shown in Fig. 11.1, together with a projection of the reconstructed blobs. The recognition results span from 60% recognition for the activities "conversation" and "take an object" to 87.5% for "presentation", with a mean correct recognition rate of 70%.

11.2 Person Activity Classification Using Gestures

Some activities of the persons in the room cannot be recognized using only the person tracking results and the 3D reconstructed objects. Human motion descriptors add the necessary information for classifying person activities that involve the motion of the body limbs. We have developed a view-independent approach to the recognition of human gestures of several people in low-resolution sequences from multiple

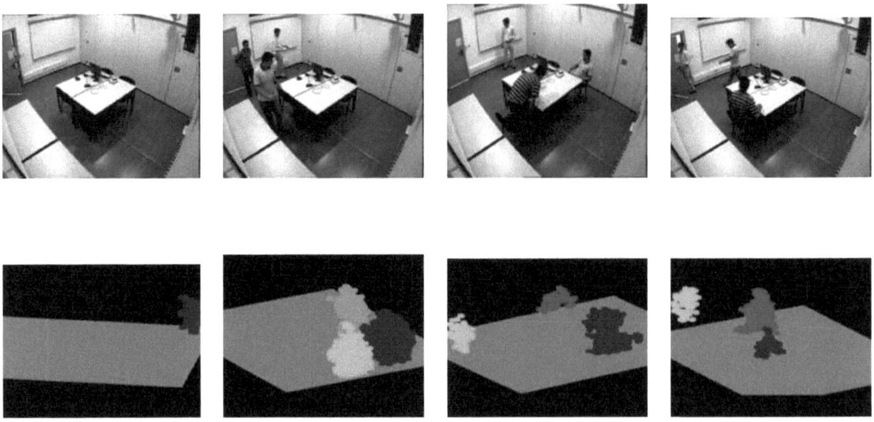

Fig. 11.1. Four frames from a presentation, showing different events that can be detected using the tracks. The first row shows the images from one camera and the second row a projection of the 3D detected blobs. The event detected in the images of the first and second columns is "person enters". The third column corresponds to "person at whiteboard" and "person sits", while in the image of the last column, a "person exits" is detected.

calibrated cameras [3]. In contrast to other multi-ocular gesture recognition systems based on generating a classification on a fusion of features coming from different views, our system performs a data fusion (3D representation of the scene) and then a feature extraction and classification. Motion descriptors introduced by Bobick and Davis. [1] for 2D data are extended to 3D and a set of features based on 3D invariant statistical moments is computed. A simple ellipsoid body model is fit to incoming 3D data to capture in which body part the gesture occurs, thus increasing the recognition ratio of the overall system and generating a more informative classification output. Classification is thus performed by jointly analyzing the motion features and the body position data obtained by fitting the ellipsoid body model. Finally, a Bayesian classifier is employed to perform recognition over a small set of actions. The actions that are more relevant to the smart-room scenario are raising hand, sitting down, and standing up. However, we have tested the system including other actions such as waving hands, crouching down, punching, kicking, and jumping. The approach taken relies on the 3D reconstruction of the detected persons. Thus, the system uses as input the same data described in Section 11.1, that is, the multiple RoIs labeled coherently along time, corresponding to the persons in the room. In the following, we describe the approach taken to analyze the person's activity using these input data.

11.2.1 Motion and Body Analysis

In order to achieve a simple and efficient low-level, view-dependent motion representation, [1] introduced the concept of motion history image (MHI) and motion energy image (MEI). We extended this formulation to represent view-independent 3D

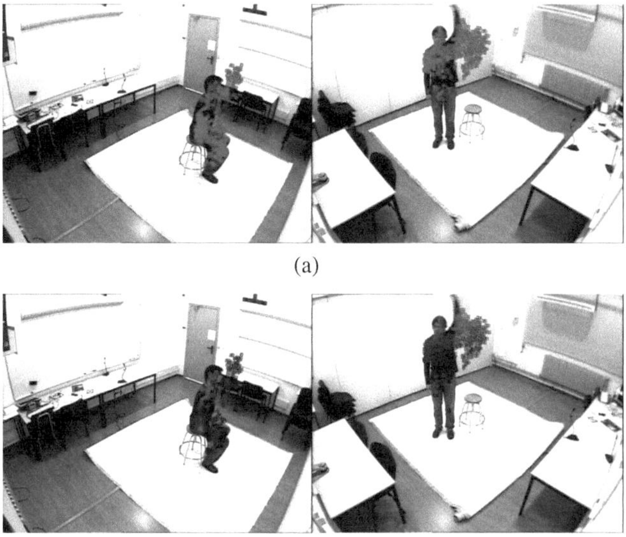

(a)

(b)

Fig. 11.2. Example of motion descriptors. In (a) and (b) are depicted the 2D projections of MEV and MHV, respectively, for *sitting down* and *raising hand*.

motion. In this way, ambiguities generated by occlusions are overcome. Analogously to [1, 2], the binary motion energy volume (MEV) $E_\tau(\mathbf{x},t)$ captures the 3D locations, where there is motion in the last τ frames. Motion detection can be coarsely estimated by a simple forward differentiation among voxel frames, still leading to satisfactory results while preserving a reduced computational complexity. Figure 11.2(a) depicts an example of MEV.

To represent the temporal evolution of the motion, we define the motion history volume (MHV) $H_\tau(\mathbf{x},t)$, where each voxel intensity is a function of the temporal history of the motion at that 3D location. An example of MHV is shown in Fig. 11.2(b).

In order to extract a set of features describing the body of a person performing an action, a geometrical configuration of the human body must be considered. An ellipsoid model of the human body has been adopted and, in spite of this fairly simple approximation compared with more complex human body models, classification results proved the validity of our assumption, as shown in Section 11.2.3.

After obtaining the set of voxels describing a given person, we fit an ellipsoid shell to model it. This information is then fed to a body-tracking module that refines this estimation by taking into account body anthropometric restrictions, imposing some motion and size constraints compatible with human bodies [4]. For example, the height of a person restricts the possible locations of arms and legs according to the average lengths of body parts. Finally, time consistency of the ellipsoid parameters is achieved by a Kalman filter.

Once the parameters of the ellipsoid representing the human body are computed, a simple body part classification can be derived. Voxels can be labeled as belonging to four categories: left/right arm/leg (see Fig. 11.3). These data will be used while classifying an action jointly with motion information.

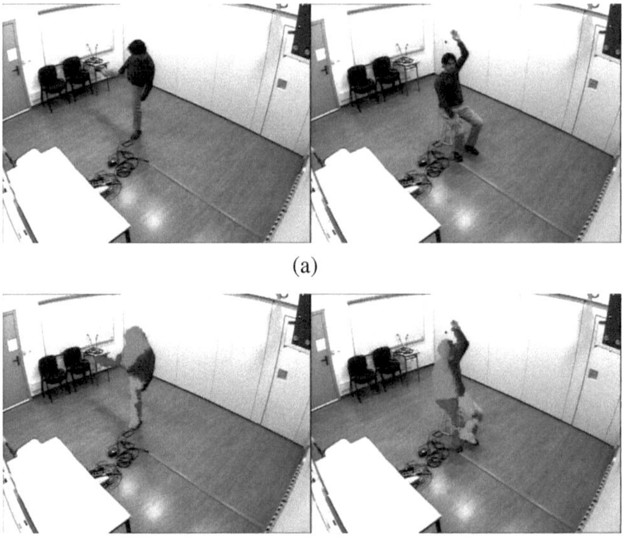

(a)

(b)

Fig. 11.3. Body analysis module output. In (a), original images for *kick* and *arise hand*. In (b), voxels belonging to the body of the person are labeled as belonging to right/left arm/leg categories.

11.2.2 Feature Extraction and Gesture Classification

Data produced by the motion and body analysis modules are processed to extract a vector of features for classification.

Informative features derived from the analyzed data (MHV and MEV in our case) are required to represent motion in a low-dimensional space. Statistical moments invariant to scaling, translation, rotation, and affine mappings were introduced by [5]. Three-dimensional invariant statistical moments [9] were used in our case. For each data set $E_\tau(\mathbf{x},t)$ and $H_\tau(\mathbf{x},t)$, two invariant moment-based feature vectors were computed, ψ_{MEV} and ψ_{MHV}, each comprising five components.

Information from body parts provided by the body analysis module can be used to generate additional features. Let ψ_{BODY} denote the four features describing the relative amount of motion voxels located in each body part. Given the computed moment-based motion features and the body features obtained for each of the actions to classify ω_j, $0 \leq j < K$, we define a full 14-dimensional feature vector $\Gamma = [\psi_{\mathrm{MEV}}\,\psi_{\mathrm{MHV}}\,\psi_{\mathrm{BODY}}]$. The dimensionality of Γ can be further reduced through principal components analysis (PCA). By analyzing the training data, we noticed

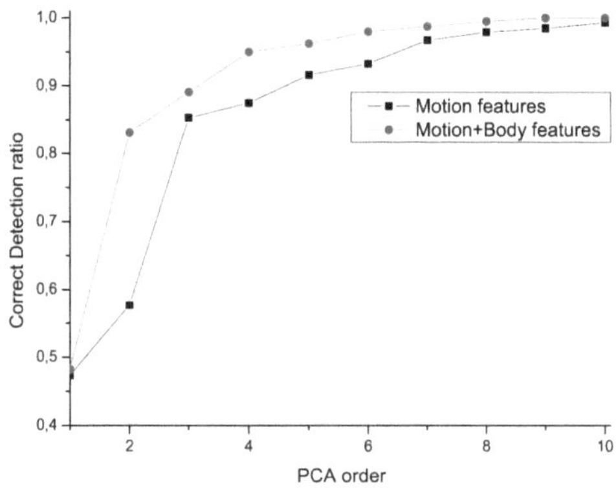

Fig. 11.4. Classifier performance evaluated with motion and body features depending on the order of the PCA analysis.

that 90% of the variance of the data was achieved by doing a dimension reduction to $d = 7$. Let us refer to the data set obtained after PCA analysis as $\hat{\Gamma}$.

The classification method is based on a Bayesian classification criterion assuming that $p(\hat{\Gamma}|\omega_j)$ is normally distributed and estimating the mean and covariance matrix of each class with the training data.

11.2.3 Experiments

In order to evaluate the performance of the proposed algorithm, we collected a set of 70 training and 30 testing multiview sequences of each action to be recognized. The gesture category set was formed by eight common actions of interest in the field of human-computer interfaces such as raising hand, sitting down, waving hands, crouching down, standing up, punching, kicking, or jumping. Moreover, to show the effectiveness of our method and its robustness against rotations, occlusions, and position, actions were recorded in different positions inside the room and facing various orientations.

In average, we got a $p(\text{error}) = 0.0154$. Experiments have been carried out with and without these features to show the influence of body part features on the overall performance. Figure 11.4 depicts the behavior of the classifier for diverse orders of the PCA analysis showing that body features increase the performance of the system. The experimental results prove the efficiency of our method, proposing an alternative to the classical methodology to multi-ocular and mono-ocular motion-based gesture analysis [1, 11, 2].

11.3 Activity Recognition and Room-Level Tracking in an Office Environment

The previously described approaches use the output from a person tracker to infer people's activities. In this work [13], we bypass the tracker and try to infer human activity directly from the camera image. This is motivated by the fact that person tracking is computationally expensive, requires a rich sensor setup, and is still not 100% reliable. Furthermore, the recognition of human activity may not depend only on the location and pose of the human body, but also on the state of objects like doors or chairs. Therefore, we are following an appearance-based approach and develop an activity recognition system that operates directly on the data from a single fixed camera and a single microphone per room.

We decompose activities in two classes, namely events and situations, both carrying a semantic meaning. In our case, events are defined to be visible or audible short-term phenomena that are spatially limited to a small area. In the presented application, we detect events like PERSON SITTING AT A DESK or PERSON ENTERING/LEAVING AN OFFICE; i.e., we focus on events that are triggered by humans. In contrast to events, we define situations to range over a longer period of time and space. Situations that are to be distinguished by our system span the entire room: MEETINGS, DISCUSSIONS, PAPERWORK, PHONE CALLS, or NOBODY PRESENT. They were chosen manually by observing the recorded data. The objective was to cover a maximum share of daily office activities in a real-world setting.

The experimental activity recognition system spans four office rooms, each occupied by one or two members of the lab, as well as the local lab room (see Fig. 11.5). Each room is equipped with a sparse sensor setup consisting of a single camera and one omnidirectional microphone.

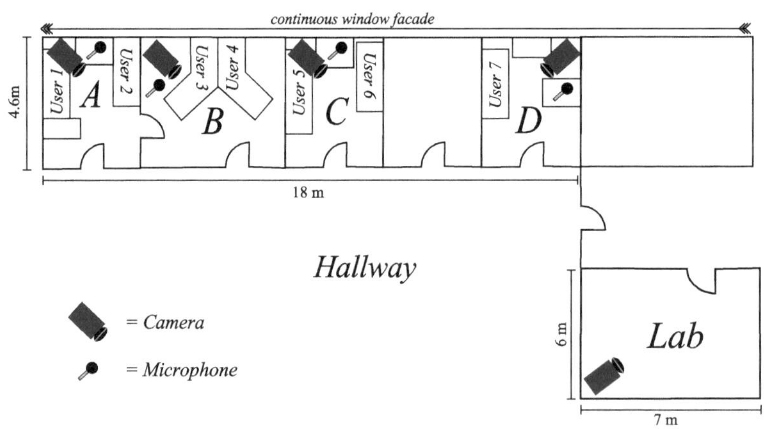

Fig. 11.5. Plan view of the rooms that were monitored for activity recognition.

Using only simple low-level features such as motion and optical flow for the video modality and signal energy, zero crossing rate, and pitch for audio, we employ a multilevel HMM activity recognition framework (see Fig. 11.6). Decomposing the parameter space into several layers reduces the amount of training data required and gives a better intuition on the learning process [10, 15]. The lower level detects events and passes them on to the situation layer. The situation layer infers room situations based on the sequence of detected events. Each layer can be trained on labeled data on its own by employing the well-known Baum-Welch parameter estimation algorithm.

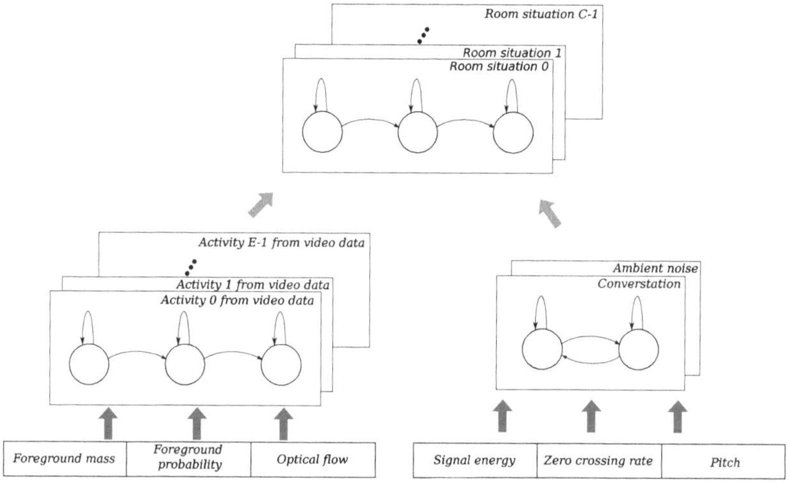

Fig. 11.6. Structure of the multilayer HMM for a single office. The lower level recognizes events, whereas the higher level represents *room situations*.

Knowing who is where on a room-level scale is the natural complement to situation recognition within the individual rooms. We track the users by inferring their locations from the sequence of events recognized by the first activity layer. We are thus exploiting common constraints like the fact that office workers tend to have a dedicated workplace that they use most of the time exclusively.

The method does not rely on conventional person identification or tracking techniques, which often pose restrictions for practical deployment due to their high sensory and computational requirements. Our approach exploits a Bayesian filter framework in a discrete state space, where the state vector contains the belief that a certain person is in a certain room or out of sight. The key problem of this multiperson tracking task is to assign observed events to the correct track, which is known as *data association* problem. A separate tracker is run for each person and data association is performed in two stages: A nearest-neighbor filter is applied to consider only the observations for the belief update that are occurring close to the highest belief state. Moreover, the observation model is designed in a way that persons can only be observed at certain places depending on the room.

It is again important to note that this tracking approach does not use appearance models to identify the individual users directly, but instead aims to infer the user's locations only from the sequence of events that are recognized by the first layer of the activity recognition model. The output of the system is the room in which the person is currently located – and not its precise location within that room.

11.3.1 Features

Due to the changing light conditions, any illumination-dependent cue such as histogram backprojection to identify skin color would be error-prone. Moreover, persons are perceived from just one camera view. Depending on their orientation, we get either frontal, side, or back views of their head, so that face detectors can hardly be employed to determine the number of persons in the room. Therefore, we are concentrating on simple but fast video features that are robust against varying lighting conditions: adaptive background subtraction and optical flow.

On the audio side, we use speech activity detection, which is an important cue to determine people's current occupation. In our office scenario, it helps, for example, to separate visually similar classes like PAPERWORK and DISCUSSIONS. In order to detect speech activity, we calculate signal power, zero crossing rate, and pitch and process them with an audio classification HMM.

The key problem is to decide which regions of the input image are relevant for certain events. In our approach, we consider the relevant foreground regions of a certain activity to be the components of a Gaussian mixture. This allows a data-driven learning approach with the well-known EM algorithm. Features are then extracted from the enclosed areas of each mixture component within three standard deviations. Together these features are capable of describing a scene by the amount of motion with the dominant direction, while preserving rough location information.

Fig. 11.7. Three Gaussian mixture components obtained from data-driven clustering. They represent areas where users often sit.

11.3.2 Experiments

For the activity recognition part, we collected data from six work days with a total length of about 34.8 hours, of which we used four days for training and two days for evaluation. To obtain ground-truth labels, the data were annotated manually. On this data set, the recognition rate of the events ranged between 63% for "visitor behind user's desk" and 100% for "somebody enters". The situations were recognized with a recognition rate of 70 – 96% depending on the type of situation; for details, see [13].

As this set of data contained only a few events of people changing offices, we recorded a second set with a scripted sequence of 44 events with a length of about one hour and we used it to evaluate room-level tracking. As the hallway was not monitored due to privacy reasons, blind gaps occurred between the cameras. On average, we could track the location of all seven people in 91.5% of the frames, and 36 of 44 transitions were correctly recognized. Table 11.1 shows ground-truth and tracking results for one of the tracked persons.

Ground Truth			Tracking Results		
Begin	*End*	*Place*	*Place*	*Begin*	*End*
0	32	Office B	Office B	0	41
46	68	Lab	Lab	49	75
77	800	Office B	Office B	81	809
809	1110	Office D	Office D	810	1102
1117	2194	Office B	Office B	1103	2197
2194	2484	Office A	Office A	2198	2496
2496	2660	Office D	Office D	2497	2671
2668	3248	Office B	Office B	2672	3263
3248	3389	Out of view	Out of view	3264	3382
3389	3685	Office B	Office B	3383	3687
3685	3699	Lab	Lab	3688	3705
3709	3719	Office A	Office D	3709	3925
3719	3926	Office B			

Table 11.1. Example trajectory for user #4 (times are given in seconds); for the sake of readability, OUT OF SIGHT is not listed for state durations of less than 30 seconds.

11.4 Conclusion

In the course of the CHIL project, three different approaches for automatic activity recognition have been implemented and evaluated. They are different in terms of the set of activities they classify, the features they use, and the actual classification method.

The first system is oriented to the recognition of room-level activities and thus uses only the detected volumes and the room configuration information as input.

It can detect interactions between people or between people and objects as well as classify the kind of activity that takes place in the smart room according to the predefined classes (meeting, conversation, etc).

The second system is oriented to the recognition of activities on a person level. For these kinds of activities, motion descriptors as well as a simple human body model are used, together with the foreground volumes. With the experiments carried out, we have concluded that a set of human body activities can be efficiently distinguished without requiring a complex human body model analysis that implies a high computational cost.

The third system works with a sparse sensor setup of one camera and one microphone per room. Events are detected based on a data-driven analysis of the sensor data. Based on the sequence of events, both office activities as well as the location of people on a room-level scale could then be inferred.

It is obviously hard, if not impossible, to find a common definition and methodology for activity recognition that fits all application domains. The current systems are dedicated to certain domains like office situations and meetings. They define a small set of activities that are specific and relevant within their domain, and they proved to be able to recognize the activities on in-domain test data. Future work, on the one hand, could try to extend the application domains, while, on the other hand, it could aim for a more detailed analysis of the activities within one domain.

One lesson learned in the CHIL project was that the development of an activity recognition system needs to be application-driven: The consumer of the information – for example, the Connector or the Memory Jog service – defines the domain and the set of meaningful activities due to its specific need. Only with that knowledge can an appropriate activity recognizer be designed.

References

1. A. F. Bobick and J. W. Davis. The recognition of human movement using temporal templates. *IEEE Transactions on Pattern Analysis and Machine Intelligence*, 23(3):257–267, 2001.
2. G. Bradski and J. Davis. Motion segmentation and pose recognition with motion history gradients. *Machine Vision and Applications*, 13(3):174–184, 2002.
3. C. Canton-Ferrer, J. R. Casas, and M. Pardàs. Human model and motion based 3D action recognition in multiple view scenarios (invited paper). In *14th European Signal Processing Conference, EUSIPCO*, University of Pisa, Florence, Italy, 4–9 Sept. 2006.
4. S. Dockstader, M. Berg, and A. Tekalp. Stochastic kinematic modeling and feature extraction for gait analysis. *IEEE Transactions on Image Processing*, 12(8):962–976, 2003.
5. M. Hu. Visual pattern recognition by moment invariants. *IEEE Transactions on Information Theory*, 8(2):179–187, 1962.
6. Y. A. Ivanov and A. F. Bobick. Recognition of visual activities and interactions by stochastic parsing. *IEEE Transactions on Pattern Analysis and Machine Intelligence*, 22:852–872, 2000.
7. J. L. Landabaso and M. Pardas. Foreground regions extraction and characterization towards real-time object tracking. In *Machine Learning for Multimodal Interaction (MLMI)*, LNCS 3869, pages 241–249. Springer, 2006.

8. K. Lari and S. Young. The estimation of stochastic context-free grammars using the inside-outside algorithm. *Computer, Speech and Language*, 4:35–56, 1990.
9. C. Lo and H. Don. 3-D oment forms: Their construction and application to object identification and positioning. *IEEE Transactions on Pattern Analysis and Machine Intelligence*, 11(10):1053–1064, 1989.
10. N. M. Oliver, B. Rosario, and A. Pentland. A Bayesian computer vision system for modeling human interactions. *IEEE Transactions on Pattern Analysis and Machine Intelligence*, 22(8):831–843, 2000.
11. R. Rosales. Recognition of human action using moment-based features. *Boston University Computer Science Technical Report, BU*, pages 98–120, 1998.
12. C. Stauffer and W. E. L. Grimson. Learning patterns of activity using real-time tracking. *IEEE Transactions on Pattern Analysis and Machine Intelligence*, 22(8):747–757, 2000.
13. C. Wojek, K. Nickel, and R. Stiefelhagen. Activity recognition and room level tracking in an office environment. In *IEEE International Conference on Multisensor Fusion and Integration for Intelligent Systems*, Heidelberg, Germany, Sept. 2006.
14. D. H. Younger. Recognition and parsing of context-free languages in time n3. *Information and Control*, 10:189–208, 1967.
15. D. Zhang, D. Gatica-Perez, S. Bengio, and I. McCowan. Semi-supervised adapted HMMs for unusual event detection. In *Computer Vision and Pattern Recognition*, pages 611–618, 2005.

12

Situation Modeling

Oliver Brdiczka[1], James L. Crowley[1], Jan Cuřín[2], Jan Kleindienst[2]

[1] INRIA Rhône-Alpes, Saint Ismier Cedex, France
[2] IBM Czech Republic, Praha, Czech Republic

CHIL services are intended to anticipate the needs of their users. An important step toward this is to model and understand human behavior. Human activity can be sensed and recognized (as described in Chapter 11). However, a higher-level representation of human actions and human relationships (social context) is necessary to effectively describe human behavior and detect human needs.

One important issue is in what detail context-aware applications can capture the complexity of social context. The majority of documented context-aware applications only use identity and location in their attempts to capture user environment changes [1, 7]. Even though these approaches produced good results, the limitation of contextual cues when modeling more complex human interactions and behavior risks reducing the usefulness of context-aware applications. Many problems can be considered as technical issues like signal processing problems or person/object recognition. However, the omission of social aspects in context modeling is a considerable problem.

Context-aware computing research typically assumes context to be a form of information that is delineable, stable, and separable from activity. Addressing the problem of social aspects in context modeling, Dourish [13] proposes an interactional view of context where context is understood as something relational, dynamic, occasioned, and arisen from human activity. Thus, context is not something that describes a setting or configuration, but it is something that people do.

This chapter defines the concepts of situation, role, relation, and situation graph to describe human behavior and relationships. An example of a lecture room illustrates a possible implementation of these concepts. Section 12.2 gives two examples that have been implemented during the CHIL project. These examples are briefly described and references to further details in this book or in the literature are given. The first example investigates the detection of different meeting states using several modalities. Different classifiers are compared and evaluated. The second example describes a probabilistic detector for interaction groups in a meeting. A hidden Markov model represents the situations corresponding to possible group configurations.

12.1 Defining Concepts: Role, Relation, Situation, and Situation Network

Dey [12] defines context as "any information that can be used to characterize the situation of an entity". An entity can be a person, place, or object considered relevant to the user and application. Loke [15] states that situation and activity are, however, not interchangeable, and activity can be considered as a type of contextual information that can be used to characterize a situation. Dey defines situation further as "description of the states of relevant entities". Situation is thus a temporal state within context. Allen's temporal operators [2, 3] can be used to describe relationships between situations. Crowley et al. [10] introduce then the concepts of role and relation in order to characterize a situation. Roles involve only one entity, describing its activity. An entity is observed to "play" a role. Relations are defined as predicate functions on several entities, describing the relationship or interaction between entities playing roles. Acceptance tests determine whether a particular entity plays a role or whether several entities are in relation. These acceptance tests associate roles and relations with relevant entities. In the following, we detail the definitions of role, relation, situation, and situation network.

Situations are a form of state defined over observations. A situation is defined using a predicate expression. The logical functions that make up this expression are defined in terms of a set of roles and relations. Situations in the context model are connected by arcs that represent events. Events correspond to changes in the assignment of entities to roles, or changes in the relation between entities.

A context [8, 10] is a composition of situations that share the same set of roles and relations. A context can be seen as a network of situations defined in a common state space. A change in the relation between entities, or a change in the assignment of entities to roles, is represented as a change in situation. Such changes in situation constitute an important class of events that we call Situation-Events. Situation-Events are data-driven. The system is able to interpret and respond to them using the context model.

The concept of *role* is a subtle (but important) tool for simplifying the network of situations. A role is an abstract entity that is able to perform certain actions. Roles are "played" by entities within a situation. The assignment of an entity to a role requires that the entity passes an acceptance test. In our framework, the *relations* that define a situation are defined with respect to roles and applied to entities that pass the test for the relevant roles.

The mapping between entities and roles is not bijective. One or more entities may play a role. An entity may play several roles. The assignment of entities to roles may (and often will) change dynamically. Such changes provide the basis for an important class of events: role-events. Role-events signal a change in assignment of an entity to a role, rather than a change in situation.

Human behavior within the environment can be described by a *script*. A script corresponds to a sequence of situations in the situation network reflecting human behavior in the scene. However, scripts in a situation network are not necessarily linear.

For example, in a lecture situation, at any instant, one person plays the role of the "lecturer" while the other persons play the role of "audience" (Fig. 12.1). "Lecturer" and "audience" share the "notSameAs" relation; i.e., the entities playing the corresponding roles are different.

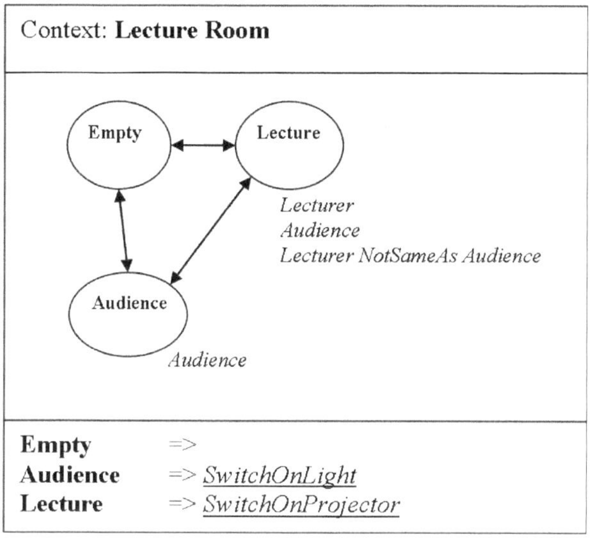

Fig. 12.1. Example of a simple context model for a lecture room. **Empty**, **Audience**, and **Lecture** are the available situations. *Lecturer*, *Audience* are the available roles and *NotSameAs* is the available relation. SwitchOnLight, SwitchOnProjector are system services.

The lecture room example can be implemented using person-entities and their location in regions of interest. We will assume that a perceptual component tracks moving objects in the video images of the scene (for example, by using image background subtraction). The resulting event stream contains the entity identifier as well as the entity position in the video images (Position *Entity X Y*). We assume that this component can determine whether or not the tracked object is a person, for example, by using skin color detection or other features (Person *Entity*).

Figure 12.2 shows an example of interest zones (in violet) for the lecture room: one interest zone next to the board ("LectureArea"), three zones around the tables and chairs ("AudienceArea"). The perceptual component tracks the entities in the scene. If the entity enters or leaves a region of interest, an abstract event of the form (enters <Region-ID> <Entity>) or (exits <Region-ID> <Entity>) is respectively sent. When the entity leaves the scene, a new event is sent to the situation model (ExitScene <Entity>). If the perceptual component loses the entity, an abstract event will be sent of the form (Lost <Entity>).

We can then define the following conditions and actions for role and relation assignment:

Fig. 12.2. Interest zones (violet) for the lecture scenario.

Role: *Lecturer*

Conditions	Action
{(Person <Entity>), (enters LectureArea <Entity>)}	*Lecturer(<Entity>))*
{(exits LectureArea <Entity>)}, {(ExitScene <Entity>)}, {(Lost <Entity>)}	*¬Lecturer(<Entity>)*

Role: *Audience*

Conditions	Action
{(Person <Entity>), ,(enters AudienceArea <Entity>)}	*Audience(<Entity>))*
{(exits AudienceArea <Entity>)}, {(ExitScene <Entity>)}, {(Lost <Entity>)}	*¬Audience(<Entity>))*

Relation: *NotSameAs*

Conditions	Action
{<Entity>≠<Entity2>}	*NotSameAs(<Entity>,<Entity2>)*
{<Entity>=<Entity2>}	*¬NotSameAs(<Entity>, <Entity2>)*

The situations in the lecture room are finally activated or deactivated by the existence or nonexistence of the necessary roles (see Fig. 12.1).

12.2 Implementations of the Situation Model

In this section, we illustrate the situation modeling approach by several examples that have been implemented during the CHIL project. These implementations focus on the detection of situations.

12.2.1 Meeting State Detection

This example illustrates a situation model trained to recognize meeting activity in smart rooms. We make no assumptions about the shape and dimensions of the meeting space; it could be a small meeting room or a large lecture hall. Similarly, we set no bounds on meeting attendance; it may have any number of participants.

The meeting recognizer works on features extracted from the stream of sensor data. We make the assumption that the smart room's video and audio sensors, through the perceptual components, provide at least one of the following streams:

1. the location of people in the room (2D or 3D coordinates);
2. the head orientation of each person in the room (angle);
3. the speech activity of a person or a group of persons (only speech vs. silence, no speech transcription).

We designed the recognizer to work with any subset or combination of these sensor streams. This is based on the practical observation that the above-described outputs are generated by independent sensors that may or may not be running at the same time. Thus, we wanted our recognizer to work in the situation where only a body tracker is running in a smart room (case 1), as well as support the case when a head pose detector and a speech activity detector are also available (streams 1+2+3).

The situations of interest are described in Table 12.1.

State	**Description**
no meeting	Empty room, people entering/leaving, drinking coffee
presentation	Presentation of lecture or seminar is going on
discussion	Discussion or around-the-table meeting

Table 12.1. Recognized meeting states.

The corpus we have used in our experiments was created by the CHIL Consortium for the purpose of the CLEAR evaluation task [16, 18]. We found that this corpus is the only data set that provides multimodal, multisensory recordings of realistic human behavior and interaction in the meeting scenario. We used five seminars prepared for the CLEAR 2007 evaluation campaign and one seminar from CLEAR Evaluation 2006. In particular, the data contain a detailed multi channel transcription of the audio recordings that include speaker identification and acoustic condition information. Video labels provide multiperson head location in the 3D space as well

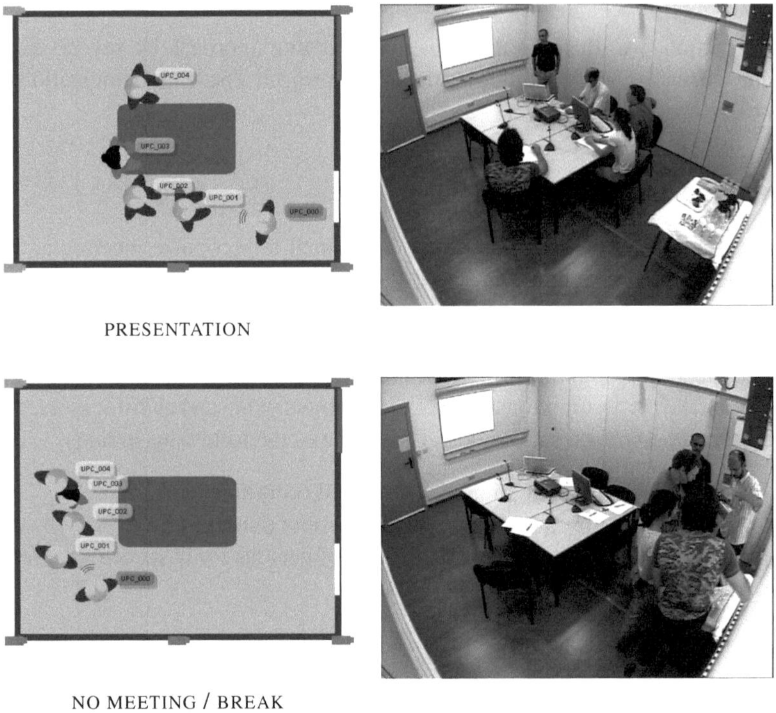

Fig. 12.3. Screenshots of presentation and break in the UPC'06 recording with SitCom's schematic visualization.

as information about the 2D face bounding box and facial feature locations visible in all camera views.

As the CLEAR recordings do not include the meeting state annotation, we asked three human annotators to label the corpus by appropriate meeting states. Their task was to identify one of the meeting states described in Table 12.1. Figure 12.3 shows a schematic visualization and corresponding pictures provided to the annotators (for presentation and break states). We can see the position and heading of each participant, visible legs for sitting persons, and highlighted bubbles for talking persons.

Table 12.2 shows the lengths of individual recordings, the number of generated feature vectors, and the data set to which they belong.

For the design, debugging, and execution of the situation model, we used the SitCom tool described in Chapter 29 of this book.

For statistical model training, tuning, and evaluation, we used the WEKA tool [19], evaluating various statistical modules for classification and feature selection.

Table 12.3 shows the percentage of correctly classified instances (10-fold cross-validation) using the selected features on training scenarios for the following classifiers: a *zero rule* classifier, a simple classifier that always predicts the most likely

	Length	Feature vectors	Data Set
AIT'07	20 min	1200	evaluation
IBM'07	20 min	1200	evaluation
FBK'07	20 min	1200	training
UKA'07	20 min	1200	training
UPC'06	13 min	780	evaluation
UPC'07	23 min	1380	training

Table 12.2. Data sizes of recordings used for training or evaluation.

meeting state, as observed in training data; a *Bayesian network* with structure trained from data; a decision-tree-based classifier (*C4.5*); and a *random forest* classifier. For details about these classifiers and their parameters, see [19].

We present the results separately for several categories of "modality", assuming that all the perceptual components or sensors may not be available in a particular room configuration. The modalities are position (**P**), heading (**H**), and speech activity (**S**).

Modality	**P**	**PH**	**PS**	**PHS**
Zero rule	72.72	72.72	72.72	72.72
Bayesian network	77.07	78.39	82.77	82.98
C4.5 classifier	77.65	78.53	82.32	83.35
Random forest	77.86	79.48	83.14	84.09

Table 12.3. Comparison of different classification methods.

We evaluate the results on the evaluation part of CLEAR data (*IBM07, AIT07,* and *UPC06* seminars) from Table 12.2 for the model built using the *random forest* classifier on the training data and applied to the evaluation data with features selected by the WEKA tool.

Both the classification methods and human annotators were uncertain in distinguishing between *presentation* and *discussion* states.

We achieved an accuracy of 76.96% when three meeting states were possible, and we found that both the classification method and the human annotator were uncertain in distinguishing between *presentation* and *discussion* states. The results for distinguishing between two states only (*meeting* and *no meeting*) reached a considerable accuracy of 95.91%.

More details about the meeting state recognition task can be found in [11, 14].

12.2.2 Interaction Group Detection

This example addresses the problem of detecting changing interaction group configurations in a smart environment. During a meeting, participants can form one big group working on the same task, or they can split into subgroups doing independent tasks in parallel. Our objective is to determine the current small-group configuration, i.e., who is interacting with whom, and thus which interaction groups are formed. As we focus on verbal interaction, one group has a minimum size of two individuals (assuming that isolated individuals do not speak). The speech of each meeting participant is recorded using a lapel microphone. An automatic speech detector parses this multichannel audio input and detects which participant stops and starts speaking. We admit the use of lapel microphones in order to minimize correlation errors of speech activity of different participants; i.e., the speech of participant A is detected as the speech of participant B.

The proposed approach is based on an HMM implementation of the context model. The observations of the HMM are a discretization of speech activity events sent by the automatic speech detector. A situation network for a meeting with four participants has been created. Each possible interaction group configuration is represented by one situation. These situations are transformed into the states of an HMM (Fig. 12.4).

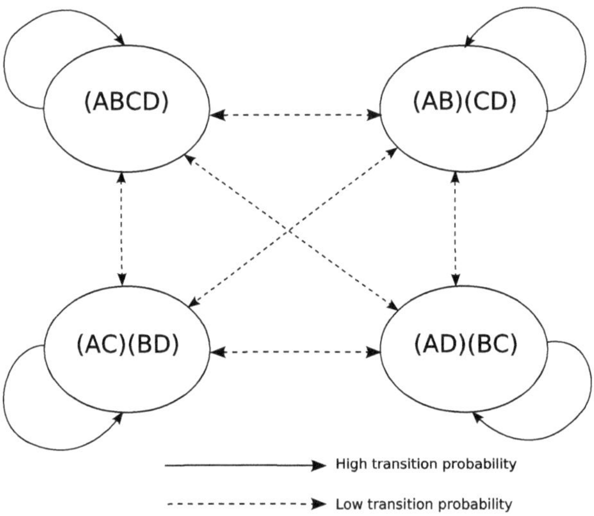

Fig. 12.4. States of the HMM implementation of the situation model for a meeting of four participants, A, B, C, D.

The probability distributions of the different states are specified based on conversational hypotheses. These conversational hypotheses assume that speech within an interaction group is more regulated than speech between distinct interaction groups. The transition probabilities between the states are set to a very low level in order to stabilize the detection of state changes, hence assuming that group changes occur in reasonable delays. To detect different group configurations, we apply the Viterbi algorithm [17] to the flow of arriving observations.

To evaluate, we recorded three small-group meetings with four participants (Fig. 12.5). Using the HMM detector, we obtained a total recognition rate for the

Fig. 12.5. Example of a configuration of two groups of two participants.

small-group configurations of 84.8% [6]. Table 12.4 shows the confusion matrix for the three experiments. This matrix indicates for each group configuration the correct and wrong detections. The lines of the matrix contain the detection results, while the columns contain the expected response.

	(ABCD)	(AB)(CD)	(AC)(BD)	(AD)(CB)
(ABCD)	0.88	0.03	0.06	0.03
(AB)(CD)	0.08	0.87	0.05	0.00
(AC)(BD)	0.22	0.01	0.77	0.00
(AD)(CB)	0.04	0.04	0.08	0.84

Table 12.4. Confusion matrix.

The results are encouraging and tend to validate the conversational hypotheses to distinguish interaction groups. The Viterbi algorithm executed on long observation sequences is quite robust to wrong detections by the speech activity detector. However, a minimum number of correct speech activity detections is necessary, as the method relies on the information of who speaks at which moment. The use of lapel microphones made it possible to limit wrong detections, as these microphones are attached to a particular person (and thus should only detect his or her speech).

12.3 Perspective: Automatic Acquisition and Adaptation of Situation Models Based on User Feedback

The situation model has been proposed as an intuitive declarative representation of context, providing both a simple means for describing human (inter)actions and a powerful tool for implementation. The previous section illustrated how to model and implement situations in different examples. Situations, roles, and relations of the model have been handcrafted by experts. However, human behavior evolves over time. New activities and scenarios emerge in an augmented environment, while others disappear. To cope with these changes, a context-aware system itself needs to evolve by adapting its contextual representation of users and environment, i.e., its context model. These adaptations can be done by experts aware of changing user needs and sensor perceptions of the system. The challenge is, however, the automatic acquisition of situations and context, and ultimately the acquisition of the entities, roles and relations from which situations emerge [9].

Context-aware systems must adapt and develop while retaining continuity and stability for users [9]. Therefore, a context model must be acquired and developed through observations and interactions with users. Recently, we investigated several methods for automatically acquiring and developing context models. The proposed methods are part of a framework for acquiring high-level context models. Figure 12.6 illustrates the different steps of the acquisition and development process. The goal is to acquire different layers of the context model with different degrees of supervision. As we consider situation to be the essential part of the context model, most acquisition methods and processes concern situation.

Further details on the framework and associated learning methods can be found in [4].

Several evaluations have been conducted based on the implementation of the proposed framework in a smart home environment. These evaluations showed an overall situation recognition rate of 88.58% and 76.48% after a situation split with feedback from the user [5]. Even though the obtained results are encouraging, error rates are still excessive. Further improvements in detection and learning algorithms are necessary in order to provide a reliable system that could be accepted and evolved by a user in his or her daily life.

12.4 Conclusion

In the chapter, we showed that modeling context is necessary when we want to sense and respond correctly to human activity in augmented environments. However, it is difficult (or even impossible) to model all possible contextual states reflecting human activity in the scene. Further, not all possible contextual cues can be sensed and integrated into a contextual model. As consequence, to realize functioning context-aware applications, we need to assume a closed world, hence modeling only human activity that is essential for correct system behavior and services. The situation model is proposed as intuitive declarative representation of context, providing both a simple

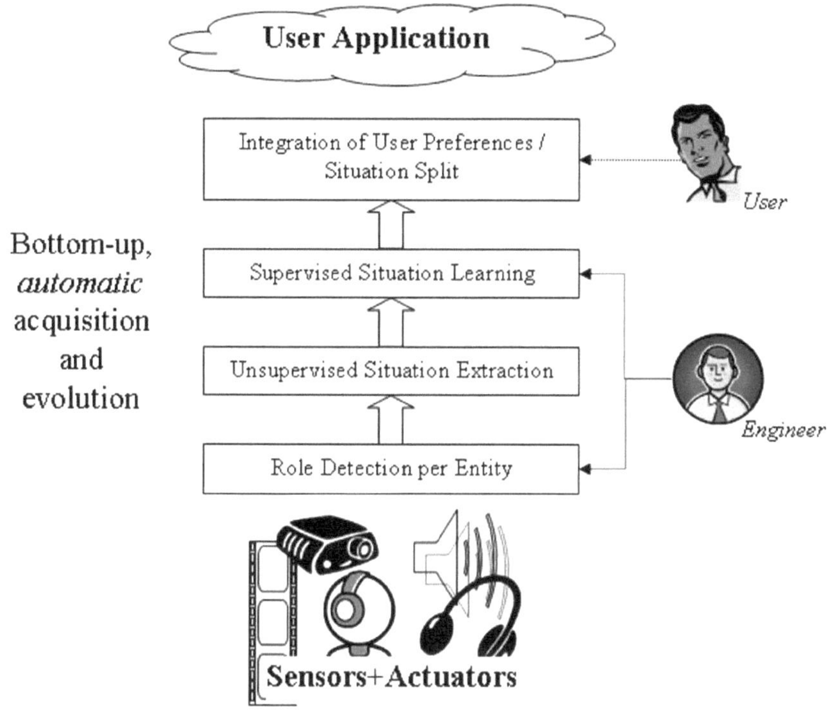

Fig. 12.6. Bottom-up automatic acquisition and adaptation of a context model.

means for describing human (inter)actions and a powerful tool for implementation. Two example implementations of the situation model have been presented and illustrated by functioning applications: meeting state detection and interaction group detection.

To conclude, we see that representing and implementing context is a nontrivial problem. Depending on application domain and sensed contextual cues, many different models and implementations are proposed. There is, of course, no general solution and framework adapted for all context-aware applications. We believe, however, that high-level models using intuitive notions of human activity are preferable, as they provide both intelligibility of human behavior in the scene and easy definition or adaptation by nonexperts.

References

1. G. D. Abowd and E. D. Mynatt. Charting past, present, and future research in ubiquitous computing. *ACM Transactions on Computer-Human Interaction*, 7(1):29–58, 2000.
2. J. F. Allen. Maintaining knowledge about temporal intervals. *Communications of the ACM*, 26(11):832–843, 1983.

3. J. F. Allen. Towards a general theory of action and time. *Artificial Intelligence*, 23(1):123–154, 1984.
4. O. Brdiczka. *Learning Situation Models for Providing Context-Aware Services*. PhD thesis, INP Grenoble, May 2007.
5. O. Brdiczka, J. Crowley, and P. Reignier. Learning situation models for providing context-aware services. In *HCI International*, July 2007. (in press).
6. O. Brdiczka, J. Maisonnasse, and P. Reignier. Automatic detection of interaction groups. In *Proceedings of International Conference on Multimodal Interfaces (ICMI)*, pages 32–36, Oct. 2005.
7. K. Cheverst, N. Davies, K. Mitchell, A. Friday, and C. Efstratiou. Developing a context-aware electronic tourist guide: some issues and experiences. In *CHI '00: Proceedings of the SIGCHI Conference on Human Factors in Computing Systems*, pages 17–24, New York, NY, 2000. ACM Press.
8. J. Coutaz and G. Rey. Foundations for a theory of contextors. In *Proceedings of the 4th International Conference on Computer Aided Design of User Interfaces*, 2002.
9. J. Crowley, O. Brdiczka, and P. Reignier. Learning situation models for understanding activity. In *Proceedings of 5th International Conference on Development and Learning, Bloomington*, May 2006.
10. J. L. Crowley, J. Coutaz, G. Rey, and P. Reignier. Perceptual components for context aware computing. In *Proceedings of UbiComp*, pages 117–134, London, 2002.
11. J. Cuřín, P. Fleury, J. Kleindienst, and R. Kessl. Meeting state recognition from visual and aural labels. In *Proceedings of 4th Joint Workshop on Multimodal Interaction and Related Machine Learning Algorithms*, Brno, Czech Republic, Jun. 2007.
12. A. K. Dey. Understanding and using context. *Personal and Ubiquitous Computing*, 5(1):4–7, 2001.
13. P. Dourish. What we talk about when we talk about context. *Personal and Ubiquitous Computing*, 8:19–30, 2004.
14. P. Fleury, J. Cuřín, J. Kleindienst, and R. Kessl. On handling conflicting input in context-aware applications. In *Workshop on Multimodal Interaction and Related Machine Learning Algorithms (MLMI)*, Edinburgh, UK, 2005.
15. S. W. Loke. Representing and reasoning with situations for context-aware pervasive computing: A logic programming perspective. *The Knowledge Engineering Review*, 19(3):213–233, 2005.
16. D. Mostefa, N. Moreau, K. Choukri, G. Potamianos, S. M. Chu, A. Tyagi, J. R. Casas, J. Turmo, L. Christoforetti, F. Tobia, A. Pnevmatikakis, V. Mylonakis, F. Talantzis, S. Burger, R. Stiefelhagen, K. Bernardin, and C. Rochet. The CHIL audiovisual corpus for lecture and meeting analysis inside smart rooms. In *Language Resources and Evaluation*. Springer, New York, 2007.
17. L. R. Rabiner. A tutorial on hidden Markov models and selected applications in speech recognition. *Proceedings of the IEEE*, 77(2):257–286, 1989.
18. R. Stiefelhagen, K. Bernardin, R. Bowers, R. T. Rose, M. Michel, and J. Garofolo. The CLEAR 2007 evaluation. In *Multimodal Technologies for Perception of Humans, Proceedings of the International Evaluation Workshops CLEAR 2007 and RT 2007*, LNCS 4625, pages 3–34, Baltimore, MD, May 8-11 2007.
19. I. H. Witten and F. Eibe. *Data Mining: Practical Machine Learning Tools with Java Implementations*. Morgan Kaufmann, San Francisco, 2000.

13

Targeted Audio

Dirk Olszewski

DaimlerChrysler AG, Research and Technology, Böblingen, Germany

Targeted audio aims at creating personal listening zones by utilizing adequate measurements. A person inside this listening zone shall be able to perceive acoustically submitted information without disturbing other persons outside the desired listening zone. In order to fulfill this demand, the use of a highly directional audible sound beam is favored. The sound beam shall be aimed at the respective listening zone target, thus implicating the expression *targeted audio*.

In general, moving from within a sound target to outside, a person will never experience an abrupt sound-level reduction from maximum to zero, similar to switching off the sound. Instead, resulting from physical laws, one has to accept a smooth sound-level decrease to certain levels outside the target. Therefore, a criterion has been evaluated first to determine the quality of targeted audio for the purpose of creating personal listening zones. It considers the audio beam directivity in terms of the sound level's decrease slope from the target to its vicinity since this influences the cross-talk ratio to the sound target's vicinity. Additionally, the audio sound level and its frequency response, which influence speech intelligibility, have been considered. Resulting from this criterion, adequate requirements for targeted audio have been defined: The audio sound level shall be reduced by 20 dB(A), respectively 10 dB at 1 kHz, at a distance of 1 m from the sound target. In several experiences, this cross-talk level reduction has been shown to be sufficient for not significantly disturbing any people nearby. The audio sound level shall achieve 80 dB(A), respectively 80 dB at 2.5 kHz, at the sound target to deliver enough headroom for speech reproduction. In order to achieve sufficient speech intelligibility, the audio sound frequency response shall cover the range between 800 Hz and 6 kHz.

A sound beam's degree of directivity in general directly depends on the sound source's physical dimensions compared to the emitted sound wavelength. Using a standard electrodynamic loudspeaker as a sound source, its dimensions have to be significantly greater than the audio wavelength and therewith roughly in the range of 1 m or more in diameter in order to achieve significant directivity at audio wavelengths. Simultaneously, sidelobes are generated due to interference between sound waves that are emitted by neighboring parts of the sound source, which lead to unwanted sound emission to the target zone's surrounding area. In contrast, a so-called parametric ultrasound loudspeaker is able to produce a highly directional audio sound beam that is virtually free from sidelobes, even when its physical dimensions are smaller than the respective audio wavelength. These facts make the ultrasound loudspeaker very suitable for applications within CHIL, since highly directional audio sound beams can be produced from very compact and therefore flexible devices.

The parametric ultrasound loudspeaker utilizes the nonlinearity of sound propagation in air to generate a highly directional audio sound beam. It therefore creates audible sound from ultrasound. The nonlinearity basically results from the air's nonlinear pressure/density relationship. For small amplitudes and therefore small sound pressure levels as they are generated in the field of everyday linear acoustics, the nonlinearity is very weak and therefore can be disregarded. At high amplitudes or high sound-pressure levels, the nonlinearity becomes significant and can be utilized for our application; see Fig. 13.1. At this point, the theories of nonlinear acoustics have to be applied.

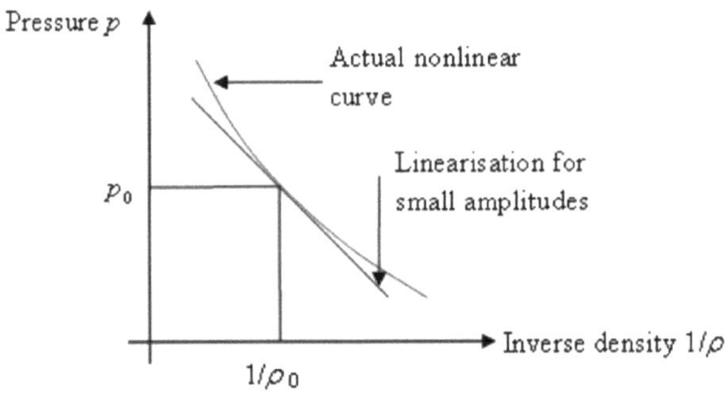

Fig. 13.1. The air's nonlinear pressure/density relationship.

If a nonlinear system is loaded with a combination of multiple signals with different frequencies at its input, the system produces additional sum and difference frequencies at its output. In the same way, the ultrasound loudspeaker uses an ultrasound carrier signal that is modulated by an audio signal and utilizes the nonlinearity of sound propagation in air to create sum and difference frequencies out of the modulated ultrasound spectrum. In an appropriate configuration, the difference frequency at the output corresponds to the modulating audio signal while all other spectral components, including the modulation spectrum and the sum frequencies, are in the ultrasound range and are therefore inaudible.

In practice, the modulated ultrasound signal is emitted at a high sound-pressure level to enable the utilization of the air's nonlinearity. Additionally, it is emitted as a narrow beam; see Fig. 13.2. This is achieved by having the ultrasound source's dimensions smaller than the audio wavelength but greater than the ultrasound wavelength, which is quite simple since the latter is in the range of only a few millimeters.

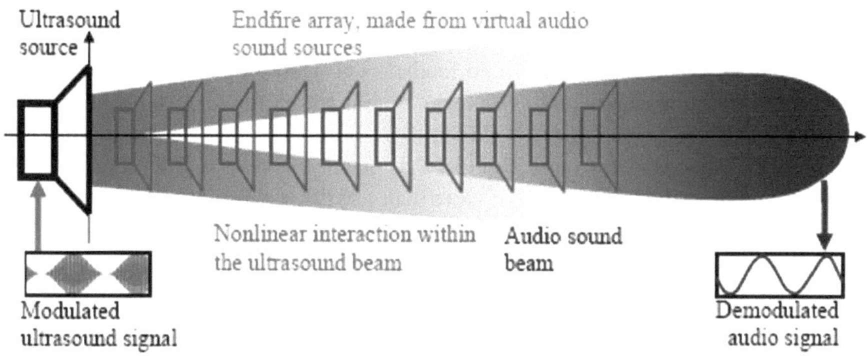

Fig. 13.2. The modulated ultrasound beam creates an endfire array that consists of virtual audio sound sources.

Due to the nonlinear interaction, as mentioned above, the audio signal is demodulated inside a column in front of the ultrasound source and forms virtual sources that form an endfire array. For the direction of the ultrasound beam, the elementary audio sources are in phase and their contributions to the overall audio sound generation add up in this direction. For all other directions, they are out of phase and their sound waves interfere destructively. Finally, the endfire array generates a highly directional audio sound beam in the direction of the inaudible ultrasound beam, although the sound source's dimensions are small compared to the audio wavelength. With increasing distance from the ultrasound source, the ultrasound level decreases – due to atmospheric absorption and spherical spreading – more and more until it becomes too weak to still utilize the air's nonlinearity to generate audio sound. This point defines the endfire array's generation length, which itself defines the maximum usable audio sound level and directivity.

Unfortunately, the ultrasound loudspeaker has two significant drawbacks: At first, the frequency response of the generated audio sound is characterized by a second-order highpass filter and therefore features a 12-dB/octave slope down to low frequencies. As a result, the system is not able to reproduce low audio such as bass frequencies with a sufficient level. At least, this lack of low-frequency reproduction can be partially compensated to achieve our quality goal mentioned above by driving ultrasound emitters with high Q-factors. By means of amplitude modulation, lower audio frequency content, which is located in the sidebands nearer to the carrier, is – due to the transducer's resonance peak – emitted with higher intensity compared to higher audio frequencies. The second drawback results from the quadratic dependence of the audio sound level from the ultrasound level. As a consequence, the demodulated audio signal contains not only the origin frequency but its first harmonic as well, which represents an unwanted distortion component. While the origin frequency level depends linearly on the modulation depth, the harmonic level depends quadratically. Therefore, by choosing an appropriate modulation index, a compromise between maximum audio level and minimum distortion has to be found.

In order to engineer an ultrasound loudspeaker according to our requirements, we first processed research studies on technologies for highly directional ultrasound beams. As a result, we decided to build ultrasound source arrays from multiple single emitters. We selected ultrasound emitters that match our design criteria as follows: high nominal sound-pressure level of more than 120 dB at a distance of 30 cm; highly resonative transfer function with Q-factors of more than 20; small size; commercial availability; high cost-performance ratio; see Fig. 13.3.

Fig. 13.3. Ultrasound transducers used for parametric ultrasound loudspeakers.

An ultrasound loudspeaker software tool, based on MATLAB®, has been adapted to simulate endfire arrays for targeted audio. It enables us to simulate beam patterns and directivities of various emitter and array geometries and designs to optimize our hardware systems.

The development of targeted audio systems within CHIL has been performed in an evolutionary manner; each step of this evolution process has been named Soundbox. The first milestone, Soundbox 0, was intended for the use with laptop computers. It is basically an array in the shape of a slim stick and is intended to be mounted on notebook computers in a way that only the person sitting in front of and working on the computer would have been received targeted audio. While it rapidly became clear that CHIL scenarios need concepts differing from this one, studies with Soundbox 0 gave us useful experience for further development steps. Since the using distance, which measures the physical distance between the user's ears and the sound source, is generally low (smaller than 50 cm), several facts limit the ultrasound loudspeaker's applicability in this case: At first, the small using distance results in a small usable generation length, limiting maximum usable audio level and directivity to insufficient values. Additionally, the small using distance causes not only the audio beam but also the ultrasound beam to reflect off the user's body. After the reflection, the latter is – due to the small distance – usually still strong enough to further generate audio sound. Therefore, audio sound is spread to a wider than desired area and the targeted audio is not really targeted any more. Finally, one has to be concerned with medical safety since high ultrasound levels have to be used to utilize the nonlinearity. To prevent damage to the human auditory system, sound exposition levels should be kept below 140 dB in the ultrasound range. At a close distance, this level might be exceeded; reducing the ultrasound levels in the near field, on the other hand, will decrease the usable audio level and directivity as well. Hence, it has to be accepted that the use of an ultrasound loudspeaker at small using distances is usually unsuccessful. Additionally, human auditory systems should be protected from being exposed to high ultrasound levels without decreasing the level itself. We therefore equipped all our subsequent prototypes with proximity sensors that turn off the ultrasound carrier when humans are detected at close distances.

The next evolution step, Soundbox I, includes signal conditioning electronics, e.g., dynamic range control to increase efficiency, an amplitude modulator, a power amplifier, and a rectangularly shaped array of piezoceramic ultrasound transducers; see Fig. 13.4.

Fig. 13.4. Soundbox I with fixed audio beam directivity.

Measurements of frequency response, directivity, audio sound-level output, and others have been done with these systems. Results are as follows: The systems are sufficient telephone-quality for speech or music at sound-pressure levels up to 80 dB (usual conversations take place at levels around 60 dB, which, for the human ear, is approximately four times weaker than 80 dB). The beam width – measured inside an anechoic chamber – is sufficiently small not to disturb other people next to a target person, but due to reflections inside usual meeting rooms (such as CHIL smart rooms), sound is scattered over a wider area than desired. Generally, our quality requirements from above are matched by this system. Figure 13.5 shows the audio frequency response.

While the theoretical audio frequency response in Fig. 13.5 shows the 12-dB/octave slope down to low frequencies as mentioned above, the measured curve shows a partially compensated frequency response due to the high Q-factor of the

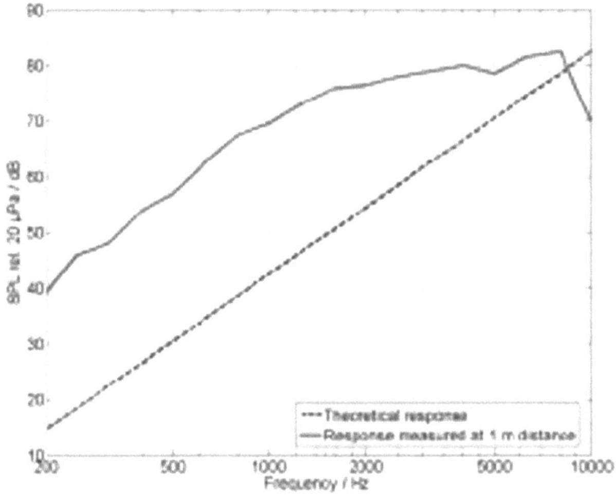

Fig. 13.5. Theoretical and measured audio frequency response.

ultrasound emitters. The next step, Soundbox II, has been developed as the first targeted audio with beam-steering capability, while all predecessors used a fixed beam direction. The system consists of a hybrid system, combining electronic (phased array technique) and mechanical (tilting) beam-steering approaches; see Fig. 13.6.

Fig. 13.6. Soundbox II shows a hybrid beam-steering concept.

Our experiments clearly show that steering of audio beams generated by ultrasound loudspeakers can be done by applying the phased array technique. Although emitters have been used that do not match criteria for phased array purposes at first sight, the hybrid approach still delivers an appropriate beam-steering performance. The system realized for our experiments mainly consists of two units: a DSP box that accounts for signal processing on four channels and an array box containing four mechanically turnable subarrays arranged as louvers. When the steering angle has to be changed, the louvers are turned into the desired direction while the DSP computes time delays matching the desired angle for each channel and passes the delayed signals to the louvers. In this way, an acoustic wavefront is emitted into the desired direction. The system allows hybrid steering of the audible beam within an angle of +/- 40° in the horizontal plane, while steering in the vertical plane over 270○ is done by turning the whole array box around its mounting handle by a stepper motor; see Fig. 13.7. Both axes are controlled via a USB interface. Steering software has been developed for Windows-based PC systems. Audio features are very similar to those of Soundbox I.

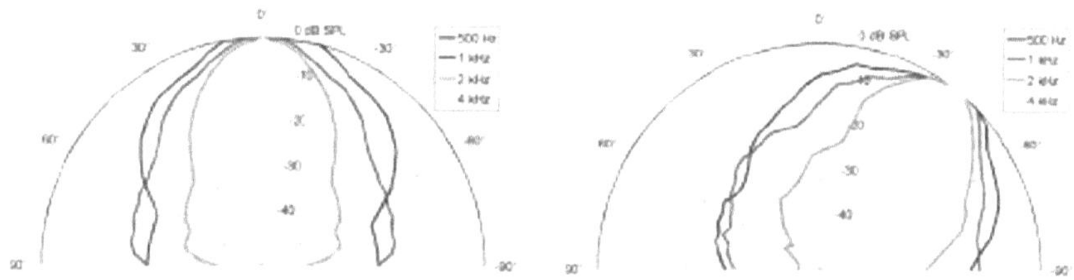

Fig. 13.7. Soundbox III audio directivity plots at different steering angles; left: 0°, right: 40°.

Fully electronic beam steering without mechanical tilting would be unsuccessful because unidirectional devices are needed. Instead, the ultrasound emitters used in the Soundbox systems feature certain directivity, as can be evaluated from Fig. 13.8.

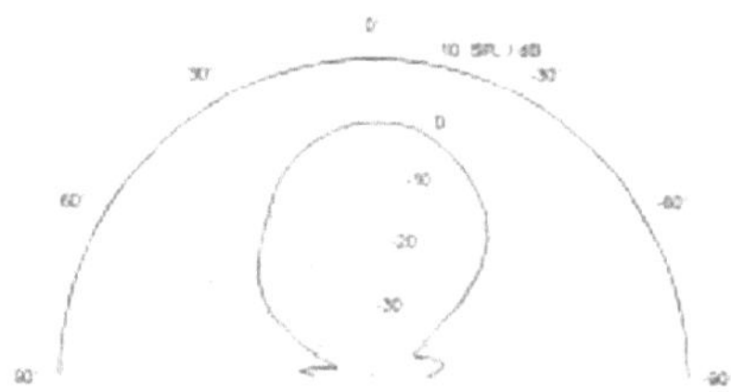

Fig. 13.8. Single ultrasound emitter directivity.

Only directional ultrasound emitters are capable of delivering high-level ultrasonic output, which is needed for the parametric sound generation. This leads to the consequence of additional steering in the way of Soundbox II or the need for a presteered profile, which we use in our final evolution step, the Soundbox III prototype. It is able to deliver up to four different audio sound beams with different content into different directions simultaneously. In contrast to Soundbox II, no moving parts are generated. Instead, it utilizes a set of 13 emitter arrays with fixed beams, pointing different directions. In contrast to phased array beam-steering methods, only those emitters needed for the desired direction are activated, thus reducing energy consumption. The disadvantage of using directional ultrasound emitters (which makes them less useful for phased array approaches) has therefore been turned into an advantage. The beams can be switched by using the hardware-software interface, thus enabling beam steering and tracking moving targets; see the example in Fig. 13.9.

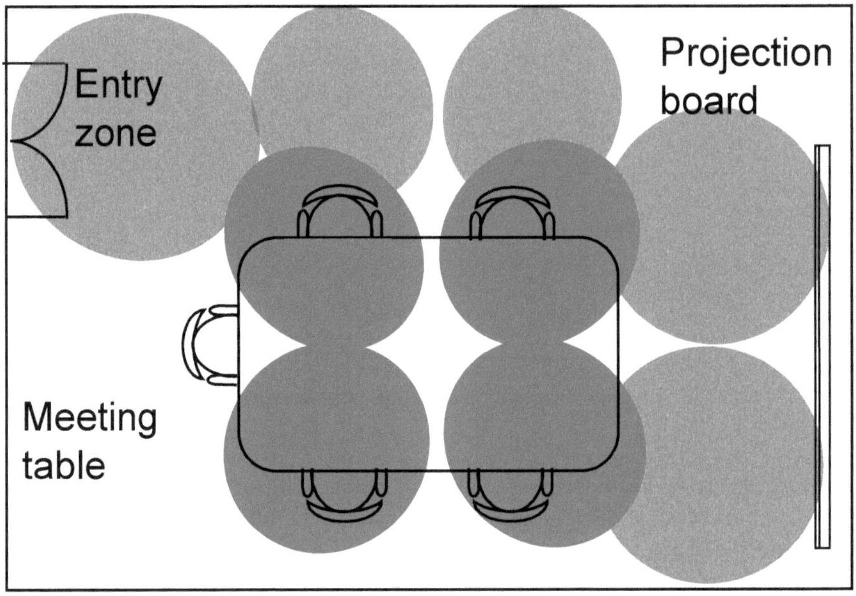

Fig. 13.9. Meeting room example; emitter device mounted under the ceiling above the table. Colored spots show audio target zones. Only 8 out of a possible 13 audio spots are illustrated.

The Soundbox III system (see Fig. 13.10) consists of two separate units: a signal processing device, which comes in a 19" rack unit, and an emitter device, which covers all 13 emitter arrays in one compact unit and is intended for ceiling mounting. The signal processing device features an interface that allows the routing process on a remote computer to be fulfilled independent of the type of operating system. This is done by applying 8-bit patterns. Additionally, the bit pattern can be set manually on the front panel to enable manual steering.

Fig. 13.10. Soundbox III hardware parts; left: emitter unit; right: signal processing unit.

Our CHIL Soundbox evolution shows that ultrasound loudspeakers can be used to deliver targeted audio to chosen areas and match our quality criteria. Systems with fixed and even steerable audio beam direction have been developed and successfully installed in CHIL smart rooms. Further details can be found in the references ([4, 2, 5, 3, 1]).

References

1. D. Olszewski. Optimum carrier frequency for ultrasound loudspeaker. In *27th Symposium on Ultrasonic Electronics (USE)*, Tsukuba, Japan, Nov. 2006.
2. D. Olszewski and K. Linhard. Highly directional multi-beam audio loudspeaker. In *Interspeech*, Pittsburgh, PA, 2006.
3. D. Olszewski and K. Linhard. Imperfections of parametrically generated sound beams caused by reflexions. In *27th Symposium on Ultrasonic Electronics (USE)*, Tsukuba, Japan, Nov. 2006.
4. D. Olszewski and K. Linhard. Messungen zur Schallerzeugung durch einen parametrischen Ultraschalllautsprecher. In *DAGA*, Braunschweig, Germany, 2006.
5. D. Olszewski and K. Linhard. Optimum array configuration for parametric ultrasound loudspeakers using standard emitters. In *IEEE International Ultrasonics Symposium*, 2006.

14

Multimodal Interaction Control

Jonas Beskow, Rolf Carlson, Jens Edlund, Björn Granström, Mattias Heldner, Anna Hjalmarsson, Gabriel Skantze

Kungl Tekniska Högskolan, Centre for Speech Technology, Stockholm, Sweden

No matter how well hidden our systems are and how well they do their magic unnoticed in the background, there are times when direct interaction between system and human is a necessity. As long as the interaction can take place unobtrusively and without techno-clutter, this is desirable. It is hard to picture a means of interaction less obtrusive and techno-cluttered than spoken communication on human terms. Spoken face-to-face communication is the most intuitive and robust form of communication between humans imaginable. In order to exploit such human spoken communication to its full potential as an interface between human and machine, we need a much better understanding of how the more human-like aspects of spoken communication work.

A crucial aspect of face-to-face conversation is what people do and what they take into consideration in order to manage the flow of the interaction. For example, participants in a conversation have to be able to identify places where it is legitimate to begin to talk, as well as to avoid interrupting their interlocutors. The ability to indicate that you want to say something, that somebody else may start talking, or that a dialog partner should refrain from doing so is of equal importance. We call this *interaction control.*

Examples of the features that play a part in interaction control include the production and perception of *auditory cues* such as intonation patterns, pauses, voice quality, and various disfluencies; *visual cues* such as gaze, nods, facial expressions, gestures, and visible articulatory movements; and *content cues* like pragmatic and semantic (in)completeness. People generally seem to use these cues in combination, and to mix them or shift between them seamlessly. By equipping spoken dialog systems with more human-like interaction control abilities, we aim to move interaction between system and human toward the intuitive and robust communication among humans.

The bulk of work on interaction control in CHIL has been focused on auditory prosodic cues, but visual cues have also been explored, and especially through the use of embodied conversational agents (ECAs) – human-like representations of a system, for example, animated talking heads that are able to interact with a user in a natural way using speech, gesture, and facial expression. ECAs are one way of leveraging the inherent abilities that we all possess in terms of decoding information in speech, visible articulation, intonation, voice quality, facial displays, gestures, and gaze and holds the potential of improving the effectiveness, robustness, and naturalness of human-computer interaction. We have also explored the usefulness of gaze direction for interaction control purposes.

This chapter describes work along a number of lines in order to multimodally capture and mimic the flow of human-human interaction. Section 14.1 describes experiments and techniques to improve the flow of the interaction by analyzing what users say and do, and Section 14.2 concerns how multimodal output – an animated talking head and prosodically as well as temporally aware speech synthesis – can be used to achieve further improvements.

14.1 Interaction Control in Spoken Dialog Systems

As mentioned, the bulk of the work on interaction control in the CHIL project has been focused around auditory, prosodic cues. The choice was made partly because prosodic cues have several advantages from a system design point of view: They are available from the speech signal alone, and furthermore it may be possible to utilize them without using ASR. Thus, they rely less heavily on other technologies than, for example, semantic completeness or gaze tracking.

14.1.1 Silence Duration Thresholds

Current spoken dialog systems commonly detect where the user ceases speaking in order to find out where they should take their turn. The method is based on the assumption that speakers have finished what they intended to say when they become silent and that these points in time are also suitable places for the system to speak. Such *endpoint detection* triggers on a certain set amount of silence, or non-speech. The end-of-utterance detectors in current automatic speech recognition typically rely exclusively on a silence threshold somewhere between 500 and 2000 ms (sic!) (cf. [14] and references mentioned therein). The method makes sense; given that a speaker is allowed to complete what she or he intends to say, the end of the utterance is likely to coincide with silence at a place where an interlocutor might take the next turn. The method segments speech into reasonably sized units, in many cases corresponding to sentences or some sentence-like units. However, spontaneous conversational speech frequently contains silent pauses inside what we would intuitively group into turns, complete utterances, or sentence-like units, and inside what are indeed semantically coherent units, and dialog systems using silence-based endpoint detection run into problems with unfinished utterances when encountering spontaneous speech [2]. Experiments presented in [11] showed that dialog systems relying on silence-based segmentation run the risk of interrupting their users in as much as 35% of all silent pauses in the kind of speech investigated.

14.1.2 Prosodic Boundaries

It follows that something is needed in addition to silence, but what should that be? A starting point is to find out what people are able to do and distinguish in terms of suitable and unsuitable places to say something in a spoken interaction. Places where speaker changes may occur (i.e., suitable places to speak) are closely linked to the notion of prosodic boundaries. Places where the speaker is allowed to finish what she is going to say by nature coincide and co-occur with prosodic boundaries. However, it is far from obvious that every kind of prosodic boundary represents a possible place for turn-taking.

Therefore, early on in CHIL, several tests were carried out to examine the relationship between prosodic boundaries and suitable places for speaker changes. Heldner et al. [18] report two of these tests, a listening test where subjects rated the appropriateness of made-up turn-takings, and a production experiment where another group of subjects was asked to indicate suitable places for turn-takings.

The stimuli used in the listening test were made up of fragments from a seminar on speech technology given in English by a lecturer of German origin. The positions to be evaluated were selected based on manual annotations of perceived prosodic boundaries using a three-level convention (strong vs. weak vs. no prosodic boundary). Each of the stimuli in the listening test consisted of a fragment from the seminar followed by a fragment of a question from somebody in the audience. These stimuli were subsequently presented to a number of subjects in a listening experiment where the task was to rate whether the questioner enters the conversation at an appropriate place. The results of the listening test showed that the listeners generally judged the turn-takings in strong boundary positions to be more appropriate than those in weak or no boundary conditions, and that weak boundaries were slightly better than no boundaries.

A subsequent production experiment was carried out to let the subjects indicate possible places for turn-takings in the same speech material that was used in the listening experiment. Here, each trial consisted of the subjects listening to a lecturer's part of the seminar and as soon as they thought it was appropriate to take the turn, they pressed a key, the sound of the lecturer stopped, and the question, *"What about <um> could you give us some <hrm> rough idea what"* was played. Subsequently, the sequence of lecturer part and question was repeated and the subjects had to rate whether the trial was successful (discard, keep), the timing of the turn-taking on a micro level (early, OK, late), as well as the politeness of the turn-taking (rude, neutral, polite). The possible places for turn-takings indicated by our subjects were analyzed in terms of whether they occurred at a weak or a strong boundary, or not at any boundary.

The production experiment supports the findings from the listening test by showing a strong preference for turn-takings at strong boundaries, although turn-takings may also occur at certain weak and no boundaries according to our subjects. Furthermore, as nearly all strong boundaries occurring in the experiment were marked as possible turn-taking positions, it is reasonable to assume that strong boundaries generally make up appropriate turn-taking positions, at least in this communicative situation.

Several additional studies highlighting various aspects of the relationships between prosodic boundaries and turn-taking have been performed. Some of these have been published in [18, 5, 17, 35].

14.1.3 Prosodic Cues

We opted for prosody, then, but there are still choices to be made. Previous work suggests that a number of prosodic or phonetic cues are associated with turn-yielding and thus potentially relevant for interaction control. These cues include, apart from silent pauses, phenomena such as various intonation patterns (rises, falls, down-steps, up-steps); decreases in speech rate; final lengthening; intensity patterns; centralized vowel quality; creaky voice quality; and exhalations. Note that both rises and falls have been associated with turn-yielding. These cues are typically located somewhere toward the end of the turn, although not necessarily on the final syllable [24, 25, 43, 15, 23, 28].

Similarly, there are studies suggesting that certain prosodic or phonetic cues are associated with turn-keeping, and these cues are, of course, also potentially relevant for the interaction control. They include phenomena such as glottal or vocal tract stops without audible release; a different quality of silent pauses as a result of these glottal or vocal tract closures; assimilation across the silent pause; and other "held" articulations (e.g., lengthened vowels, laterals, nasals, or fricatives) [23] [28]. In particular, level intonation patterns in the middle of the speaker's fundamental frequency range have been observed to act as turn-keeping cues in several different languages. For example, Duncan [8] reported that any pattern other than a level tone in the speaker's mid-register (a 2 2 | pattern in the Trager -Smith prosodic transcription scheme [40]) signals turn-yielding in English. Thus, the mid-level pattern acts as a turn-keeping signal, although Duncan did not use that term. Similarly, Selting [30] reported that level pitch accents before a pause are used to signal a turn-holding (or turn-keeping) in German; Koiso et al. [22] observed that flat, flat-fall, and rise-fall intonation patterns tended to co-occur with speaker holds (i.e., turn-keeping) in Japanese; and in another study on Japanese conversations Noguchi, and Den [27] reported that flat intonation at the end of pause-bounded phrases acts as an inhibitory cue for backchannels. Furthermore, in a study of final pitch accents and boundary tones in the turn-taking system of Dutch, Caspers [6] identified two intonation patterns that seem to be associated with turn-keeping: an accent lending rise followed by level high pitch (H* %) used for bridging syntactic breaks between utterances; and a filled pause with a mid-level boundary tone (M %) for bridging hesitations within syntactic constituents. However, Caspers could not find any intonation patterns clearly associated with turn-yielding. This observation led her to conjecture that turn-changing is the unmarked case and that only the wish to keep the turn needs to be marked with specific intonation patterns.

Based on this, we decided to see if this mid-level pattern could be automatically extracted and used to inform the interaction control of a spoken dialog system. A number of requirements must be placed on such an extraction if it is to be of any practical use.

14.1.4 `/nailon/` – Online Automatic Extraction of Prosodic Cues

In order to use additional information, including prosody, to improve interaction control in practical applications such as in spoken dialog systems, the information needs to be made available to the system, which places special requirements on the analyses. First of all, in order to be useful in live situations, all processing must be performed automatically, in real time, and deliver its results with minimal latency (cf. [31]). Furthermore, the analyses must be online in the sense of relying on past and present information only, and cannot depend on any right context or look ahead. There are other technical requirements: The analyses should be sufficiently general to work for many speakers and many domains, and they should be predictable and constant in terms of memory use, processor use, and latency.

In order to meet these requirements, **`/nailon/`**, a software package for online, real-time prosodic analysis, was developed. **`/nailon/`** is based on the Snack software library [34] and provides a number of analyses. Originally, the choice of methods to include in the analyses was in part ad hoc, in part based on literature. During development and testing, most of the ad hoc methods were underpinned. A more detailed description of the workings of **`/nailon/`** can be found in [12].

In order to extract data that can be used for making decisions about interaction control – that is, pitch patterns preceding silence –**`/nailon/`** does the following, in turn:

Voice/speech activity detection (VAD/SAD). The first step is to decide whether or not there is speech. The VAD built into **`/nailon/`** is trivial and intended for testing purposes only. VAD/SAD is a research area in its own right, and for real use, methods for plugging in external VAD/SAD are provided. Note also that channel separation is not provided by **`/nailon/`**, which expects channel-separated input – one speaker per channel. The VAD/SAD is performed on the frame level, but the results may be smoothened by subsequent processes.

Extraction of pitch, intensity, and voice. Next, an ESPS extraction of pitch, intensity, and voicing [39] is done for each frame judged to be speech. This extraction is modified to run repeatedly over a small window to run at real time with a latency of one frame.

Online speaker normalization. The pitch values are normalized against an incremental model of the speaker's range to provide range relative values in terms of standard deviation [42], which in turn gives an idea of "high", "medium", and "low" pitch, from the speaker's perspective. The method has been validated in [13].

Pause detection. **`/nailon/`** flags consecutive silence of a certain length. The length used in the tests has been 300 ms, but the system works with lengths below 200 ms on a normal desktop computer. The length chosen, plus a few milliseconds for processing, constitutes the smallest amount of silence needed in the speaker channel for the system to detect a suitable place to speak. It should be compared to the 500-2000 ms used by most current systems.

Convex hull extraction. Whenever silence of a certain length is found, a modified convex hull extraction [26] is used to find the last syllable-like entity – pseudo-psyllable or psyllable – preceding the silence. The use of convex hulls was inspired by Nick Campbell and colleagues, who used convex hulls for automatic classification of phrase final tones in the (JST/CREST ESP Project) [20].

Finally, the decision of whether or not a given silence is a suitable place to speak is made not by **`/nailon/`**, but by the interaction control software, which may be trained or rule-driven. In either case, the normalized pitch pattern over the last psyllable preceding the silence is the input used for such a decision.

The data extracted by **`/nailon/`** have been used to perform interaction control studies. In [11] it was shown that the number of incorrect turn-taking decisions can be reduced substantially by combining standard silence-based endpoint detection with an automatic classification of intonation patterns. In the process, it is also possible to decrease the length of the required silence without any loss in performance. This can be used to make a conversational computer more responsive by allowing it to reply faster without simultaneously making it more obtrusive.

Automatically classified level intonation patterns in the middle of the speaker's fundamental frequency range were found to act as turn-keeping cues, and may thus be used to avoid interrupting human interlocutors with high precision. Although there are several observations of the function of these mid-level intonation patterns, to our knowledge, they have never been used to avoid interrupting users of spoken dialog systems before.

14.1.5 Interaction Model

Before we turn to the production, or output, side of spoken human-machine interaction, we need a model of its flow that caters to simultaneous speech, multiple speakers, etc. For this purpose, we have developed an interaction model [10] (see Fig. 14.1) that can be viewed as an extended version of the AVTA system [21] but differs in that it models the situation subjectively and separately for each participant, while it lacks a combined, objective "God's view" of the interaction. Although we have not exploited this property within the CHIL project, it is worth noting that this subjectiveness is useful for modeling interaction under transmission latency, since it can capture the fact that the effect of round-trip latency is perceived differently by different subjects.

The model is computationally simple yet powerful. It consists of three parts: a state derived directly from each participant's speech activity, a state derived from the speech activity of all participants, and events representing changes in these states.

The first state (SPEECH/SILENCE) continuously models speech/nonspeech as a binary state on a per-participant level. At any given point in time, each participant may be either speaking or not speaking. The only input the model takes is speech/nonspeech decisions from each participant's VAD.

The second part of the model is a four-way decision of the communicative state (SELF/OTHER/NONE/BOTH), again repeated for each participant. These states are derived from the SPEECH/SILENCE state of each participant. From participant P's point

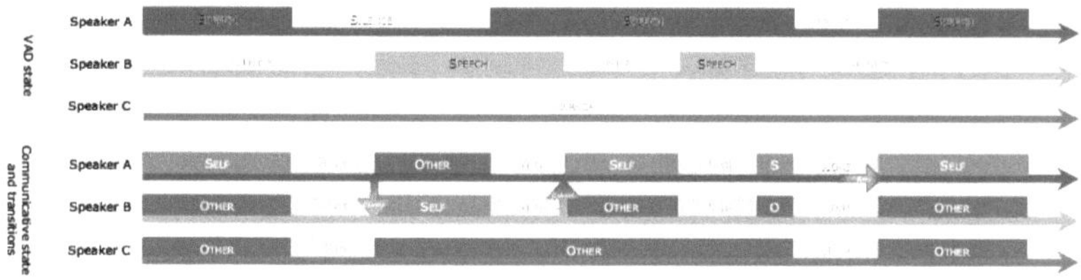

Fig. 14.1. Model of spoken interaction flow.

of view, the state is NONE if none of the participants is speaking. It is SELF if P is speaking but no one else is. If one or more other participants are speaking and P is silent, it is OTHER; finally, if both P and some other participant are speaking, it is BOTH.

Finally, the model includes transitions from one communicative state to another for each participant. If P is in state NONE and someone else starts speaking, P goes from NONE to OTHER and the participant who started speaking goes from NONE to SELF.

The system is modeled as any other participant in the model. At any given point in time, the current state (SELF/OTHER/NONE/BOTH) from the system's point of view can be read. If the system is silent and has something to say, either it may wait for the NONE state to appear, or it may decide to barge in. In making these decisions, other factors than the state of the interaction model can be weighted. Typically, factors such as how urgently the system wants to say something (e.g., if the message is that the fire alarm has been set off, the system should disregard any politeness considerations); what type of utterance the system wants to make (feedback and backchannels, for example, may be easier to fit into other people's speech and also cause less frustration if slightly mismatched); and, of course, cues such as intonation and gaze, as discussed above. Other contextual features matter as well, for example, the emotional state of the current speaker(s) (see Chapter 10).

14.1.6 Other Modalities: Eye Gaze

As widely noted in the literature and above, speaker changes in conversation correlate with a great many things that are not prosody. Among the more prominent ones is eye gaze. In the CHIL project, a study using a Tobii x50 eye tracker (see Fig. 14.2 and 14.3) corroborates the statement that eye gaze behavior is relevant for interaction control. The study shows that people tend to look more at each other when they listen than when they speak, as has been reported in many previous studies. The results also show that interlocutors tend to glance at each other around the time of speaker changes, presumably to ensure understanding as well as a smooth speaker change [19].

Fig. 14.2. Experimental setup.

Fig. 14.3. Tobii x50 eye tracker.

14.2 Multimodal Output and Interaction Control

Many of the interaction experiments in CHIL incorporate the use of an animated talking head. We now turn to provide a general overview of the talking head animation system used in this research, and specifically the generation of expressive visual speech animation, which is a prerequisite for reaching the type of conversation on human terms that we have targeted in the project. An experiment to assess the speech intelligibility gain provided by a talking head when used in combination with the targeted audio system that is described in Chapter 13 is also included, as it relates to interaction control in that the aim is not to disrupt the flow of a meeting, for example.

14.2.1 Animated Talking Head

Our talking head is based on the MPEG-4 Facial Animation standard [29]. It is a textured 3D model of a male face comprised of approximately 15,000 polygons. The mesh has been parameterized to allow for realistic deformation, using a framework based around a combination of professional 3D modeling tools and in-house custom algorithms, and a flexible animation engine.

In order to render visual speech movements, we employ a data-driven methodology. We start from a time-aligned transcription of the speech to be synthesized. The time-aligned transcription can be obtained from a text-to-speech system, if we are synchronizing with synthetic speech, or it can be produced by a phoneme recognizer (as in the Synface system [3]) or a phonetic aligner [33]. Next, an articulatory control model is applied to convert the time-aligned phonetic transcription of the utterance into control parameter trajectories to drive the articulation of the talking head model. In order to produce convincing and smooth articulation, the articulatory control model has to account for coarticulation, which refers to the way in which the realization of a phonetic segment is influenced by neighboring segments.

14.2.2 Expressive Speech

The articulatory control model has been trained to reproduce the facial movements of a real speaker. While previous work on visual speech synthesis typically has been aimed at modeling neutral pronunciation, this model is also capable of synthesizing expressive speech. We used an opto-electronic motion tracking system to collect a multimodal corpus of acted emotional speech. The system allows the dynamics of emotional facial expressions to be captured by registering the 3D coordinates of a number of reflective markers at a rate of 60 frames/second. For this study, our speaker, a male native Swedish amateur actor, was instructed to produce 75 short sentences with the six emotions of happiness, sadness, surprise, disgust, fear, and anger, plus neutral, yielding 7 times 75 recorded utterances. A total of 29 IR-sensitive markers were attached to the speaker's face, four of which were used as reference markers (on the ears and on the forehead). The marker setup largely corresponds to MPEG-4 feature point (FP) configuration.

The recorded corpus with expressive speech was then used to train articulatory control models based on PCA analysis [7] for the five emotions happy, angry, surprised, sad, and neutral. To provide suitable training data, the recorded marker positions were first encoded as MPEG-4 FAPs, and the audio track was phonetically labeled using a forced alignment system. The resulting models can later be used to synthesize articulatory movements for novel arbitrary Swedish speech, thereby modeling expression and articulation in an integrated fashion. Figure 14.4 shows snapshots from resulting animations of the same utterance for the four emotions happy, angry, surprised, and sad. See [4] for further details.

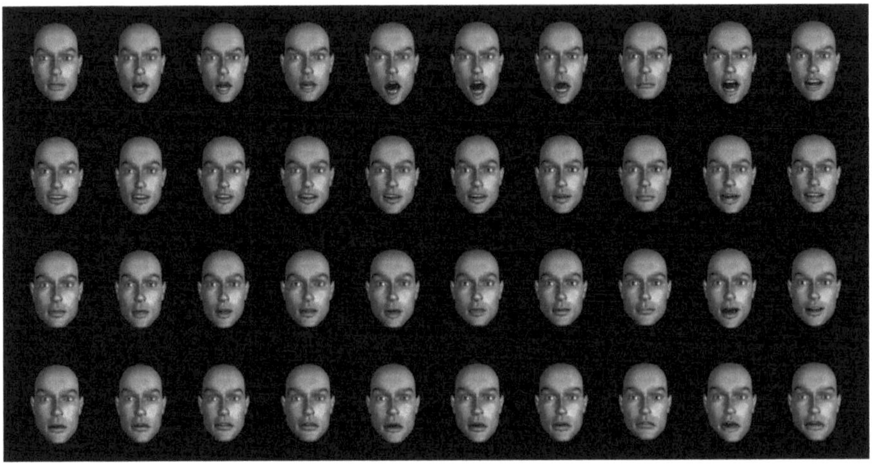

Fig. 14.4. Snapshots taken every 0.1 s from the animation of the fragment "jag ska köpa..." ("I will buy..."), synthesized with four different expressive speech models: happy (top row), angry (second row), surprised (third row), and sad (bottom row).

14.2.3 Unobtrusive Speech

While properly synchronized and articulated visual speech synthesis is an important property of animated talking agents for multimodal interaction that improves realism, it also adds to the intelligibility of the speech output [32]. This property may be exploited for spoken interaction in noisy environments. It may also be used, for example, in quiet meeting-room scenarios to deliver spoken messages to individuals without disturbing the meeting, since it allows the volume to be kept to a minimum. A talking head combined with directed audio makes it possible to achieve even less disturbance for meeting partners.

An experiment was conducted to test how the intelligibility of speech is affected by the listener's position relative to the audio beam and to measure the possible augmentation of intelligibility that the use of a talking head can provide for this type of application. Two different listening positions were tested: one where the subject was positioned in the audio beam (0°) and one where the loudspeaker was rotated 45° away from the subject. For each position, both an audio-only condition and a condition where a talking head was displayed on the screen were evaluated.

The distance between the subject's head and the loudspeaker was approximately 80 cm. To better simulate a real environment, no specific means were taken to ensure that the subject remained still during the experiment. Due to problems with sound reflection, and in order to minimize disturbance outside the audio beam, a low-output sound level was used. The task consisted of listening to nonsense VCV (vowel-consonant-vowel) words and identifying the consonant in each word. The seven consonants to be identified were [f, s, m, n, k, p, t] uttered in an [a_a] context, produced by a male speech synthesis voice. Each of these seven VCV words occurred a total of four times in each condition, and the presentation order was randomized. The order in which the different conditions were presented was also rotated to avoid having possible learning effects affect the overall result. The answering sheet consisted of a forced choice among the seven consonants in the test.

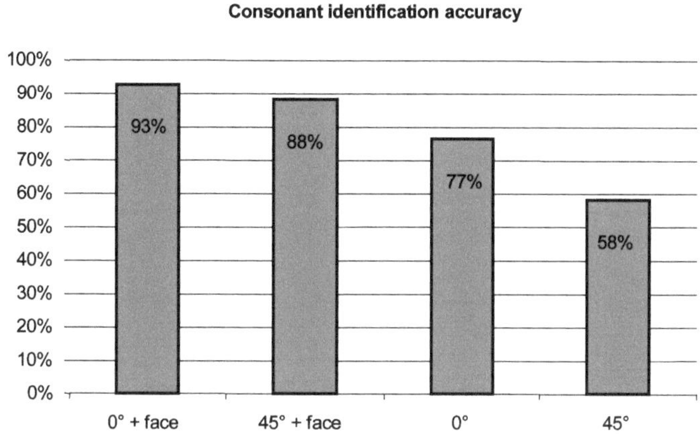

Fig. 14.5. Recognition rates for nonsense words with targeted audio, with and without talking head support.

The results of the test are shown in Fig. 14.5. The highest recognition accuracy, 93%, was obtained for the condition where the subjects were positioned in the audio beam and the talking head was used (0° + face condition). For the 45° + face condition, the recognition rate was 88%. When only the sound was available, 77% recognition accuracy was reached for the 0° condition, and 58% for the 45° condition. The result of this last condition (audio-only, 45°) was significantly ($p < 0.05$) from the other three conditions. The differences in accuracy for the 0° condition, between having access to a talking face or not (i.e., 0°+audio-only vs. 0°+face), was also significant ($p < 0.05$). For further details, see [38].

14.2.4 MushyPeek – An Experimental Framework

Another aspect affecting the decision, from the system's point of view, of whether to speak or remain silent, is that it is not a binary choice. The system can also opt to make "feelers" to see whether it is given the floor or not. It may do this vocally, for example, by "clearing its throat" softly, but it may also use gestures in an animated talking head representing the system. In [10], we present MushyPeek, an experimental framework inspired by [16], in which listeners can hear the speakers' real voices while watching what they are told to be graphic representations of the speakers and their gestures on monitors. The framework can be seen as a special case of the *transformed social interaction* discussed by Bailenson et al. [1]. In MushyPeek, the participants are placed in separate rooms, and each participant is equipped with a headset connected to a Voice-over-IP call (http://www.skype.com/). On both sides, the call is enhanced with SynFace [3] – a lip-synchronized animated talking head representing each participant. As both talking heads represent real persons (the participants), we refer to them as *avatars* in the following. This basic setup constitutes the communicative backbone of the framework. In addition, the framework contains experiment-specific components for voice activity detection (VAD), interaction modeling and control, gesture realization, and logging. All components communicate over TCP/IP connections. The framework is symmetrical in that both participants have the same setup. The general layout is shown in Fig. 14.6.

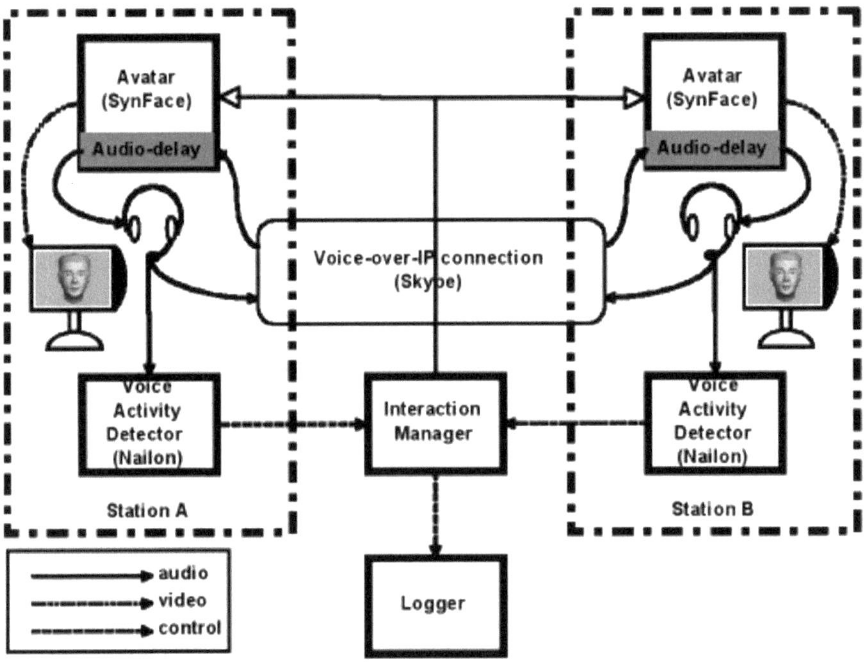

Fig. 14.6. The MushyPeek experimental framework.

A first experiment in MushyPeek using the interaction model and /nailon/ prosodic analysis, together with turn-taking gestures in an animated talking head, showed that it is indeed possible to unobtrusively affect people's willingness to release the floor in this manner [9].

14.2.5 Short Feedback Utterances

A final aspect that is crucial to creating an unobtrusive and smooth flow in human-machine interaction has to do with the manner in which the system says things. Within CHIL, KTH has investigated the use of small responsive backchannels [41] as well as unobtrusive and responsive feedback and clarification requests [37, 36]. These aspects have also been implemented in many demo applications, such as *the Hummer*, which provides backchannels such as "uh-huh" and "OK" at appropriate places, and an animated talking head doing narration while taking the listener's level of attention into consideration. The Hummer has been demonstrated in plenum at several conferences, and the narrating talking head has been demonstrated at technology days and at IST Helsinki 2007.

References

1. J. N. Bailenson, A. C. Beall, J. Loomis, J. Blascovich, and M. Turk. Transformed social interaction: Decoupling representation from behavior and form in collaborative virtual environments. *Presence: Teleoperators & Virtual Environments*, 13(4):428–441, Aug. 2004.
2. L. Bell, J. Boye, and J. Gustafson. Real-time handling of fragmented utterances. In *Proceedings of NAACL Workshop on Adaptation in Dialogue Systems*, pages 2–8. Carnegie Mellon University, Pittsburgh, PA, 2001.
3. J. Beskow, I. Karlsson, J. Kewley, and G. Salvi. *SYNFACE – A Talking Head Telephone for the Hearing-Impaired*, pages 1178–1186. Springer-Verlag, New York, 2004.
4. J. Beskow and M. Nordenberg. Data-driven synthesis of expressive visual speech using an MPEG-4 talking head. In *Proceedings of Interspeech 2005*, Lisbon, Sept. 2005.
5. R. Carlson, J. Hirschberg, and M. Swerts. Cues to upcoming Swedish prosodic boundaries: Subjective judgment studies and acoustic correlates. *Speech Communication*, 46:326–333, 2005.
6. J. Caspers. Local speech melody as a limiting factor in the turn-taking system in Dutch. *Journal of Phonetics*, 31:251–276, 2003.
7. M. M. Cohen and D. W. Massaro. *Modelling Coarticulation in Synthetic Visual Speech*, pages 139–156. Springer Verlag, Tokyo, 1993.
8. S. Duncan, Jr. Some signals and rules for taking speaking turns in conversations. *Journal of Personality and Social Psychology*, 23(2):283–292, 1972.
9. J. Edlund and J. Beskow. Pushy versus meek – using avatars to influence turn-taking behaviour. In *Proceedings of Interspeech 2007 (ICSLP)*, pages 682–685, Antwerp, Belgium, 2007.
10. J. Edlund, J. Beskow, and M. Heldner. Mushypeek – an experiment framework for controlled investigation of human-human interaction control behaviour. In *TMH-QPSR Vol. 50: Proceedings of Fonetik 2007*, pages 65–68, Stockholm, 2007.
11. J. Edlund and M. Heldner. Exploring prosody in interaction control. *Phonetica*, 62(2-4):215–226, 2005.
12. J. Edlund and M. Heldner. /nailon/ – software for online analysis of prosody. In *Proceedings of the Ninth International Conference on Spoken Language Processing (Interspeech 2006 - ICSLP)*, Pittsburgh, PA, 2006.

13. J. Edlund and M. Heldner. Underpinning /nailon/: Automatic estimation of pitch range and speaker relative pitch. In C. Müller, editor, *Speaker Classification*. Springer/LNAI, forthcoming.
14. L. Ferrer, E. Shriberg, and A. Stolcke. Is the speaker done yet? faster and more accurate end-of-utterance detection using prosody in human-computer dialog. In *Proceedings of the Seventh International Conference on Spoken Language Processing (ICSLP 2002)*, volume 3, pages 2061–2064, Denver, CO, 2002.
15. C. E. Ford and S. A. Thompson. Interactional units in conversation: syntactic, intonational, and pragmatic resources for the management of turns. In E. Ochs, E. A. Schegloff, and S. A. Thompson, editors, *Interaction and grammar*, pages 134–184. Cambridge University Press, Cambridge, 1996.
16. J. Gratch, A. Okhmatovskaia, F. Lamothe, S. Marsella, M. Morales, R. J. van der Werf, and L.-P. Morency. Virtual rapport. In *Proceedings of 6th International Conference on Intelligent Virtual Agents*, Marina del Rey, CA, US, 2006.
17. M. Heldner, J. Edlund, and T. Björkenstam. Automatically extracted f0 features as acoustic correlates of prosodic boundaries. In *Proceedings of Fonetik 2004*, pages 52–55. Department of Linguistics, Stockholm University, 2004.
18. M. Heldner, J. Edlund, and R. Carlson. Interruption impossible. In M. Horne and G. Bruce, editors, *Nordic Prosody: Proceedings of the IXth Conference, Lund 2004*, pages 97–105. Peter Lang, Frankfurt am Main, 2006.
19. V. Hugot. *Eye gaze analysis in human-human communication*. Master thesis, KTH Speech, Music and Hearing, 2007.
20. C. T. Ishi. Perceptually-related f0 parameters for automatic classification of phrase final tones. *IEICE Transactions on Information and Systems*, E88D(3):481–488, 2005.
21. J. Jaffe and S. Feldstein. *Rythms of dialogue*. Personality and Psychopathology. Academic Press, New York, 1970.
22. H. Koiso, Y. Horiuchi, S. Tutiya, A. Ichikawa, and Y. Den. An analysis of turn-taking and backchannels based on prosodic and syntactic features in japanese map task dialogs. *Language and Speech*, 41(3-4):295–321, 1998.
23. J. K. Local and J. Kelly. Projection and "silences": Notes on phonetic and conversational structure. *Human Studies*, 9:185–204, 1986.
24. J. K. Local, J. Kelly, and W. H. G. Wells. Towards a phonology of conversation: turn-taking in Tyneside English. *Journal of Linguistics*, 22(2):411–437, 1986.
25. J. K. Local, W. H. G. Wells, and M. Sebba. Phonology for conversation: Phonetic aspects of turn delimitation in London Jamaican. *Journal of Pragmatics*, 9(2-3):309–330, 1985.
26. P. Mermelstein. Automatic segmentation of speech into syllabic units. *Journal of the Acoustical Society of America*, 58(4):880–883, 1975.
27. H. Noguchi and Y. Den. Prosody-based detection of the context of backchannel responses. In *Proceedings of the Fifth International Conference on Spoken Language Processing (ICSLP'98)*, pages 487–490, Sydney, Australia, 1998.
28. R. Ogden. Turn transition, creak and glottal stop in finnish talk-in-interaction. *Journal of the International Phonetic Association*, 31(1):139–152, 2001.
29. I. S. Pandzic and R. Forchheimer. *MPEG-4 Facial Animation – the Standard, Implementation and Applications*. John Wiley & Sons, Chichester, 2002.
30. M. Selting. On the interplay of syntax and prosody in the constitution of turn-constructional units and turns in conversation. *Pragmatics*, 6:357–388, 1996.
31. E. Shriberg and A. Stolcke. Direct modeling of prosody: An overview of applications in automatic speech processing. In *Proceedings of Speech Prosody 2004*, pages 575–582. Nara, Japan, 2004.

32. C. Siciliano, G. Williams, J. Beskow, and A. Faulkner. Evaluation of a multilingual synthetic talking face as a communication aid for the hearing impaired. In *Proc of ICPhS, XV International Conference of Phonetic Sciences*, pages 131–134, Barcelona, Aug. 2003.
33. K. Sjölander and M. Heldner. Word level precision of the nalign automatic segmentation algorithm. In *Proceedings of the XVIIth Swedish Phonetics Conference, Fonetik 2004*, pages 116–119, Stockholm University, may 2004.
34. K. Sjölander. The Snack sound toolkit, 1997. `http://www.speech.kth.se/snack/`.
35. G. Skantze and J. Edlund. Robust interpretation in the higgins spoken dialogue system. In *COST278 and ISCA Tutorial and Research Workshop (ITRW) on Robustness Issues in Conversational Interaction*, Norwich, UK, 2004.
36. G. Skantze, D. House, and J. Edlund. Grounding and prosody in dialog. In *Working Papers 52: Proceedings of Fonetik 2006*, pages 117–120. Lund University, Centre for Languages & Literature, Department of Linguistics & Phonetics, Lund, Sweden, 2006.
37. G. Skantze, D. House, and J. Edlund. User responses to prosodic variation on fragmentary grounding utterances in dialogue. In *Proceedings Interspeech 2006*, pages 2002–2005, Pittsburgh, PA, 2006.
38. G. Svanfeldt and D. Olszewski. Perception experiment combining a parametric loudspeaker and a synthetic talking head. In *Proceedings of Interspeech*, pages 1721–1724, 2005.
39. D. Talkin. A robust algorithm for pitch tracking (rapt). In B. Kleijn and K. Paliwal, editors, *Speech Coding and Synthesis*, pages 495–518. Elsevier, New York, NY, 1995.
40. G. L. Trager and H. L. Smith. *An Outline of English Structure*. American Council of Learned Societies, Washinton, DC, 1957.
41. Å. Wallers, J. Edlund, and G. Skantze. The effects of prosodic features on the interpretation of synthesised backchannels. In E. André, L. Dybkjaer, W. Minker, H. Neumann, and M. Weber, editors, *Proceedings of Perception and Interactive Technologies*, pages 183–187. Springer, Kloster Irsee, Germany, 2006.
42. B. Welford. Note on a method for calculating corrected sums of squares and products. *Technometrics*, 4(3):419–420, 1962.
43. B. Wells and S. MacFarlane. Prosody as an interactional resource: Turn projection and overlap. *Language and Speech*, 41(3-4):265–294, 1998.

15

Perceptual Component Evaluation and Data Collection

Nicolas Moreau[1], Djamel Mostefa[1], Khalid Choukri[1], Rainer Stiefelhagen[2], Susanne Burger[3]

[1] Evaluations and Language Resources Distribution Agency, ELDA, Paris, France
[2] Universität Karlsruhe (TH), Interactive Systems Labs, Fakultät für Informatik, Karlsruhe, Germany
[3] Carnegie Mellon University, interACT, School of Computer Science, Pittsburgh, PA, USA

Systematic evaluation is essential to drive the rapid progress of a broad range of audiovisual perceptual technologies. Within the CHIL project, such evaluations were undertaken on an annual basis, so that improvements could be measured objectively, and different approaches compared and assessed.

CHIL organized a series of technology evaluations with subsequent evaluation workshops, during which the systems and obtained results were discussed in detail. A first project-internal technology evaluation was held in June 2004. Here baseline results of available technologies used in the real-life lecture scenarios CHIL initially focused on were established. Twelve evaluation tasks were conducted, including face and head tracking, 3D person tracking, face recognition, head pose estimation, hand tracking and pointing gesture recognition, speech recognition (close-talking and far-field), acoustic speaker tracking, speaker identification, acoustic scene analysis, and acoustic event detection.

Then, in January 2005, a more formal evaluation was organized, which now also included multimodal evaluation tasks (multimodal tracking, multimodal identification). External research groups were also invited to participate. Also in 2005, CHIL research teams took part in the NIST (National Institute for Standardization, USA) Rich Transcription (RT) Meeting Evaluation 2005. This annually RT evaluation series focuses on the evaluation of content-related technologies such as speech and video recognition. CHIL contributed test and training data sets in the lecture room domain and participated in the Speaker Location Detection task.

Many researchers, research labs, and, in particular, a number of current major research projects worldwide are working on technologies for analyzing people's activities and interactions. However, common evaluation standards, metrics, and benchmarks for such technologies, as, for example, for person tracking, are missing. It is therefore hardly possible to compare the developed algorithms and systems. Hence, most researchers rely on using their own data sets, annotations, task definitions, and evaluation procedures. This again leads to a costly multiplication of data production and evaluation efforts for the research community as a whole. In order to overcome this situation, we decided in 2005 to completely open up the project's evaluations by creating an open international evaluation workshop called CLEAR, Classification of Events, Activities, and Relationships [17, 5], and transferred part of the CHIL technology evaluations to CLEAR. CLEAR's goal is to provide a common international evaluation platform and framework for a wide range of multimodal technologies and to serve as a forum for the discussion and definition of related common benchmarks, including the definition of common metrics, tasks, and evaluation procedures. The creation of CLEAR was possible by joining forces with NIST, who, among many other evaluations, also organizes the technology evaluation of the U.S. Video Analysis Content Extraction (VACE) program [19]. As a result, CLEAR could be jointly supported by CHIL, NIST, and the VACE program. The first CLEAR evaluation was conducted in the spring of 2006 and was concluded by a two-day evaluation workshop in the UK in April 2006. CLEAR 2006 was organized in cooperation with the NIST RT 2006 evaluation [12]. This additionally allowed for sharing data between both evaluations by also harmonizing the 2006 CLEAR and RT evaluation deadlines. For example, the speaker localization results generated for CLEAR were also used for the far-field speech-to-text task in RT 2006. The CLEAR 2006 evaluation turned out to be a big success. Overall, around 60 people from 16 different institutions participated in the workshop, and nine major evaluation tasks, including more than 20 subtasks, were evaluated [5]. Following the success of CLEAR 2006, CLEAR 2007 took place in May 2007, in Baltimore, Maryland, again organized in conjunction and co-located with the NIST RT 2007 evaluation [15, 16].

An important aspect of the technology evaluations was using real-life data representing situations for which the applications and technologies in CHIL were developed, and annotated with the necessary information for various modalities. Therefore, a number of data sets were created, which enabled the development and evaluation of audiovisual perception technologies. The data were collected inside so-called smart rooms at five different locations. All of these rooms were equipped with a broad range of cameras and microphones.

After an initial data set collected in 2004 for CHIL's first internal technology evaluation, new data sets in slightly modified scenarios were collected in 2005 and 2006. These sets constituted the main test and training data sets used in the CLEAR 2006 and 2007 evaluations. The audio modalities (close-talk and far-field) of the 2005, 2006, and 2007 data sets were also part of the RT 2005, 2006, and 2007 evaluations. Additionally, this audio portion was employed in a pilot track in the Cross Language Evaluation Forum CLEF 2007 [6].

The utilization of the CHIL data sets in such high-profile evaluation activities demonstrates the state-of-the-art nature of the corpus, and its contribution to the development of advanced perception technology. Moreover, the numerous scientific publications produced as results of the evaluations are further enhancing the importance of the corpus. The CHIL data sets are publicly available to the community through the language resources catalog [7] of the European Language Resources Association (ELRA).

The next two sections describes how the CHIL data sets were collected (Section 15.1) and annotated (Section 15.2).

The last section of this chapter (Section 15.3) details the various evaluation activities conducted during CHIL internally, RT and CLEAR.

15.1 CHIL Data Overview

Data were collected throughout the four years of the CHIL project: in several modalities, in real-life situations under different conditions, in various scenarios and locations, using and testing a large amount and variety of sensors. The resulting CHIL corpus has the potential to drastically advance the state-of-the-art in the field by providing numerous synchronized audio and video streams of 86 real lectures and meetings, captured at five recording sites. Although not the first corpus to address the meeting or lecture scenarios, it significantly overcomes deficiencies of other existing data sets. The vast majority of such publicly available corpora focuses on the audio modality alone, mainly aiming at speech technology development, such as the ICSI [8] and ISL [4] meeting corpora. Other data sets recently collected by NIST [13] and the AMI project [2] exhibit many similarities to the CHIL corpus by providing multimodal and multichannel data sets inside smart rooms. However, they are either limited to a single data collection site or contain scripted and somewhat constrained static interaction among the meeting participants. Therefore, to our knowledge, the CHIL corpus is the only data set that provides multimodal, multisensory recordings of realistic human behavior and interaction in lecture and meeting scenarios, with a desirable data variability due to the numerous recording sites and sessions. These are key attributes that allow researchers to evaluate algorithmic approaches on real data, breaking the toy-problem barrier, to test generalization to mismatched data, and to experiment with channel and modality fusion schemes.

The CHIL corpus is accompanied by rich manual annotations of both its audio and visual modalities. In particular, it contains a detailed multichannel verbatim orthographic transcription of the audio modality that includes speaker turns and identities, acoustic condition information, and named entities for part of the corpus. Furthermore, video labels provide multiperson head locations in the 3D space as well as information about the 2D face bounding boxes and facial feature locations visible in the camera views. In addition, head pose information is provided for part of the corpus.

15.1.1 Data Collection Setup

Five smart rooms were set up as part of the CHIL project and utilized in the data collection efforts. These rooms were located at the following partner sites:

- the Research and Education Society in Information Technologies at Athens Information Technology, Athens, Greece (AIT),
- the IBM T.J. Watson Research Center, Yorktown Heights, NY (IBM),
- the Centro per la ricerca scientifica e tecnologica at the Bruno Kessler Foundation, Trento, Italy (FBK-irst),

Fig. 15.1. Example camera views recorded at the five CHIL smart rooms during lectures (upper row) and interactive seminars (meetings) (lower row).

- the Interactive Systems Labs of the Universität Karlsruhe, Germany (UKA),
- and the Universitat Politècnica de Catalunya, Barcelona, Spain (UPC).

These five smart rooms were medium-sized meeting or conference rooms with a number of audio and video sensors installed, and with a supporting computing infrastructure. The multitude of recording sites provides a desirable variability in the CHIL corpus, since the smart rooms obviously differed from each other in their size, layout, acoustic and visual environment (noise, lighting characteristics), as well as sensor properties (location, type); see also Fig. 15.1. Nevertheless, it was also crucial to produce a certain degree of homogeneity across sites to facilitate the technology development and evaluations, which is why a minimum common hardware and software setup was specified regarding the recording sensors and resulting data formats. All five sites complied with these minimum requirements, but often added additional sensors. A minimal setup consisted of:

- A set of common audio sensors, namely:
  - a 64-channel linear microphone array;
  - three four-channel T-shaped microphone clusters;
  - three tabletop microphones;
  - close-talking microphones worn by the lecturer and each of the meeting participants.
- A set of common video sensors, which include:
  - four fixed cameras located at the room corners;
  - one fixed, wide-angle panoramic camera located under the room ceiling;
  - one active pan-tilt-zoom camera.

This set was accompanied by a network of computers to capture the sensory data, mostly through dedicated data links, with data synchronization realized in a variety of ways. A schematic diagram of a smart room including its sensors is depicted in Fig. 15.2.

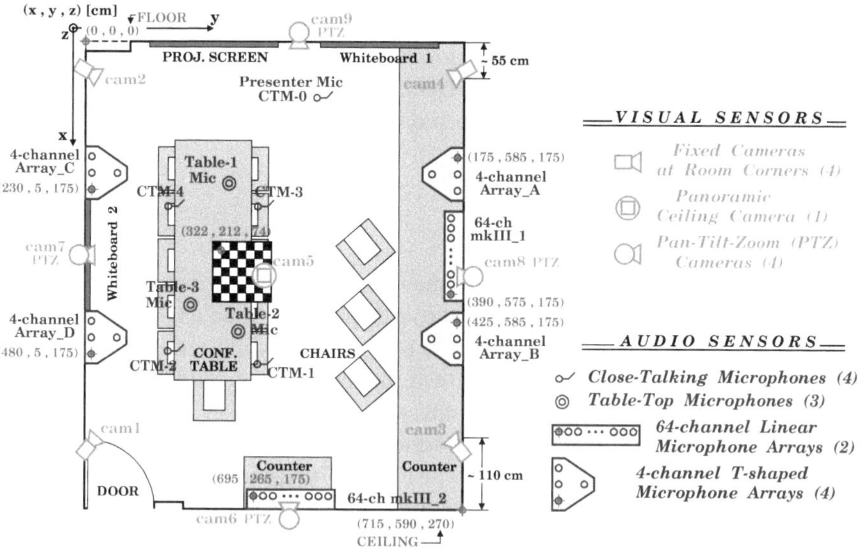

Fig. 15.2. Schematic diagram of the IBM smart room, one of the five installations used for recording the CHIL corpus. The room is approximately $7 \times 6 \times 3 \times m^3$ in size and contains nine cameras and 152 microphones for data collection.

Audio Sensor Setup

Each smart room contained a minimum of 88 microphones that captured both close-talking and far-field acoustic data. In particular, at least one 64-channel linear microphone array was used for the far-field audio recording. Here, all five sites decided to work with the Mark III array developed by NIST [11]. It was placed on the smart room wall opposite the speaker area. Using a microphone array allowed audio beamforming for speech recognition and speaker localization. The microphone array was accompanied by at least three additional microphone clusters located on the room walls, each consisting of four microphones organized in an inverted "T" formation of known geometry to allow far-field acoustic speaker localization. Additional far-field audio is collected by at least three tabletop microphones. These are positioned on the meeting table, distributed an appropriate distance from each other in an arbitrary formation. As a contrast to the far-field audio condition, close-talking microphones were used to record the lecture presenter and, in the case of interactive seminar recordings, each seminar participant. At least one of these microphones was wireless to allow the presenter to freely move around. The five recording sites were able to add more sensors if they wished, as long as they provided the minimum setup. For example, the IBM smart room contained two NIST Mark III arrays; the FBK-irst room used seven T-shaped arrays.

For audio data capture, all microphones beside the NIST Mark III microphones were connected to a number of RME Octamic eight-channel pre-amplifiers/digitizers. The pre-amplifier outputs were sampled at 44.1 kHz and 24 bits per sample, and were recorded to a computer in WAV format via an RME Hammerfall HDSP9652 I/O card. The 64-channel NIST Mark III data were similarly sampled and recorded in SPHERE format, but were fed into a recording computer via an Ethernet connection in the form of multiplexed IP packets.

Video Sensor Setup

Four of the fixed video cameras were mounted close to the corners of the room, also close to the ceiling, with significant overlapping and wide-angle fields-of-view. They were adjusted such that any person in the room would always be visible by at least two cameras. The fifth camera was mounted on the ceiling, directly facing downwards. It used a fish-eye lens that covered the entire room. The type of cameras installed varied among the sites, being either firewire or analog, providing images in resolutions ranging from 640×480 to 1024×768 pixels, and frame rates from 15 to 30 fps. All fixed cameras were calibrated with respect to a reference coordinate frame, with both extrinsic and intrinsic information provided in the corpus. In addition to the fixed cameras, at least one active pan-tilt-zoom (PTZ) camera was available in all five smart-room setups. Its purpose was to provide close-up views of the presenter during lectures or meetings. Significant differences existed in the PTZ camera setups among the recording sites in

- the number of cameras used,
- their type (analog, digital),
- their control medium (serial, network),
- the control mechanism (human operator, automatic).

An example of smart-room camera views is depicted in Fig. 15.3.

A number of dedicated computers were used to capture the data. All video streams were saved as sequences of JPEG-compressed images. This allowed an easy non-linear access to the frames, as well as an exact absolute timestamping. It is also worth mentioning that most meeting recordings were accompanied by brief video sequences that contain empty room images, immediately followed by capturing all participants entering the room. These were provided to assist background modeling in video processing algorithms.

15.1.2 Quality Standards

In order to have every site producing the same quality of data, CHIL internally developed quality standards for all the sensors.

Fig. 15.3. Sample synchronous images captured at the IBM smart room during an interactive seminar (meeting). The five fixed camera views are depicted in the upper rows, while pan-tilt-zoom camera views are shown in the lower row (see also Fig. 15.2).

Video Quality Standard

The minimum frame rate of the four angle cameras and the central ceiling fish-eye camera was set to 15 frames per seconds (fps). The data streams were saved as sequences of JPEG images in a fixed name standard: seq_xxxxx.jpg, with *xxxxx* being the frame number. A specific file called seq.index contained the table of correspondence between the frame and its associated timestamp. A file called seq.ini stored all the camera-related information.

The maximum possible desynchronization between the five cameras for the entire length of a recording was set at 200 ms.

The deviation of the synchronization was measured by introducing a distinct and well-observable audiovisual signal at the beginning and end of each recording. The decision of how to realize this was left to the recording site. However, using a clapboard was recommended (as used in movie studios) since such a board also provides a good method for synchronizing the audio and video channels.

Microphone Array Quality Standard

The fourth channel of the 64 channels was extracted for each of the Mark III recordings. A specific file called timestamps.ini was created to store the timestamp of an eventual packet loss. The maximum desynchronization due to packet loss during one recording was set at 200 ms. If more package loss occurred, the recording had to be redone.

Hammerfall Quality Standard

In the case of the remaining 20 microphones captured by the Hammerfall sound card, including the T-shaped arrays, the tabletop microphones, and the close-talking microphones, a timestamps.ini file was also created. Here, the maximum desynchronization due to packet loss during one recording was set at 50 ms. If more package loss occurred, the recording had to be redone.

Additional Information

A specific information directory was included in each recording so that every site provides the same information in a structured manner. This contained (1) a directory for calibration information with 10 pictures per camera and the camera's calibration results and (2) a directory for background information with pictures of the background before the meeting, when the room was still empty.

A seminar data sheet was also required for each recording. It mainly contained information about the attendees: a picture showing the identity tags; which microphones corresponded to the respective attendee; etc. Also included were the presentation slides. This information was mainly collected to support the transcription and annotation work and to make it more reliable.

15.1.3 Data Collection Scenarios

The CHIL corpus comprises two types of scenarios: lectures and meetings, also referred to as "noninteractive seminars" and "interactive seminars", respectively. Both are similar in that a presenter gave a talk in front of a seminar audience; both significantly differ, however, in their degree of interactivity between the audience and the presenter, as well as in the number of participants. The topics of the collected seminars were quite technical in nature, spanning the areas of audio and visual perception technologies, but also biology, business, and so on. The language spoken was English, while most subjects exhibited strong nonnative accents, such as Italian, German, Greek, Spanish, Indian, Chinese, and so on.

Noninteractive Seminars (Lectures)

In this scenario, the presenter talked in front of an audience of typically 10 to 20 people. There was little interaction between the presenter and the audience, usually in form of a question-and-answer section toward the end of the presentation. As a result, there was not much activity in the audience, and the respective recorded data were often of low quality with lots of background noise. Hence, the lecture analysis concentrated on only the presenter. The lecture dataset of the CHIL corpus consequently contains only transcriptions of the presenters' speech, and also the respective evaluation tasks were mainly focused around the presenter. Examples of noninteractive seminars are depicted in the upper row of Fig. 15.1. The noninteractive seminar data were recorded at UKA in 2003, 2004, and 2005, and at FBK-irst in 2005. A total of

46 noninteractive seminars became part of the CHIL corpus (see also Table 15.1); most are between 40 and 60 minutes long.

Site	# Seminars	Year	Type	Evaluations
UKA	12	2003/2004	Lectures	CHIL Internal Evals., CLEF07
UKA	29	2004/2005	Lectures	CLEAR06, RT05s, RT06s, CLEF07
FBK-irst	5	2005	Lectures	CLEAR06, RT06s
UKA	5	2006	Meetings	CLEAR07, RT07s
FBK-irst	5	2006	Meetings	CLEAR07, RT07s
AIT	5	2005	Meetings	CLEAR06, RT06s
AIT	5	2006	Meetings	CLEAR07, RT07s
IBM	5	2005	Meetings	CLEAR06, RT06s
IBM	5	2006	Meetings	CLEAR07, RT07s
UPC	5	2005	Meetings	CLEAR06, RT06s
UPC	5	2006	Meetings	CLEAR07, RT07s

Table 15.1. Details of the 86 collected lectures/noninteractive seminars (upper part) and meetings/interactive seminars (lower part) that comprise the CHIL corpus. The table depicts the recording site, year, type, and number of collections, as well as the evaluations where the data were used.

Interactive Seminars (Meetings)

The number of participants of interactive seminars was small, between three and five people. The audience was seated around a table and wore close-talking microphones. The presenter either also sat at the table or moved around and was equipped with a wireless close-talking microphone. The audience was able to interrupt at any time by asking questions or making comments or suggestions. This frequently led to discussions among all participants, resulting in a significantly higher interaction between presenter and audience than in the noninteractive seminars.

To ensure a broad variety of events and interactions, the seminar participants were asked to possibly include the following situations in their meetings:

- participants enter or leave the room;
- some attendees stand up and go to the whiteboard;
- discussions occur among the attendees;
- participants stand up for a short coffee break;
- during and after the presentation, there are questions from the attendees with answers from the presenter.

In addition, to allow a more meaningful evaluation of the corresponding technology, the participants were encouraged to generate a series of acoustic events:

- sounds when opening and closing the door;
- interruptions of the meeting due to ringing mobile phones;

- attendees coughing and laughing;
- attendees pouring coffee in their cup and putting it on the table;
- attendees playing with their keys;
- keyboard typing, chair moving, etc.

Since all participants actively contributed to the meeting, all were of interest to the meeting analysis, and hence, all speakers were transcribed. Examples of interactive seminars are depicted in the lower row of Fig. 15.1. The interactive seminars were recorded at AIT, IBM, and UPC in 2005, as well as at all five sites during 2006. A total of 40 interactive seminars is part of the CHIL corpus; most are approximately 30 minutes in duration. Table 15.1 summarizes the recorded data sets, when and where they were recorded, and in which evaluation campaigns they were used.

Seminar Participants

The seminar participants were mainly staff and students from the respective recording sites, but national and international visitors participated also. This led to a broad variety in participants' nationalities and to many different native and nonnative accents in English in the collected data. As an example, the CHIL RT 2007 evaluation data set [3] with 71 individual speakers, including only five female voices, comprises speakers originating from 17 different countries, with the biggest groups being Spaniards (23%), Italians (15%), and Greeks and Germans (each 14%). Figure 15.4 shows the distribution of the speakers' country of origin for the CHIL 2007 test and development set speakers.

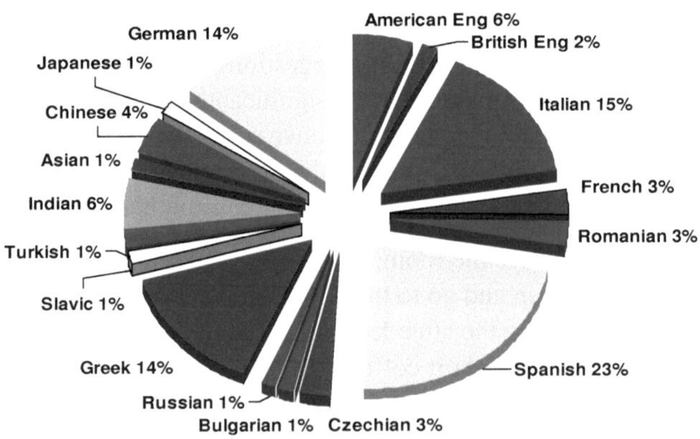

Fig. 15.4. Speaker accents: distribution of countries of origin in the CHIL 2007 development and test sets.

15.2 CHIL Corpus Annotations

The CHIL Consortium devoted significant efforts in identifying useful and efficient annotation schemes for the CHIL corpus and providing appropriate labels in support of technology evaluation and development. As a result, the data set contains a rich set of annotations in multiple channels of both audio and visual modalities. Table 15.2 gives the total duration of annotated data (lectures and meetings) used in each of the CHIL evaluation campaigns.

Evaluation Campaign	Development Data	Evaluation Data
CHIL Internal Evals.	2h 20min	1h 40min
CLEAR06	2 h 30 min	3 h 10 min
CLEAR07	2 h 45 min	3 h 25 min

Table 15.2. Total duration of annotated data for development and evaluation purposes, in each of the CHIL evaluation campaigns.

15.2.1 Audio Channel Annotations

The data collection in the CHIL smart room resulted in multiple audio signals recorded through close-talking microphones as the near-field condition and, in parallel, through tabletop microphones, T-shaped clusters, and the Mark III microphone array as the far-field condition. The recorded speech as well as environmental acoustic events were carefully segmented and annotated by human transcribers at two locations, the European Language Resources Distribution Agency (ELDA) and the interACT Center at Carnegie Mellon University (CMU).

Orthographic Transcriptions

The transcriptions were done by native English speakers. The transcribers followed detailed transcription guidelines containing conventions for the transcription of spontaneous speech and labels for the annotation for nonlexical speech events (e.g., filled pauses, coughing). The manual transcription process started by transcribing the speaker contributions of all recorded near-field channels on orthographic word level. The start and end of the contributions were manually segmented. The transcription of the near-field condition was then compared and adapted to one of the far-field channels. Nonaudible events were removed and details recorded by only the far-field sensors were added. The effort put into the transcription process was time-intensive; it took about 30 times of the real-time duration of a recorded speech signal, including several review passes.

As mentioned above, in the case of the noninteractive seminars, the near-field condition of only the presenter's contributions was transcribed since nobody else

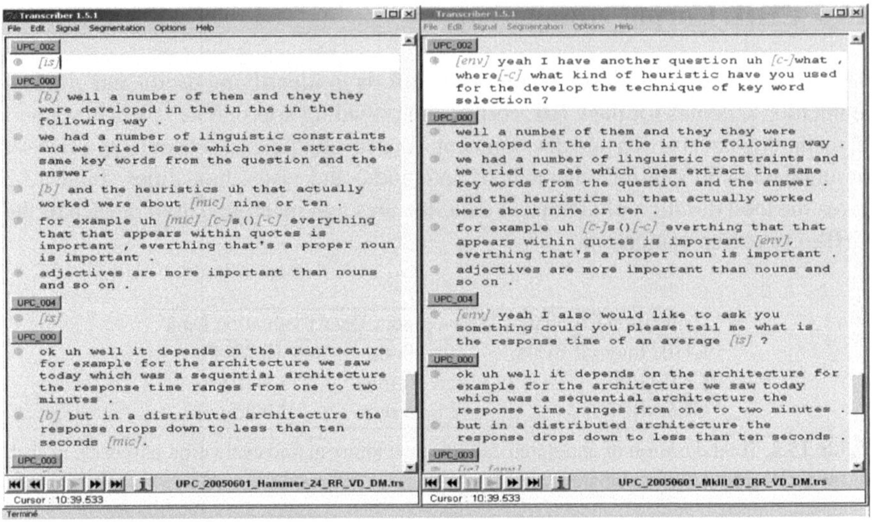

Fig. 15.5. Transcription of a lecture segment using the Transcriber tool. Near-field transcription (left side) versus far-field transcription (right side) is shown.

wore close-talking microphones. The annotators used the Transcriber tool [18]. Potential contributions from the audience as well as environmental noises were transcribed as the far-field version using a channel of the far-field recording.

An example of the resulting transcripts is depicted in Fig. 15.5. The left window shows the near-field transcription of several turns in a lecture, while the right window depicts the transcription of the far-field condition of the same lecture. It can be seen that speaker "UPC_002", a participant sitting in the audience, was not audible in the near-field condition where his utterances were tagged as "inaudible speech (is)". However, the same utterances were clearly understood in the far-field condition and are therefore included in the transcription.

In contrast to the noninteractive seminars, all participants in the interactive seminars (meetings) wore individual close-talking microphones. These conversations contain overlapping speech and discursive turns reacting to each other (question/answer, argument/counter argument), similar to a multiparty meeting conversation. To display all recorded speech signals in a multitrack parallel view, transcribers worked with TransEdit, a tool originally developed for the transcription of multiparty meetings. The near-field transcription was again adapted to the far-field condition, using only one channel of the Mark III channels. On a side note, many of the segmentation boundaries of the far-field segmentation of the utterances needed to be adjusted because reverberation in the far-field recordings resulted in shifted segment boundaries.

Annotation of Acoustic Events

Following the orthographic transcription of close-talking and far-field audio, the environmental acoustic events were annotated as a third level of annotation. These annotations were used as reference and training material supporting the "acoustic event detection and classification" task in the CLEAR evaluations.

Acoustic events describe all audible events in a recording. Accordingly, speech contributions were also considered to be acoustic events but were only categorized by the label SPEECH, instead of lexically transcribed as unique words. The remaining set of labels for acoustic events consisted of DOOR SLAM, STEP, CHAIR MOVING, CUP JINGLE, APPLAUSE, LAUGH, KEY JINGLE, COUGH, KEYBOARD TYPING, PHONE RINGING, MUSIC, KNOCK (door, table), PAPER WRAPPING, and UNKNOWN.

The annotation of acoustic events was carried out as an independent additional labeling process using the Annotation Graph Tool Kit (AGTK). Similar to TransEdit, AGTK enables the annotation of multiple overlapping events [1].

Acoustic events were labeled on several different types of data sets: acoustic events that occurred in the CHIL seminar corpus, and recordings of series of artificially produced events. The first set of data was labeled listening to the fourth channel of the Mark III microphone array. The artificially produced acoustic events were recorded without temporal overlap in two data sets in the FBK-irst and UPC smart rooms, used as a quiet environment.

15.2.2 Video Channel Annotations

Facial Features and Head Location Information

Video annotations were manually generated using an ad hoc tool developed at the University of Karlsruhe, modified by ELDA. The tool displayed one picture per second, in a sequence, for all camera views. To generate the labels, the annotator performed a number of clicks on the head region of the person of interest. This was only the lecturer in the case of the noninteractive seminars, but these were all participants in the case of the interactive seminars. In particular, the annotator first clicked on the head centroid (e.g., the estimated center of the person's head), then on the left eye, followed by the right eye, and finally on the nose bridge (if it was visible). In addition, the annotator delimited the person's face with a bounding box. The 2D coordinates of the marked points within the camera plane were consequently saved to a corresponding label file. This procedure allowed the computation of the 3D head location of the persons of interest inside the room, based on the camera calibration information. Figure 15.6 depicts an example of video labels. It shows the head centroid (white), the left eye (blue), the nose bridge (red), the right eye (green), and the face bounding box.

Head Pose Annotations

In addition to the 2D face and 3D head location information, parts of the lecture recordings were also labeled with gross information about the lecturer's head pose. In particular, only eight head orientation classes were annotated. This limitation was still deemed to be a feasible task for human annotators, considering the low-resolution captured views of the lecturer's head. The head orientation label corresponded to one of eight discrete orientation classes, ranging from a 0° to a 315°

Fig. 15.6. Example of video annotations for an interactive seminar in the UPC smart room. Face bounding boxes and facial feature annotations are depicted for two camera views.

angle, with an increment of 45°. Overall, 19 lecture videos were annotated with such information. These videos were used in the CLEAR head pose technology evaluation.

15.2.3 Validation Procedures

The video annotations were validated internally. Each annotation file was automatically scanned using a tool developed by ELDA. This tool automatically detected most of the potential annotation errors, i.e., permuted right and left eye labels or missing labels. During a subsequent second validation pass, a human operator manually checked and corrected the video labels, supported by the error listings produced by the automatic scanning tool. It was ensured that the person who checked a given seminar was different from the one who initially labeled it.

In the same way, each orthographic transcription was validated by a human transcriber, different than the original first-pass transcriber. In a final consistency pass, one person listened to all transcribed data once again, and ran additional automatic check procedures such as spell checks, lexicon consistency checks, and transcription format checks.

Moreover, a further cross-validation check of the video labeled data (at UKA) and of the audio transcriptions (between ELDA and CMU) was done.

15.3 CHIL Evaluations Overview

The CHIL corpus has been the cornerstone data set in the evaluation of a multitude of audiovisual perception technologies for human activity analysis during lectures and meetings. This section gives a brief overview of the evaluation paradigms for the various perception and content extraction technologies addressed in the evaluation campaigns where the CHIL corpus served as test and training data.

15.3.1 Evaluation Tasks

The following tasks were evaluated on CHIL data sets during CHIL-internal evaluation campaigns and international evaluations:

- **Person tracking**: the tracking of one person or all persons in the scene (3D) or in all images (2D). Subtasks included acoustic, visual, and audiovisual tracking. The developed approaches and results are discussed in Chapter 3.
- **Person identification**: the identification of all persons in the scene. Subtasks included the acoustic, visual, and audiovisual identification, as well as different lengths of audiovisual segments for training and testing (see Chapter 4).
- **Head pose estimation**: the estimation of the 3D head orientation of a selected person (e.g., the lecturer) by using all camera views (see Chapter 5.
- **Speech activity detection and speaker diarization**: the correct detection of speech segments. In the case of the speakers' diarization, the identities of speaker voices had to be determined (the question of "who spoke when"). The related systems and results are discussed in Chapter 6.
- **Automatic speech recognition (ASR)**: the production of a transcription of the speaker's speech. Subtasks included different microphone conditions, such as ASR from close-talking (CTM), tabletop (TT), or far-field microphones (FF). The related systems and results are discussed in Chapter 6.
- **Acoustic event detection**: the detection and identification of a number of acoustic events in meeting or lecture data (see Chapter 7).
- **Question answering (QA) and summarization (SA)**: the automatic answering of user questions by means of extracts from speech transcripts. The answers could contain exact and specific facts, such as names of persons and organizations, or summaries (see Chapter 8).

Table 15.3 summarizes the tasks evaluated on the CHIL corpus and also shows in which evaluation campaigns these tasks were evaluated. Further details of the evaluation tasks, including detailed evaluation procedures and metrics, can be found in [9, 14, 15]. Various systems that were used in these evaluations are discussed in Chapters 3 to 8.

15.3.2 Evaluation Packages

The CHIL evaluation resources are publicly available to the academic and commercial communities as a set of "evaluation packages" through the ELRA General Catalog [7].

Generally, an evaluation package consists of a complete documentation (including the definition and description of the evaluation methodologies, protocols, and metrics) along with the data sets and software scoring tools necessary to evaluate developed systems in a given technology. The evaluation packages will therefore enable external participants to benchmark their own systems and compare their results to the results obtained during the official and original evaluation.

The CHIL evaluation packages, in particular, consist of the following:

Tasks	Evaluation
Person tracking (2D,3D,A,V,AV)	CHIL internal, CLEAR'06, CLEAR'07
Person identification (A,V,AV)	CHIL internal, CLEAR'06, CLEAR'07
Head pose estimation	CHIL internal, CLEAR'06, CLEAR'07
Acoustic event detection	CHIL internal, CLEAR'06, CLEAR'07
Speech recognition (CT,FF,TT mics)	RT'05, RT'06, RT'07
Speech activity detection	CHIL internal, RT'05, RT'06, RT'07
Speaker diarization	RT'07
Question answering	CHIL internal, CLEF'07
Summarization	CHIL internal

Table 15.3. Overview of the tasks and evaluation workshops, which used parts of the CHIL corpus.

- a document describing in detail the content of the package, as well as the respective evaluation (tasks, metrics, participants, results, etc.);
- the raw audio recordings of the seminars (Hammerfall, close-talking microphones, and microphone array channels);
- the raw video recordings of the seminars (streams of the four corner cameras and ceiling camera);
- the video annotations and audio transcriptions of the seminars;
- useful information about each seminar (attendees, slides, calibration information, background pictures);
- additional databases specific to some evaluation tasks (head pose, pointing gestures, isolated acoustic events).

In addition, a range of specific data is provided for each evaluation task, allowing the package users to reproduce an evaluation under the same conditions and to compare their results with those of the original evaluation participants:

- documentation about the evaluation procedure (metrics, submission format, etc.);
- the input data, as received by the participants during the evaluation;
- the participants' submissions;
- the reference labels;
- the scoring tools;
- the participants' results.

Several evaluation package "suites" have been produced so far: The first is composed of evaluation packages related to the CHIL-internal evaluation of January 2005 [10]. A description of the data, tools, and results is available in a public document [9]. The second suite covers the CLEAR 2006 evaluation [14]. The third and last suite covers the CLEAR 2007 evaluation [15, 16] (see also Table 15.1).

15.4 Conclusions

In this chapter, we have presented an overview of the CHIL corpus, a one-of-a-kind audiovisual database of lectures and meetings held inside smart rooms equipped with a multitude of sensors. The corpus has been collected as part of the CHIL project, aimed at developing and evaluating audiovisual perception technologies analyzing and supporting human activity and interaction during lectures and meetings. The resulting data set has largely contributed to advancing the state-of-the-art in the field by providing numerous synchronized audio and video streams of 86 real lectures and meetings, captured in five recording sites over several years. Significant effort has been dedicated to accompany the recorded data with rich multichannel annotations in both audio and visual modalities. The CHIL corpus has already been utilized in international evaluations and is publicly available to the research community. We therefore strongly believe that it represents an important contribution and a resource crucial to the development of robust perception technologies for the analysis of realistic human interaction.

We have also given an overview of the perceptual technology evaluations and evaluation workshops that have been organized and supported by the CHIL project. These were held on a yearly basis and covered a wide range of perceptual technology evaluation tasks, such as person tracking, person identification, head pose estimation, acoustic event detection, speech recognition, speech activity detection, speaker diarization, question answering, and summarization.

An important contribution of the CHIL project was the creation of an international evaluation workshop called CLEAR (Classification of Events, Activities and Relationships) in which part of the CHIL perceptual technology evaluations took place. CLEAR was successfully held in 2006 and 2007, and provided an until then missing international evaluation forum for the multimodal perception of humans.

References

1. The AGTK Annotation Tool, `http://agtk.sourceforge.net`.
2. AMI – Augmented Multiparty Interaction, `http://www.amiproject.org`.
3. S. Burger. The CHIL RT07 evaluation data. In *Multimodal Technologies for Perception of Humans, Proceedings of the International Evaluation Workshops CLEAR 2007 and RT 2007*, LNCS 4625, Baltimore, MD, 2007.
4. S. Burger, V. McLaren, and H. Yu. The ISL meeting corpus: The impact of meeting type on speech style. In *Proceedings of the International Conference on Spoken Language Processing*, Denver, CO, 2002.
5. Classification of Events, Activities, and Relationships Evaluation and Workshop: `http://www.clear-evaluation.org`.
6. The CLEF Website, `http://www.clef-campaign.org/`.
7. ELRA Catalogue of Language Resources, `http://catalog.elra.info`.
8. A. Janin, D. Baron, J. Edwards, D. Ellis, D. Gelbart, N. Morgan, B. Peskin, T. Pfau, E. Shriberg, A. Stolcke, and C. Wooters. The ICSI meeting corpus. In *Proceedings of the International Conference on Acoustics, Speech, and Signal Processing (ICASSP)*, Hong Kong, 2003.

9. D. Mostefa, K. Bernardin, H. K. Ekenel, N. Gourier, D. Macho, R. Malkin, F. Marques, R. Morros, K. Nickel, M. Omologo, G. Potamianos, R. Stiefelhagen, and J. Turmo. CHIL Public Deliverable D7.6: Exploitation Material for CHIL Evaluation Campaign 1. http://chil.server.de/servlet/is/8063/.
10. D. Mostefa, M.-N. Garcia, and K. Choukri. Evaluation of multimodal components within CHIL. In *Proceedings of the 5th International Language Resources and Evaluations Conference (LREC)*, Genoa, Italy, 2006.
11. The NIST MarkIII Microphone Array, http://www.nist.gov/smartspace/cmaiii.html.
12. The Rich Transcription 2006 Spring Meeting Recognition Evaluation website, http://www.nist.gov/speech/tests/rt/rt2006/spring.
13. The NIST Smart Space project. http://www.nist.gov/smartspace/.
14. R. Stiefelhagen, K. Bernardin, R. Bowers, J. Garofolo, D. Mostefa, and P. Soundararajan. The CLEAR 2006 evaluation. In *Multimodal Technologies for Perception of Humans, Proceedings of the First International CLEAR Evaluation Workshop, CLEAR 2006*, LNCS 4122, pages 1–45, Southampton, UK, Apr. 6-7 2006.
15. R. Stiefelhagen, K. Bernardin, R. Bowers, R. T. Rose, M. Michel, and J. Garofolo. The CLEAR 2007 evaluation. In *Multimodal Technologies for Perception of Humans, Proceedings of the International Evaluation Workshops CLEAR 2007 and RT 2007*, LNCS 4625, pages 3–34, Baltimore, MD, May 8-11 2007.
16. R. Stiefelhagen, R. Bowers, and J. Fiscus, editors. *Multimodal Technologies for Perception of Humans, Proceedings of the International Evaluation Workshops CLEAR 2007 and RT 2007.* LNCS 4625. Springer, Baltimore, MD, May 8-11 2007.
17. R. Stiefelhagen and J. Garofolo, editors. *Multimodal Technologies for Perception of Humans, First International Evaluation Workshop on Classification of Events, Activities and Relationships, CLEAR'06.* LNCS 4122. Springer, Southampton, UK, Apr. 6-7 2006.
18. The Transcriber Tool home page, http://trans.sourceforge.net.
19. VACE – Video Analysis and Content Extraction. https://control.nist.gov/dto/twiki/bin/view/Main/WebHome.

Part III

Services

16

User-Centered Design of CHIL Services: Introduction

Fabio Pianesi[1] and Jacques Terken[2]

[1] Foundation Bruno Kessler, irst, Trento, Italy,
[2] Technische Universiteit Eindhoven, The Netherlands

Services, that is, organized sets of functionalities targeting users, are at the core of the CHIL enterprise. It is at this level, in fact, that the general vision – putting the computer in the loop of human interaction – is made concrete and enjoyable to the user; it is services that users see, interact with, and exploit to better achieve their objectives. In the conception of this book, services are neither simple collections of technologies, nor showcases aimed to concept-proof technological advances, nor integration add-ons. Rather, they are the visible outcome of an organized effort towards:

- Understanding what the metaphor of 'the computer in the loop of human communication' really amounts to. Because of the novelty of the field, the only way to give concreteness to the metaphor is to deal with the task of specifying, designing, implementing and evaluating actual service, this way providing insights that can be brought to bear on future efforts.
- Going in depth into the human (social and cognitive) factors that must be considered to bring about the vision's added-value, avoiding the 'yet another clumsy technology' effect. As stated again and again in this book, the overall vision of CHIL starts from the consideration of human communication needs and the idea that the computer must support them. This is tantamount to requiring that the services be aware of the social environment in which they operate, be capable of adapting to varying cognitive and interaction needs, and, first and foremost, that their designers have a clear understanding of these issues and know how to deal with them.
- Articulating the multidisciplinary nature of this domain, understanding the contributions of, and mutual influence among, fields such as multimodal scene analysis, human-computer interaction, social and cognitive psychology. It is a straightforward consequence of it goals and objectives that CHIL's endeavor is a multidisciplinary one. The crucial consequence is that the pursuance of those goals require a level of integration and contaminations across the contributing disciplines and research field that is unprecedented: The solution to the "human in the loop' problem cannot be attained through the straightforward integration of separate components contributed by them.

In the end, the term services as used here indicates both the system prototypes providing a set of functionalities that give substance to the 'CHIL' vision, and the very effort, including all the user-centered issues, behind their design and development.

The services we are going to discuss in this book all address social situations at large, and group interaction and dynamics in particular. Within this vast domain, a number of issues have been isolated along with functionalities meant to provide an answer to them. These areas span three different aspects of group interaction: (a) Managing tasks, in particular the group access to, and creation and manipulation of, information; (b) Managing connections between the people in a meeting and the outside world; (c) Managing the social dynamics in groups. Before turning to give some details about these different aspects and the services addressing them, let us emphasize once again that the primary purpose of developing CHIL services was to contribute to the understanding of the metaphor of the 'computer in the loop of human communication'. Thus, we were not so much interested in optimizing concrete applications, but rather in investigating the interplay between design choices, design process, and users' relevant psychological and social dimensions for this new class of services. In order to investigate these aspects we needed to develop concrete concept proposals that people could experience with, for us to be able to measure their reactions, and use the insights so gained into a better understanding of the 'computer in the human loop' metaphor.

- Face-to-face interaction and collaboration in meetings can often be difficult, little productive, and frustrating. Working groups often have a hard time in setting up and respecting agendas, allocating time to the topics to discuss, collecting the material produced, and working out agreed upon minutes. A solution that will be explored below is to provide working groups with Shared Workspaces (SWs) that support them in their need to keep focalization on the crucial issues, keep track of, and manage, time, etc. SWs are infrastructures for encouraging cooperation among participants, whereby the system provides a multimodal interface for entering and manipulating contributions from different participants, such as the use of the agenda and the discussion of outcomes notes, and allowing people to share information and ideas and accomplish common tasks. By merging multimodal and computer supported collaborative work (CSCW) technologies [17], a SW offers groups a table-top shared environment in which the members can edit documents and draw diagrams; pass their products over to other members; manage the agenda and keep track of the passing of time; access an automatically produced meeting report collecting all the material produced. The system, in turn, helps users to manage and organize space, also based on its awareness of people's position around the table.
- Another problematic area in social context, and in particular meetings, is the timely availability of information that the single members of the group, or the group as whole, need in order to better pursue their goals. The Memory Jog service is a non-obtrusive assistant for meetings, lectures and presentations settings, aiming at providing pertinent information and memory aids. It acts both proac-

tively, when pertinent and socially appropriate, and reactively when requested by the user, providing information about past events (what happened while I was away?) and participants (who's that person?).

- A particular type of social situation is that established between two (or more) people by means of a (cell) phone. Crucial issues in this context are the availability of the callers, and the appropriateness of the time and place for the call to be accepted. Most cell phone owners have received interrupting phone calls at inconvenient moments; but what if a rejected phone call was of unexpected importance? What if the call is closely related to the current task? At the same time, on the caller's side, it can become quite annoying to lose time playing phone tag trying to reach others, leaving streams of voice mails, asking third persons about the whereabouts of the original contactee, not being available oneself for returning calls, using less appropriate alternative communication media, and so on. The connector service is a context-aware communication application that intelligently connects people. It maintains an awareness of its users' activities, preoccupations, and social relationships and mediates a proper connection at the right time between them. In addition to providing users with important contextual cues about the availability of potential callers, this information is used to adapt the behavior of the contactee's device automatically in order to avoid inappropriate interruptions. Moreover, the connector can use any available output device to deliver information to users in the most unobtrusive way possible.
- Professionals agree that as much as 50% of meeting time is unproductive, and that up to 25% of meeting time is spent discussing irrelevant issues [10]. Often, the origin of these problems is in the dysfunctional behavior of meeting participants, and external interventions such as facilitators and training experiences are commonly employed in order to improve performance of meetings. Facilitators participate in the meetings as external elements of the group, their role being to help participants to maintain fair and focused behavior, as well as directing and setting the pace of the discussion. Training experiences aim at increasing the relational skills of individual participants by providing online or offline (with respect to meetings) guidance or coaching so that the team is eventually able to overcome or cope with its dysfunctionalities. Multimodal analysis and interaction techniques can be exploited to, at least partially, automate team facilitation and coaching. Below we describe the Relational Cockpit and the Relation Report, two services that provide online and offline feedback, respectively, about people's social behavior, in order to guide them toward more functional behavior.

  The Relational Cockpit monitors low-level correlates of the social dynamics of the conversation, such as eye gaze and speaking time of individual participants, and presents feedback about these behavioral properties during the meeting through a peripheral display (a projection on the meeting table).

  The Relational Report service provides offline guidance. It monitors groups and generates individual reports about the addressed participant's behavior. The system observes the meeting as a coach would do, and generates reports that are not minutes; they do not address the content of the interaction, but present a more qualitative, meta-level interpretation of what happened in the social dynamics of the group, focusing on the addressee. The reports are delivered privately to each participant after the meeting, and their purpose is that of informing her about her behavior rather than evaluating it. In doing so, it provides an opportunity to the addressee to think reflectively [1], bringing in a different perspective on the group relationship and own behavior [2], much as a human coach would do.

16.1 Methodology

A key premise of the CHIL effort is that computers should meet the needs and goals of users. Unfortunately, decisions about how technologies should be implemented are highly complex, as indicated by the extremely high failure rate for most innovations [5, 13, 16]. Some of the key causes of these failures are the following: (1) people often cannot articulate their needs and goals; introspection is unreliable [15]; (2) people are very poor predictors of their behavior [9]; (3) people and designers often cannot predict actual use as opposed to intended use [7, 11, 14]; (4) needs or desires may be in conflict [3]; (5) differences among people, such as gender or culture, can be critical determinants of technological preferences [4, 18]; (6) there are multiple stakeholders in any technology decision, many of whom are not directly assessed [12]; and (7) an enormous set of strongly held norms and expectations, many of them culturally dependent, have evolved to guide face-to-face and small teamwork interactions [6]; systems aiming to support or otherwise play a role in group settings might clash with established practices, or modify equilibria in unpredictable manners. To address these difficulties in moving from technological innovations to desirable multimodal services for social interaction, a mix of qualitative and quantitative methods was adopted. Keywords in concept development were "user-centered" and "iterative".

- We began by studying the context of use by ethnographically informed methods, interviewing and observing office workers in strategic meetings and student teams in weekly team meetings, discussing their progress, and planning their subsequent activities. These observations, and interviews resulted in the identification of significant moments during meetings, and illuminated how meetings relate to the larger pattern of team activities. Also, this work made clear that a narrow focus on productivity would be useless, as it neglects the social function of meetings both for the participating individuals and their organizations.
- Next, group interviews were conducted through focus group meetings in which people discussed their views on problems associated with meetings. These sessions confirmed what had already been stated in the literature about main problems with meetings: lack of preparation (e.g. no agenda); lack of agenda discipline and time management (digressions and long discussions); dominance, resulting in some participants monopolize the discussion, lack of participation and, ultimately, disengagement by other participants. Furthermore, when asking for valuable tools, automatic minutes and related means for getting access to the content of the meeting were frequently mentioned items.

- The outcomes of the previous stages were used to derive initial requirements for our services, which were then turned into hypotheses about desirable functionalities. According to the user-centered design philosophy, the functionalities were submitted to formative evaluation by means of mock-ups, low-fidelty prototypes, and Wizard of Oz studies. The purpose was to enable people to get an impression of what it would be like to employ the CHIL services in this way providing feedback information to the design team to further refine the design. The outcomes of these initial tests led to adjustments for the next cycle, in which working prototypes of the services were implemented, incorporating actual perceptual components.
- In a final round, for some of the services summative evaluations were conducted in order to address specific research questions, such as
  - Do CHIL services enhance productivity in the sense that people employing them achieve better results, and/or in a shorter time?
  - What about the user experience about employing CHIL services? Do they want to use them in the future, and can we predict their intention to use from information about aspects such as perceived usefulness, ease of use, control, intrusiveness, etc., which have been widely discussed by many technology acceptance models as Davis' TAM [8]?
  - What are people's opinions about salient characteristics of CHIL services such as autonomy and proactivity?

  Again, this final round of evaluations involved a combination of qualitative and quantitative methods, with a slight prevalence of the former.

Besides those prescribed by the user-centered design approach, a number of specific studies were conducted to investigate peculiar issues raised by the users services relationships. For instance, upon observing the users' behavior with the SW we realized that psychological ownership could affect their feelings toward sharing material, and accessing documents available in the common space. Hence a targeted experimental study was designed and conducted to elucidate the role of psychological ownership in shared workspaces [19]. Similarly, it seemed important to understand whether the way the relational report conveys its information can affect its acceptance, and for this case too a specific study for was designed and executed.

16.2 Methodological Issues

Before turning to the chapters dealing with the individual services and the user studies conducted on those services, we briefly address a number of methodological issues that we ran into while developing and testing the services. In the first place, while there was a general wish and sometimes a right out necessity to move from mock-ups to actual prototypes including operational perceptual components, often there was a considerable gap between the capabilities of the technologies and the needs of services. For instance, the Memory Jog enabled people to get easy access to recordings of earlier parts of a meeting through automatically generated annotations.

However, the speech technology needed to provide the annotations was not sufficiently advanced, so that the actual annotations were fairly limited from a service point of view. As a result of the technological limitations, in several instances primary user needs could not be addressed and instead only rather marginal user needs could be addressed by the prototypes.

Secondly, the ultimate proof for services for supporting (co-located) collaboration is their performance and ultimate adoption in field studies, where people can experiment with the services over longer periods of time and adapt them to their needs and working practices, and/or accommodate their working practices to the opportunities provided by the new technologies. Instead, due to the lack of robustness of perceptual component evaluations had to be conducted primarily in the laboratory. Furthermore, due to the nature of the questions asked, especially questions concerning productivity gains, control had to be exerted over the events during evaluations, so that we had to resort to scripted tasks instead of allowing people to experience the employment of CHIL services for their naturally occurring working practices. Finally, as is the case with most lab studies, only initial use was investigated, so that people's judgments were based on limited experience with the services. It needs to be pointed out, though, that where technology allowed it, field studies were run (cf. the chapters for the Shared Workspace and the Connector). A further complication that was observed in those instances is that it takes considerable effort to find suitable participants for field studies, precisely because of the fact that the participants have to work with new technologies while performing their naturally occurring activities, taking extra effort that may hamper their performance.

Finally and of a lesser importance, conducting experiments on services for supporting collaboration and group communication requires larger numbers of subjects, so that conducting experiments takes more effort in collecting the required number of subjects. As a result, in several cases the experiments involved fewer subjects than desirable.

For these reasons, the conclusions derived from the service evaluations must be considered tentative and require further investigation. In particular, the acceptance of the services and the adoption to naturally occurring work practices require further development of services and further user studies. However, we believe that the studies reported in the next chapters provide useful first steps and have provided meaningful initial results that will help to guide further development.

16.3 Overview of Part III

The following chapters each discuss one of the services by first providing details about the architecture and functionalities of the final versions, and then by resorting to typical scenarios and use cases. The second part of each chapter presents the salient steps in the UCD cycles, emphasizing the new knowledge acquired that can be generalized to other multimodal services. Particularization of the methodological issues that pertains to the single service will also be discussed here. Chapters 17 and 18 present services that support the management of content in meetings. Chapter 17 presents the Collaborative Workspace. Chapter 18 presents two versions of the Memory Jog service. One focuses on annotating meeting data with information about the Who and What of the meeting in order to support meeting participants in retrieving information about episodes of earlier meetings of earlier parts of the same meeting. The other Memory Jog service focuses on exploring the information push-and-pull aspects of intelligent information services. Chapter 19 presents the Connector services which helps people better manage the risk of untimely interruptions in an "always-on" world, where we can reach each other anytime, and anyplace. The Connector does so by representing the receiver's availability to the caller, so that the caller can take a more informed decision on how and when to initiate contact. Moreover, the Connector empowers the receiver by augmenting the communication modality itself, by offering novel forms of answering (previously unanswerable) calls. Chapters 20 and 21 address the management of social relations in a meeting. Chapter 20 presents the Relational Cockpit, a service that provides feedback on people's social behavior in meetings during the meeting. Chapter 21 presents the Relational Report, a service that provides feedback on people's social behavior in meetings after the meeting.

References

1. C. Andersen. A theoretical framework for examining peer collaboration in preservice teacher education. In *Proceedings of the 2000 Annual International Conference of the Association for the Education of Teachers in Science*, 2000.
2. G. Bloom, C. Castagna, and B. Warren. More than mentors: Principal coaching. *Leadership*, 32(5):20–23, 2003.
3. S. Brave and C. Nass. *Emotion in Human-Computer Interaction*, pages 81–96. Lawrence Erlbaum Associates, Inc., Mahwah, NJ, 2003.
4. L. Cherny and E. Weise. *Wired Women: Gender and the New Realities of Cyberspace*. Seal Press, Seattle, WA, 1996.
5. C. Christensen. *The Innovator's Dilemma*. Harper Business, New York, 2003.
6. H. Clark. *Using Language*. Cambridge University Press, 1996.
7. A. Cooper and P. Saffo. *The Inmates Are Running the Asylum*. Macmillan, Indianapolis, IN, 1999.
8. F. Davis. User acceptance of information technology: System characteristics, user perceptions and behavioral impacts. *International Journal of Man-Machine Studies*, 38(3):475–487, 1993.
9. T. Dockery and A. Bedeian. Attitudes versus actions: Lapiere's (1934) classic study revisited. *Social Behavior and Personality*, 17(1):9–16, 1989.
10. M. Doyle and D. Straus. *How To Make Meetings Work*. The Berkley Publishing Group, New York, NY, 1993.
11. C. Fisher. *Studying Technology and Social Life*. Sage, Beverly Hills, CA, 1985.
12. B. Friedman, editor. *Human Values and the Design of Computer Technology*. Cambridge University Press/CSLI, New York, NY, 1999.
13. G. Moore and R. McKenna. *Crossing the Chasm*. Harper Business, New York, NY, 2002.
14. D. Norman. *The Design of Everyday Things*. Currency Doubleday, New York, NY, 1988.

15. B. Reeves and C. Nass. *The Media Equation: How People Treat Computers, Television, and New Media Like Real People and Places.* Cambridge University Press, New York, NY, 1998.
16. E. Rogers. *Diffusion of Innovations.* Free Press, New York, NY, fifth edition, 2003.
17. S. Scott, K. Grant, and R. Mandryk. System guidelines for co-located, collaborative work on a tabletop display. In *ECSCW'03: Proceedings of the 2003 Eighth European Conference on Computer-Supported Cooperative Work*, 2003.
18. G. Tan, S. Brave, C. Nass, and M. Takechi. Effects of voice vs. remote on U.S. and Japanese user satisfaction with interactive HDTV systems. In *CHI'03: Extended Abstracts on Human factors in Computing Systems*, pages 714–715, New York, NY, 2003. ACM Press.
19. Q. Wang, A. Batocchi, I. Graziola, F. Pianesi, D. Tomasini, M. Zancanaro, and C. Nass. The role of psychological ownership and ownership markers in collaborative working environment. In *ICMI'06: International Conference on Multimodal Interfaces*, 2006.

17

The Collaborative Workspace: A Co-located Tabletop Device to Support Meetings

Chiara Leonardi, Fabio Pianesi, Daniel Tomasini, Massimo Zancanaro

Foundation Bruno Kessler, irst, Trento, Italy

In this chapter, we describe the design and development of co-located multi user collaborative displays that aim to support effective social interactions within a small group of professionals engaged in meeting activities. Several studies have investigated and tried to go beyond the design paradigm developed for single-user/single-desktop interaction in order to support face-to-face collaborative activities. Several helpful design principles have emerged from research on shared workspaces and co-located technologies.

As to the design dimensions, Scott and colleagues [21] pointed out some challenges faced by tabletop systems such as support for natural interaction, transitions between activities and between personal and group work, transitions between the tabletop and the external world, the use of physical objects, access to shared physical and digital objects, flexibility in users' arrangement, and simultaneous user interactions. Other studies, for example, Kruger and colleagues [8], pointed out the issues related to the spatial organization of the information on shared surfaces and the role of orientation. Regard to social aspects, social protocols and coordination policies in accessing shared surfaces have been also considered as important aspect that must be addressed while designing co-located interfaces [11]. Furthermore, Rogers and Lindley [16] compared horizontal and vertical workspace suggesting that the former are more effective in promoting group cooperation and in encouraging participant to switch between roles and explore ideas.

Suggestions have also been formulated concerning the criteria to be employed in the formative and summative evaluation of co-located cooperative systems. Classical dimensions and levels of analysis used in HCI such as usability and efficiency should be reconsidered and extended when coping with the challenge of supporting groups and communities with cooperative technologies. Group performance, level of collaboration, shared understanding of problems, and quality of interaction are all relevant, though they are difficult to evaluate and measure. Indeed, some studies tried to identify metrics of collaborative work by considering, e.g., the time for task/activity completion or the level of users' engagement in common tasks. Difficulties have emerged though, above all those concerning the interpretation and the reliability of results (see for instance [5]).

A possible alternative to the attempt at measuring the outcomes of the encounters between groups and co-located systems consists in attempting to understand the processes through which technologies become meaningful for people, whereby the dimensions for both the design and the evaluation emerge through the analysis of the social processes of using technologies in real contexts of use. With respect to understanding how technological artifacts are turned into meaningful instruments during peoples courses of action, we tried to take advantage of the research concept of appropriation. The appropriation of technology has been defined as the mutual adaptation between technology and user's practices. It is a twofold process by which a technology is adopted which entails modifying users' activities to fit system requirements - and adapted that is, accommodated into the actual and situated working practices ([1, 4]). Studies dealing with appropriation are consequently long-term, longitudinal studies conducted in ecological settings that take into account people using fully functional and robust systems in their daily activities. In the context of iterative prototyping, appropriation should be considered that a research concept which drives our understanding of the relationships between technology, which doesn't have a pre-established meaning independent from its use, and users, who are not merely passive recipients of new technology [22].

The focus of our study is on the initial encounters between groups of people and technologies and on the ongoing invisible work people accomplish to fit technologies in their courses of action through a sense-making process. As suggested by Carroll [1], considering user appropriation of technology as part of the design process entails that users' needs are expressed through action as people interact with the technologies thus supporting the design of malleable and flexible technologies and finally encouraging a critical reflection on the relationships between the technology as designed and technology in use. We exploited these ideas in the design of a co-located interactive table top device to support small group meetings, by means of an iterative design approach involving the design and use of both low- and high-fidelity prototypes. Initial insights were first derived, and then prototyped and further refined by observing groups of workers using the technologies during their meeting activities. The chapter is organized as follows: We first introduce the motivations and general scenario for the design of our co-located interface. We then present the research areas that influenced our approach and design methodology. Then, we present the design process of the system, starting from the analysis of teams using low-fidelity mock-ups, up to studies of teams using a functioning release of the system in their meeting activities. Finally, we describe the last prototype developed.

17.1 Related Work

In the last years, the very concept of meeting has widely changed, as have the ways to support this activity. Early attempts in the direction of the latter include the design of Electronic Meeting Systems (EMS), which are part of the broader class of Group Support Systems (GSS). EMS [12] consisted of networked computers and other audio visual supports like whiteboards and projectors. Rogers and Rodden [17] pointed out the costs and limitations of such complex technologies; for instance, the need of a facilitator who leads the meeting, as well as the rigid and unnatural structure of the interaction imposed by the system.

Another class of facilities that aim at supporting meetings consists of the Group Decision Support Systems (GDSS). Their main goal is to support individuals in formulating and communicating decisions during face-to-face meetings and to facilitate the group decision-making process.

The concept of meeting lying behind both EMS and GDSS is that of a well-planned, collective, and goal-directed activity, that cannot but take advantage of sets of well defined procedures to be more effective. Meetings, however, often feature a smoothly organized activity that only seldom takes the form of formal sessions. Several studies conducted in the field of CSCW (see for example, [18, 19]) emphasize the integration of meetings into heterogeneous arrangements of cooperative activities, shifting the focus onto their spatially and temporally distributed character, and on the coordination and articulation work needed to deal with the interdependencies among workers.

Capitalizing on these insights, more recent works have explored the use of large interactive displays and multimodal devices in the design of ubiquitous workspaces: for example, the iROS room-based environment designed at Stanford University [6]. Yet, despite the keep it simple principle that inspired the design, this meeting room often still needs a facilitator to manage the activity flow.

Overall, these studies suggest that flexible and open technologies offering opportunities rather than constraining interaction are needed to meet the requirement of supporting face-to-face collaboration in an unobtrusive way. They also demonstrate the importance of simplicity in the design of meeting supportive technologies and the number of social and psychological challenges groups face when engaged in computer-mediated activities.

Recently, co-located collaborative work around shared surfaces has become a major topic in the research agenda. In a seminal work, Tang [23], discussing the analysis of activity around a shared drawing surfaces proposes the following design implications: (1) the need to consider hand gestures as central communicative resources, (2) the mediating role of the shared drawing space, (3), the structuring role of spatial orientation among members.

More recently, Scott and colleagues [21] capitalizing on existent researches on co-located interaction around shared surfaces, proposed the following set of guidelines concerning how to best support collaborative work while limiting disruptions and overhead:

- *natural interpersonal interaction*, at the heart of each collaborative activity, should be supported;
- the technology should not require users to excessively switch between activities when interacting around a table, rather *fluid transition between activities* should be allowed;
- technology must support users in maintaining and relying on both personal and group areas by fostering *transitions between personal and group work*;

- access to the outside world from the meeting and vice versa must be supported by allowing *transitions between tabletop collaboration and external work*;
- technology should be able to accommodate and support the *use of physical objects* (e.g., pens, sheets of papers, etc.) during the meeting;
- *shared access to physical and virtual objects* must be granted to all participants;
- the design of a digital tabletop should consider the physical properties of the table, such as the size, the physical movement of people and allow *flexible user arrangements*;
- the tabletop device must permit *simultaneous user interaction*, that is, it should be usable by more than one participant at the same time. Other issues addressed in the field of co-located collaborative work encompass:
- the use and partition of territory on shared interfaces. They distinguish three main areas on the tabletop surfaces that play an indirect role in mediating and facilitating group interaction: *group areas*, defined as the workspaces that each participant can access; *personal areas* (often the areas directly in front of the relevant person), which individual members use to perform own independent activities (e.g., reading and writing); *storage areas*, where task and non task resources are organized;
- the role of social protocols defined as standards and polite behaviors to coordinate the actions of users accessing a groupware. Morris and colleagues [11] discuss the incorporation of coordination policies (mechanisms for avoiding conflicts) in digital interfaces. They observe that they may be useful when the size of the group and the amount of documents increase and, consequently, when it is difficult for users to monitor and coordinate the activities.

Other studies have dealt with the socio cognitive aspects of integrating shared displays into existing settings, investigating the negotiation taking place around their access and exploitation, and the complex interplay between the setting and the technology [13]. Rogers and Rodden [17], while studying the relationships between the physical affordances of the display and the social and cognitive behaviors it supports or constrain, singled out a number of potential problems:

- when interacting with a public surface, uncertainty can increase because of the lack of knowledge about, e.g., what people have accessed, or who is controlling what;
- people need to continue relying on the artifacts they are used to, such as pens, paper, communication devices or even a cup of coffee, when interacting with shared technologies;
- tensions can emerge when the display contravenes the norms of social acceptability and physicality; e.g., being forced to stand close, side by side in front of a vertical display can cause awkwardness and the feeling of invading each other's personal space.

17.2 User-Centered Design of a Tabletop Interface

The target population of our study were professionals in a working environment. This category of users is well known to often form working teams to achieve specific goals (for instance, software design), to conduct specific activities (organize a conference), or planning activities (debriefing and project meetings); moreover, they are used to being involved in a plurality of teams at the same time. Teams engaged in our study were seldom long-lasting groups, but were rather networks of researchers and developers working on many projects at the same time, and participating in several teams. These teams are often heterogeneous and ramified arrangements of people and complex interdependencies among the members exist. Such a fluid structure occasionally led to practical problems for the longitudinal observation of the same group. Teams participating in the study were also characterized by an informal organization of work and, consequently, of meetings too. Although most of them were FBK-irst researchers and collaborators, none of the individuals involved in our study was participating in the CHIL project, and none of their members worked in fields related to human computer interaction.

The design process consisted of three phases involving prototypes of increasing complexity (one low-fi and two hi-fi prototypes) as depicted in Fig. 17.1.

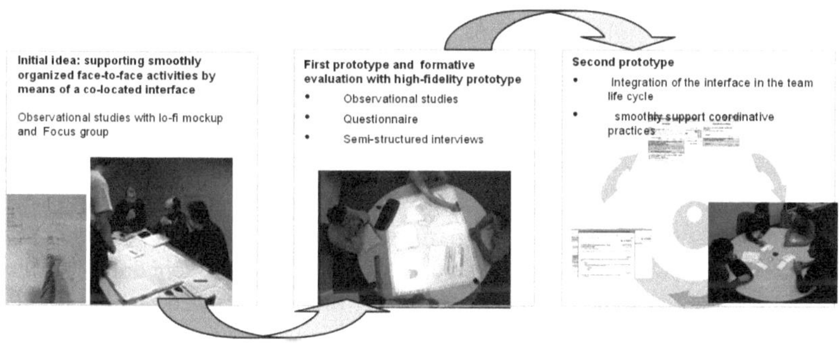

Fig. 17.1. Phases of the design process and methodologies used to collect data.

17.3 Initial User Study: Whiteboard as Mock-up

The initial requirements were elicited by observing small groups conducting meetings around a traditional whiteboard placed horizontally as a low-fi mock-up. The whiteboard was used to simulate the future interactive table, working as a "triggering artifact" [2] to promote the analysis of current practices. Participants' attitudes and experience were further investigated by means of focus groups. This first user study was designed to investigate how small groups of people would use shared horizontal surfaces as resources to coordinate their meeting activities. Four markers were also made available since we were particularly interested in observing to what extent participants used the whiteboard simultaneously.

Several project managers were contacted and asked to conduct their meetings in this setting, according to their own schedule and agenda. Five sessions of natural occurring meetings of four different groups were video taped with two or three cameras, one placed above the whiteboard and the other two facing the participants. Groups ranged in size from three to six participants and meetings lasted from 45 to 90 minutes.

Observations were conducted on the videos, focusing on the use of space on the whiteboard and the coordination of activities around it. Three researchers first analyzed the videos autonomously. Insights and hypotheses were then discussed during debriefing sessions.

The meetings were of three different types: design, debriefings and presentation meetings. In design meeting decentralized communication processes prevailed: The main activities performed were brainstorming, exploring and negotiating ideas, and problem solving. As already pointed out by Olson and colleagues [14] problem-solving activities more or less chaotic are frequent when groups engage in discussion about technical features of objects sustaining their activity by means of diagrams and sketches. In these meetings, the equal access to communication is paralleled by an equal access to the whiteboard. A substantial sharing of the four markers was observed. The whiteboard was mainly employed as group space for communicating and negotiating ideas, and the positions of the participants, who typically sat in a circle around the table, reflected the structure of the communication process. Debriefing (or project) meetings relied on unilateral communication processes. Their main goal was to inform the group, and in particular the leader (who usually drove the meeting for most of its duration), about individual work carried out by the members, and plan future work accordingly. In this case, meetings were seen as resources to articulate teamwork: Goals were identified, each participant clarified the state-of-the-art to the others, the situation was jointly interpreted, other meetings were scheduled, and tasks were assigned. The available material resources, the four markers, were often kept by the leader who used them to summarize key points. Presentation meetings were characterized by the interlacing of activities of presentation and discussion. One of our examples, for instance, started with a PhD student presenting his work. At the end of the presentation, the student's work was discussed by the student himself, his advisors and the other four members of the group. During the presentation, the communication was unilateral and the whiteboard was used by the student to communicate ideas, while his advisor took notes (still on the whiteboard but in a sort of personal space). In a second phase, the communication process was decentralized and characterized by an alternation of questions and answers, and discussion moments. Each participant accessed the whiteboard, mainly referring to the diagrams previously sketched by the student. The nature of the meeting was reflected in the positioning of the participants, with the student standing on one side of the table, and the other participants sitting at the opposite side [see figure 17.2. (c)]. It is worth noticing that while in the debriefing meetings the asymmetric access to the whiteboard is related to the very structure of the group (that is, the existence of a leader), in the presentation meetings it is more related to the task (presentation). The communication structures observed in our meetings are consistent with those reported by Leavitt [9] in his bipartite classification of the communication process in small group meetings into decentralized (circle model, peer-to-peer) and unilateral (asymmetric) ones. In the former, all the members take an active part in the discussion and no one formally leads the interaction; in the latter, one of the members rules the discussions directing herm or his speech toward the other participants who, in turn, direct their answer, ask questions, and propose suggestions to the leader.

In the meetings observed, no specialized tools (calendar, traditional paper-based agendas or laptop to write the minutes) were brought in and used by the participants. The activities were coordinated mainly verbally, with the whiteboard working as a support for making verbal information persistent e.g., by sketching agendas and by drawing to-do lists summarizing the outcomes and decisions taken. Planning and storage activities were the two main activities carried out to coordinate group actions.

Planning through agenda. Groups relied mainly on informal, non-material agendas for their meetings. Both in debriefing and in design sessions, items to discuss were usually specified at the beginning of the meeting by the project manager, on the basis of informal schedules agreed upon by participants beforehand, usually through e-mail or verbal discussions. At the beginning of the meetings, the list of issues to deal with or goals to achieve was verbally presented, and often summarized on the whiteboard in the form of an agenda. The latter would then remain on the surface until the end, with other contributions being organized around it (see Fig. 17.2 (a)).

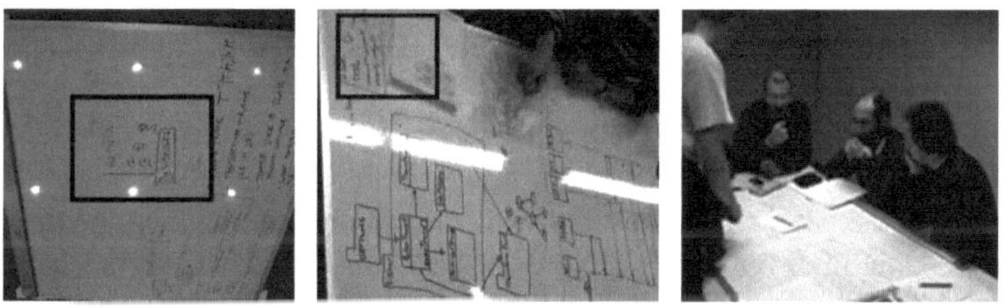

Fig. 17.2. (a) the agenda produced at the beginning of a meeting, and (b) the summary of the discussion written on the whiteboard in another meeting, and (c) the participants' physical configuration during a presentation meeting.

Storage activities. Meetings were also characterized by the attempts to summarize the decisions taken and to keep track of the documents produced as both long-term and short-term memory of the group processes. Personal notes were written on sheets of paper, while group-related ones were made public by writing them on the whiteboard. In some cases, it was difficult to distinguish the latter from personal notes, as was the case with the PhD advisor mentioned above, who used the whiteboard to note the question he was going to ask at the end of the presentation. Henceforth, we will reserve the term "storing activities" to those meant to produce group-related records. A typical sequence observed during a design meeting is when a consensual vision on a problem is reached e.g., after the group has discussed a design problem using a diagram which is then written nearby an existing diagram (see Fig. 17.2 (b)). In one case the project manager took a picture of the whiteboard and then distributed it to each participant.

Regarding the use of the space on the table and on the whiteboard, our observations confirmed the tripartite organization suggested by Scott and colleagues [20]. Group, personal and storage areas were created and used to organize individual and group work around the shared whiteboard.

Group areas were used mainly for jointly contributed material: explanations, build-up and diagrams. As said above, they also served as a basic resource upon which tools for meeting management were created: agenda and minutes. Although group areas were free and public spaces accessible to everyone, very rarely a participant deleted, corrected or overwrote other members' contributions. In other words, the actual operations on group areas were read and write, not delete, correct or overwrite. In very few cases the whiteboard was simultaneously accessed by two participants to sketch a diagram and to take personal notes. In all other cases participants tended to take turns in using the shared surface by minimizing simultaneous use.

Personal areas, usually outside the whiteboard, were used to carry out personal work, such as writing personal notes, reading photocopies and copying diagrams written and discussed by the group. Personal areas were only seldom created on the shared surface, suggesting that the whiteboard was mainly conceived as a common resource. In all cases, the creation of personal spaces on the whiteboard was limited to group members with a high enough level of participation in the interaction. This was the case of the PhD advisor discussed above who, while the student was talking, created a personal space on the whiteboard where he noted the questions he would ask at the end of the presentation.

Storage areas were used to physically store personal documents, photocopies, agendas and even objects not directly related to the meeting (e.g., watches and drinks). Since the table was almost completely occupied by the whiteboard, participants were sometimes obliged to store documents on the surface. This lack of space was particularly evident when the meeting was primarily based on the discussion of documents.

17.3.1 Focus Groups

After the sessions with the mock-up, four focus groups were organized with the participants of the teams involved in the study. Short video clips of the meetings were shown and then participants were invited to discuss the experience of working around a horizontal surface.

The focus groups confirmed the importance of the different meeting typologies in the use of the resources, supporting the idea that a tabletop system should be flexible enough to accommodate them. According to some participants, the tabletop devices, and more generally, working around a horizontal surface, would be more useful for meeting with a decentralized communication process. Some participants mentioned that keeping track of the information produced during meetings may support groups in better managing the discussion and that it also may be useful in creating a long-term memory available for after-meeting activities. Participants suggested that the data produced during the meeting could be organized either on a temporal basis, reflecting the temporal order in which they are produced, or according to thematic criteria.

Regarding the use of the shared whiteboard, an important issue that emerged from the discussion was the lack of a shared perspective which seemed to discourage and limit the interaction because of the delay in reading and understanding what was written on the table. Many participants also lamented the lack of space on the horizontal surfaces which was felt to be the more of a limiting factor the more the use of diagrams was required by the meeting. A specific problem in this respect was the difficulty of keeping track of, and reusing information contained in diagrams that had to be erased to free space. Finally, the lack of space and the difficulty of giving an internal structure to the group space create confusion as a participant put it "we wrote wherever there was enough space, and this made it difficult to keep track of the discussion."

The focus groups regarded the coexistence of group and private areas as an important requirement for a tabletop device. Personal spaces, in fact, would give everyone the chance actively take part in the discussion by privately modifying contents and publicly submitting the results to the group. At the same time, participants emphasized the need for supports accompanying the transition of information from the private to the public status.

Finally, the preference for paper-based resources was confirmed and motivated by the tangibility "you can touch and feel the paper", the versatility "you can bring them into your office or home"; also, the symbolic value of the paper emerged. According to participants, paper permits both sharing information with co-workers, and taking personal notes. Interestingly, some participants complained that the whiteboard laid on the table left too small a space for individual uses of paper. Laptops, in turn, were described as potentially distracting the attention from the meeting (e.g., checking e-mails, writing documents not related to the meeting) and not offering more advantages than paper for personal notes. Moreover, they can seriously damage the communication flow and the visual contact, because of the vertical screens that obstruct the visual interaction. As a participant says "a laptop disturbs the interaction, I can't see the other people's eyes". Some participants also referred to the greater naturalness of handwriting for taking personal notes: "Handwriting is better because it is more natural for people to think and write down thoughts than typing the same things on the keyboard".

17.3.2 Initial Requirements for a Co-located Interface

From the information discussed above, the following requirements were derived for the first design of the tabletop device:

- allow empty space to accommodate bloc notes and other personal objects;
- allow an efficient use of the available space;

- allow the participants equal access to the information;
- support the coordination and storage activities, making it possible to handle an agenda and save the documents produced during the meeting.

17.4 The Collaborative Workspace: First Design

Starting from the above requirements, we decided to adopt a paper-based metaphor for the design. The system was designed as a shared space top-projected onto a standard table where virtual sheets of paper can be created, edited, rotated, and shared by the participants. These objects are acted upon by an electronic pen. A commercial tool that allows only one pen, which emulates a mouse was used. A first mock-up was prepared in Flash and a more robust prototype was then implemented in MS C#.

Free-form documents were designed to sustain communication and sharing of information among participants. Virtual sheets can be created, oriented, moved, and erased by participants by means of the electronic pen. Texts can be edited through a wireless keyboard or the pen; diagrams and sketches can be drawn using the pen (see Fig. 17.3).

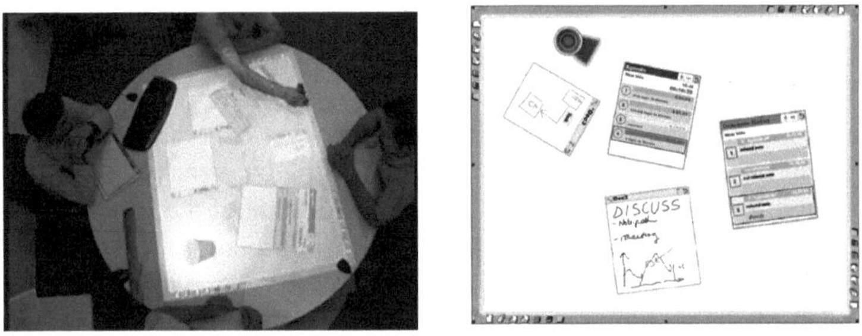

Fig. 17.3. Left, the Collaborative Workspace. Right, a snapshot of the interface

Use of space. The interface has a rectangular shape and is projected onto a circular table. This allows for four empty areas on the table that can be used as personal areas, permitting the storage and use of one's paper documents, photocopies, or laptops.

The interface space consists of the projected area with the virtual documents in it. Thats with the pens, they can be opened, shrunk to save space, rotated, and passed on to other participants to make the content accessible to other members. They can be saved and retrieved, or removed from the desktop by dragging them into the trash bin. To permit an equal access to the shared workspace, a toolbar is reproduced on each side of the interface. It enables users to open new documents (as blank sheets), select the pen modality (writing or erasing), import/save a document, and choose the pen stroke (three combinations of color and stroke: red/thin; blue/medium; gray/large).

Coordination tools. An electronic agenda was designed in the form of a specialized sheet of paper, to provide support for the scheduling and management of meetings. Agenda items are displayed in it, and they can be made active/inactive by clicking with the pen. To help improve the temporal organization of the meeting, a timer is displayed on the active issue, which pauses when the latter is made inactive. The agenda is dynamic: Items can be deleted or added during the meeting.

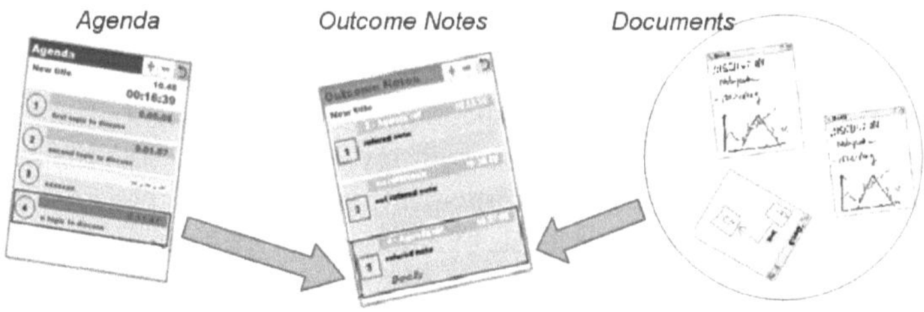

Fig. 17.4. The active item of the Agenda (The Item 4) is automatically associated with the Outcome Notes List. Documents produced can be stored into the appropriate item of the Outcome Notes through drag'and'drop.

Storage and track-keeping tools. A special document, the Outcomes Notes List, is made available to facilitate the organization of the meeting outcomes. It provides a space where participants can keep track of the decisions taken writing into the boxes with the keyboard and a repository to store the documents produced by associating them to the appropriate outcome item. Each entry is automatically associated with the current agenda issue; documents can be dragged-and-dropped to be indexed and saved as snapshots of the original one. As with the Agenda, each item can be added, removed, or sorted (see Fig. 17.4). Underlying the design of the Agenda and of the Outcome Notes List was the assumption that groups would first schedule their activities, then produce and work on documents of various type, which would eventually be stored into the Outcome Notes List. As we will see in the next section, observations on actual usage of the tools challenged two assumptions.

17.5 The Second User Study

A user study involving six small groups was organized in the spirit of a formative evaluation. The study was meant to inform the design of the next implementation and to analyze both the individual and the social processes involved in the use of a co-located interface.

Both video-based observations and interviews acknowledged the high understandability and effectiveness of the paper metaphor. The openness, under specificity, and versatility of the virtual sheets together with the functionalities associated to them (i.e., the possibility of drawing with the pen and typing with the keyboard) led groups to widely use them according to specific needs and tasks to be performed. The general setup appeared flexible enough to accommodate the diverse needs of different groups. Thus in design meetings, sketching and drawing diagrams were much more common than text writing; documents were often enlarged to cover almost the whole surface, realizing large rotatable and adjustable surfaces where diagrams and schemes were sketched, in this way providing for strong attention foci around which the discussion was organized. The lack of a common visual perspective did not emerge as an important limit: Apparently, participants could easily refer to the document content through pointing.

In debriefing meetings, text editing was more typical, with texts written using the keyboard to keep track of decisions and to write to-do lists for organizing future activities; in design meetings, drawing diagrams was predominant. The lack of a shared perspective caused more problems with text editing than with diagrams. As already observed in [21] and [24], the participants in front of the person editing the content showed a reduction in their levels of attention and participation. Finally, the Collaborative Workspace was underused in presentation meetings and sometimes completely ignored when the meeting turned into a seminar with slides projected on the wall.

Comparing the horizontal with the vertical whiteboard setting, a project manager reported during an interview that "the very advantage of the device is that it simplifies the interaction with other members of the group ..... The relationship among members becomes more egalitarian. It's very different from the school-like blackboard model". Particularly appreciated was the share ability of the documents during and after the meeting. The possibility of having multiple copies of them and sharing them via mail after the meeting was acknowledged as a major asset of the tabletop system for meetings (it is worth noting that the prototype did not support the after-meeting sharing of the documents; it just allowed them to be saved). Even if the paper metaphor was appreciated, however, a few problematic aspects emerged. In particular, the ways documents are moved, rotated, resized, and stored were sometimes problematic. The direct manipulation of documents suffers from some major limitations:

- In order to share a document with others, participants had to first rotate the document and then move it. This is different from what happens with traditional sheets of papers which can be rotated while moved.
- The mechanism implemented to manage the resizing of windows was quite misleading: When the participants enlarged a window to provide more space for content editing, the latter was resized as well.
- Document saving was an easy task since it required just clicking on the toolbar the "save" icon; yet it was not intuitive how to retrieve the content of the saved windows. Mainly for this reason, Outcome Notes were preferred to stored documents.

In general, participants felt that virtual documents should fulfill two requirements: They should be as intuitive as an ordinary sheet of paper; and they should provide all the digital functionalities provided by text editors and sketching applications.

Regarding the Agenda and the Outcome Notes List, the analyses of the use of the two coordination tools revealed some problems in the design, mostly due to inadequacies of the assumptions about the work flow. In particular, the sequential accomplishment of actions required by the system (activate Agenda Item to automatic establish a relation with the Outcome Notes List) did not match the actual practices observed. Above all, people found it counterintuitive that documents had to go through the Outcome Notes for storage, and most participants tried to directly associate the documents to the relevant item in the agenda. The consequences were a sense of frustration and time spent understanding the relationship between the Agenda and the Outcome Notes List. Regarding the Agenda, the main advantages acknowledged by participants were the possibility to keep the attention focused, and the ability to manage the meeting via a public schedule. Compared with "normal" meetings, however, the initial phase of the construction of the digital agenda was seen as an additional task requiring the specification, translation, and renegotiation of previous (in)formal agreements into a formal one. A smoother and less effort-consuming transition from previous informal or semiformal agreements to a formal digital agenda would be more appropriate. Outcome Notes were largely used to store the documents produced during the meeting (but usually missing its relationship with the Agenda tool), as both a short- and a long- term memory of group outcomes, and participants recognized this as a major feature of the system. In design meetings, the Outcome Notes List was often used as a general indexing tool, to organize and reorganize items to be discussed, to temporarily store documents (mainly diagrams and sketches), often with the goal of saving/freeing space on the surface. As we said above, saving and retrieving documents using the bottom "save" on the toolbars was not intuitive. The Outcome Notes List allowed, on the contrary one, to visualize the documents saved and to retrieve them through drag-and-drop. Especially in debriefing meetings, the Outcome Notes List mainly worked as a repository for group notes and minutes, written using the keyboard. Two different processes were exploited to this end. In the first, a participant was more or less formally appointed the task of providing the minutes/notes. The alternative consisted of collective processes during which different interpretations were first proposed and discussed until a shared view was reached, and then written down (often under dictation). The first strategy had clear benefits for the group but presented a number of disadvantages for the individual in charge of the task; he or she felt the burden of writing the right (shareable) interpretation, feeling jeopardized in his or her active participation in other meeting activities (see below for what concerns role distribution). With the second strategy, the collective responsibility was a clear advantage for a successful decision-making process; however, the process was more time- and effort-consuming, and often required groups to reconsider their working practices and routines. The collective strategy just discussed, exemplifies an interesting function of the two coordination tools, namely as a support toward a public and shared representation of the situation. Another example in this direction is provided by the Agenda timer: The display of the time spent in the discussion of a given item often prompted groups to reflect on the temporal aspects of the meetings, in some cases stimulating them to plan the following meetings differently. Such attempts toward a use of the system as a means to consciously improve the group's activities depended on the group's intrinsic motivation. For instance, one of our groups, Group 3, used the system five times, for a total of more than six hours;

in the interview the push toward improving group work emerged clearly from all the participants, and the project manager clearly stated that they had already tried to exploit structured methodologies to improve their efficiency. From the observations, this group appeared to adopt a more systematic use of the timer than the other groups. We tried to more deeply investigate the value of having a document, a sort of report, available after the meeting for each of the participants, to sustain a shared representation of the meeting among group members. During interviews, participants stated that having such a report to be accessed after the meeting would be important in terms of the problem of keeping a long-term memory of group meetings. They also suggested that group-related documents are rarely recovered and are usually lost. While speaking about their own attempts to structure meeting activities, members mentioned Group 3 also reported that a shared folder was created to store group documents but that, after a period, the group abandoned this habit because of the time and effort required to manage this process.

Traditional and electronic tools coexisted, and their use was negotiated their space with the functionalities provided by the Collaborative Workspace. Pre-existing artifacts tended to complement system functionalities, mainly along three directions: communication among members in time-pressing situations and during presentations, extension of the physical surface, and by providing links between the work carried out during the meeting and related work conducted before and after it. The resort to traditional media in particular paper was higher in the presence of time and/or performance pressure; in some of these cases, the tabletop device ended up being totally encircled by "real" sheets of papers, and mostly used as a normal table. This result was partly motivated by the limited accuracy of the digital pen, which forced people to write more slowly than with a traditional pen. On the other hand, the resort to the virtual sheets increased whenever saving contents and making them available after the meeting became important. That is, virtual sheets were preferred to real ones whenever the need for reliably storing material surpassed the time/performance pressure.

17.6 Re-Thinking the Collaborative Workspace

From the information discussed above, the following requirements were derived for the second design of the tabletop device:

- sharing of meeting's results should be better supported;
- rotation and movements of documents should be better managed;
- the Agenda and the Outcome Notes List should be re designed.

Other aspects related to limitations of the particular hardware used were not addressed in the second prototype.

17.7 The Collaborative Workspace: Second Design

In order to have a more natural way of rotating documents, the Rotate and Translate algorithm proposed by Kruger and colleagues [7] was implemented. This technique allows the rotation and movement of the window with a single gesture by acting on the window border.

Concerning the coordination of activities through open tools, the functionalities of the Agenda and Outcome Notes List were collapsed into a single functionality. This new tool looks like the former agenda and allows the participants using it to keep track of the time spent on each item. Furthermore, by dragging documents on an item, it stores them for further use. The agenda tool was also augmented with a tool for the automatic generation of the minutes of the meeting. In its simplest form, the minutes contains just the indication of the time spent and the documents dragged. A more complex use of the agenda tool, producing a more elaborate report, is by using a special type of documents, called the Notes. They look pretty similar to textual documents, but they also have space to store other documents as attachments. When dragged into an Agenda tool's item, however, the text Notes becomes part of the report. A careful use of the Notes, therefore, allows the groups to generate rich reports during the meetings.

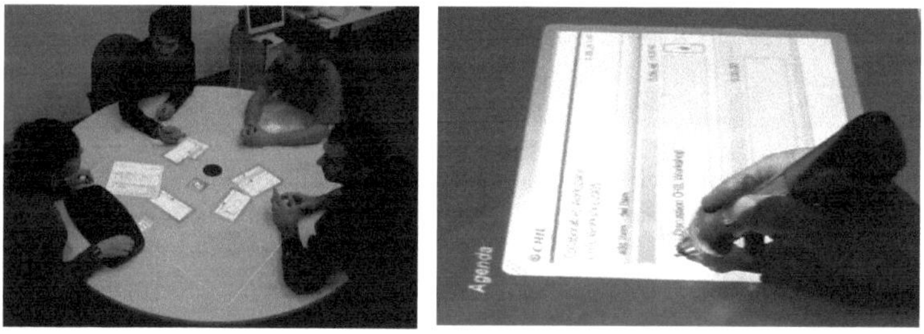

Fig. 17.5. Left, the Collaborative Workspace. Right, a snapshot of the interface.

Finally, the system was augmented by a Web application meant to be used before and after a meeting. The application allows the teammates to schedule a meeting, upload documents, and preset the agenda. After the meeting, the group will find online the minutes of the meeting in either in the shorter form or the longer form discussed above. We expect that this functionality will lead to a better use of the report functionalities after some use.

The second prototype was implemented using the WPF capabilities of the new .NET3 framework available as part of Microsoft's MS Vista operating system.

17.8 The Third User Study

The third study focused on user acceptance of the Collaborative Workspace application. As suggested by Dillon [3], there is no complete theory that predicts acceptance of a system. There is, however, a growing understanding of the key factors that affect acceptability. Several studies demonstrated that usability is a necessary but a sufficient factor to ensure acceptability of a technology. According to Rogers [15], acceptance of a system is a complex phenomenon that is related, among other factors, to the compatibility of the new technology with pre-existing working practices and norms.

Three teams working in the area of IT were involved in our study. Eight meetings were video taped by two cameras and table microphones for a total of about 14 hours of system usage. We observed from two to four meetings for each team. Although each team consisted of a small number of core people (three or four), the number of participants in the meetings varied. The length of meetings varied from half an hour to two hours (with a particular case of 4 hours). The methodology used was again video observation and interviews.

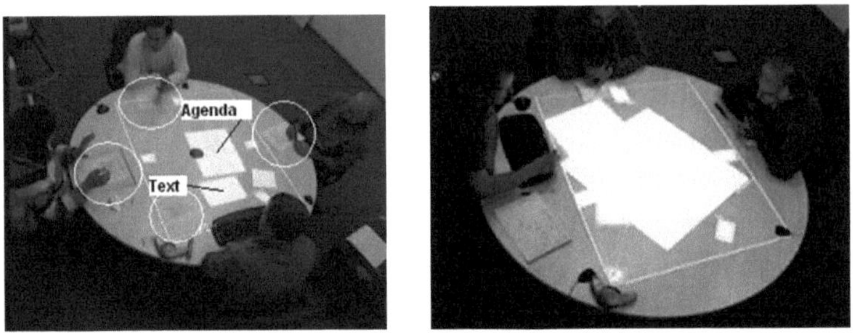

Fig. 17.6. Left. Organization of space and task allocation among group members during the first meeting: participant A takes notes (in his personal area); the Agenda he manages but placed it in the public area. Right In the second meeting, the team used the CW documens almost exclusively, which are enlarged and placed in the public area, to support the technical discussion.

The CW proved to be flexible enough to make it possible for each group to accommodate it into their work practices while not preventing the teams from continuing to use their normal tools (from sheets of papers to personal laptops). Yet, the introduction of the system required the groups to engage in a sense-making process that was sometime disruptive for the cohesion and the efficiency of the meeting (for details, see [10]).

The three teams attributed different meanings and values to the system and they exploited the opportunities offered differently. One team regarded the system as an organizational tool. They recognized and used the possibility of setting the agenda before the meeting, and of keeping track of the time spent on the different items during the meeting. Moreover, the system stimulated participants' reflection about meeting organization. Another team was more motivated by the possibility of keeping track of changes made to documents. The agenda and the possibility to produce the meeting reports semi-automatically were not recognized as relevant for them.

In general, the effectiveness of all three teams seemed to be enhanced after of the system was used a few times. In particular, the team recognized the following advantages at a group level:

- Efficient document management: The digital contents allow for a smooth transition of data throughout the team lifecycle (premeeting, meeting, postmeeting).
- Support for time management during meeting. One of the major complaints of groups is the waste of time during traditional meetings. Timing mechanisms provided by the CW allow teams to be aware of the resources used and proved to be a support for group process restructuring.
- Support for group memory. The report was considered a key element for efficient teamwork. The possibility of comparing the initial "desiderata" (the information entered through agenda, topics to be discussed and expected time) with the outcomes of the report (how many topics were actually discussed? How much time was spent on each item?) was recognized as an important factor for a more efficient way of working.

The distribution of roles among members emerged as a key factor to understanding acceptability of the system. The availability of new tools the website, the agenda, and the report at the end of the meeting as well as the features of the system (that assume a collaborative use of the functionalities) demanded that groups reconsider and formalize some of the practices they informally carry out. This in particular led to a restructure of the roles and responsibilities among members. The work of some group members rose in complexity for example, all three groups delegated the responsibility of updating the website to one member. Other roles, were instead simplified, and some tasks decreased in complexity: For example, the note-keeper could share his or her notes with the rest of the group during the meeting itself.

Even if group efficiency can improve with the introduction of teamwork facilities, acceptability ultimately depends on perceived costs and benefits for individuals, and on the value attributed to the changes introduced by the system. In our case, if advantages at a group level are more or less clear, benefits at the individual level remain problematic. Asymmetrical task distribution led some individuals to feel frustrated by the workload imposed by the system. In particular, taking minutes caused major difficulties. While the advantages of having a public report are widely recognized by all members, the practical accomplishment of this task seemed problematic. All three groups found it difficult to efficiently organize this activity and tried more than one strategy to deal with it. Appointed individuals lamented the difficulty in following the meeting discussion, the embarrassment coming from the public nature of the writing, the difficulty in selecting the right information and transcribing it, and the fear of giving a personal interpretation rather that an 'objective' one. In some way, the system causes a reduction in the steps usually taken to write a final report, thus augmenting the complexity of the task. Finding a balance between the accomplishment of group objectives and the individual ones seems a key aspect. These individual difficulties should be taken seriously inasmuch as for groupware to be accepted the whole group is expected to respond positively, to find it useful and acceptable. The literature on groupware acceptance clearly shows that in several cases the rejection of groupware was due to the refusal of technology by a few individuals despite the clear advantages for the whole group.

References

1. J. Carroll. Completing design in use: Closing the appropriation cycle. In *Proceedings of ECIS 2004*, Turku, Finland, Jun. 2004.
2. A. Crabtree. *Designing Collaborative Systems. A Practical Guide to Ethnography.* Springer-Verlag, 2003.
3. A. Dillon. User acceptance of information technology. In W. Karwowski, editor, *Encyclopedia of Human Factors and Ergonomics*. Taylor and Francis, 2001.
4. M. Huysman, C. Steinfield, C.-Y. Jang, K. David, M. H. I. T. Veld, J. Poot, and I. Mulder. Virtual teams and the appropriation of communication technology: Exploring the concept of media stickiness. *Computer Supported Cooperative Work*, 12(4):411–436, 2003.
5. K. Inkpen, R. Mandrik, J. M. DiMicco, and S. S. Scott. Methodologies for evaluating collaboration in co-located environments. In *Proceedings of Workshop on Methodologies for Evaluating Collaboration in Co-located Environments at CSCW*, Chicago, 2004.
6. B. Johanson, A. Fox, and T. Winograd. The interactive workspaces project: Experiences with ubiquitous computing rooms. *IEEE Pervasive Computing Magazine*, 1(2), Apr.,Jun. 2002.
7. R. Kruger, M. Carpendale, S. Scott, and A. Tang. Fluid integration of rotation and translation. In *Proceedings of the ACM Conference on Human Factors in Computing Systems (CHI 2005)*, Portland, OR, 2005.
8. R. Kruger, M. S. T. Carpendale, S. Scott, and Greenberg. Roles of orientation in tabletop collaboration: Comprehension, coordination and communication. *Journal of Computer Supported Collaborative Work*, 13(5-6):501–537, 2004.
9. H. J. Leavitt. Some effects of certain communication patterns on group performance. *Journal of Abnormal and Social Psychology*, 46:38–50, 1951.
10. C. Leonardi and M. Zancanaro. A trouble shared is a troubled halved: Disruptive and self-help patterns of usage for co-located interfaces. In *Presented at the Abuse-Workshop held in conjunction with INTERACT05*, Rome, Italy, 2005.
11. M. R. Morris, K. Ryall, C. Shen, C. Forlines, and F. F. Vernier. Beyond social protocols: Multi-user coordination policies for co-located groupware. In *Proceedings of CSCW 2004*, 2004.
12. J. A. Nunamaker, A. R. Dennis, J. S. Valacich, R. Vogel, and J. F. George. Electronic meeting systems to support group work. *Communications of ACM*, pages 42–61, 1991.
13. K. O'Hara, M. Perry, E. Churchill, and D. Russel, editors. *Public and Situated Displays*. Kluwer, 2004.
14. G. Olson, J. S. Olson, M. R. Carter, and M. Storrosten. Small group design meetings: An analysis of collaboration. *Human-Computer Interaction*, 7(4):347–374, 1992.
15. E. Rogers. *Diffusion of Innovations*. Free Press, New York, NY, fifth edition, 2003.
16. Y. Rogers and S. Lindley. Collaborating around vertical and horizontal displays: which way is best? *Interacting With Computers*, 16:1133–1152, 2004.
17. Y. Rogers and T. Rodden. Configuring spaces and surfaces to support collaborative interactions. In K. O'Hara, M. Perry, E. Churchill, and D. Russel, editors, *Public and Situated Displays*, pages 45–79. Kluwer, 2004.
18. K. Schmidt and L. Bannon. Taking CSCW seriously: Supporting articulation work. *Computer Supported Cooperative Work*, 1(1-2):7–40, 1992.

19. K. Schmidt and C. Simone. Coordination mechanisms: Toward a conceptual foundation of CSCW system design. *Computer Supported Cooperative Work*, 5(2-3):155–200, 1996.
20. S. D. Scott, M. S. T. Carpendale, and K. M. Inkpen. Territoriality in collaborative tabletop workspaces. In *Proceedings of the ACM Conference on Computer-Supported Cooperative Work, CSCW 2004*, Chicago, IL, 2004.
21. S. D. Scott, K. D. Grant, and R. L. Mandryk. System guidelines for co-located, collaborative work on a tabletop display. In *Proceedings of the European Conference on Computer-Supported Cooperative Work CSCW2003*, Helsinki, Finland, 2003.
22. L. Suchman. Located accountabilities in technology production. *Scandinavian Journal of Information Systems, 2002, 14(2)*, 14(2):91–105, 2002.
23. J. C. Tang. Finding from observational studies of collaborative work. *International Journal of Man-Machine Studies*, 34(2):143–160, 1991.
24. D. Wigdor and R. R. Balakrishnan. Empirical investigation into the effect of orientation on text readability in tabletop displays. In *Proceedings of the European Conference on Computer Supported Cooperative Work*, 2005.

18

The Memory Jog Service

Nikolaos Dimakis[1], John Soldatos[1], Lazaros Polymenakos[1], Janienke Sturm[2], Joachim Neumann[3], Josep R. Casas[3]

[1] Athens Information Technology, Attiki, Greece
[2] Technische Universiteit Eindhoven, Eindhoven, The Netherlands
[3] Universitat Politècnica de Catalunya, Barcelona, Spain

The CHIL Memory Jog service focuses on facilitating the collaboration of participants in meetings, lectures, presentations, and other human interactive events, occurring in indoor CHIL spaces. It exploits the whole set of the perceptual components that have been developed by the CHIL Consortium partners (e.g., person tracking, face identification, audio source localization, etc) along with a wide range of actuating devices such as projectors, displays, targeted audio devices, speakers, etc. The underlying set of perceptual components provides a constant flow of elementary contextual information, such as "person at location x_0,y_0", "speech at location x_0,y_0", information that alone is not of significant use. However, the CHIL Memory Jog service is accompanied by powerful situation identification techniques that fuse all the incoming information and creates complex states that drive the actuating logic. Essentially, it tries to identify meaningful situations in cases when people interact, and improve the overal experience for the current event.

We consider scenarios such as meetings, lectures, presentations, etc, in which people interact continuously and do not appreciate being "disturbed" by a computer service. The CHIL Memory Jog service operates in the background and provides pertinent information to the participants in a nonobtrusive way. In this chapter, we elaborate on the service system specifics, the functionality and features of the application and the evaluation process, which was organized to determine how appealing the service was to a wide range of people. Athens Information Technology (AIT, Greece) and the Universitat Politècnica de Catalunya (UPC, Spain) developed two services, presented in the next sections, and the Technische Universiteit Eindhoven supervised the service evaluation process.

18.1 The AIT Memory Jog Service for Meeting, Lecture and Presentation Support

The Memory Jog service is a ubiquitous human-centric, context-aware service that facilitates human human collaboration and human computer interaction during meetings, presentations, and lectures by providing pertinent information to the participants in a nonobtrusive way. It benefits from the CHIL Reference Architecture, presented in Chapter 22, which acts as a breadboard for context-aware services [13] and exploits the infrastructure along with its associated middleware components, in order to

- Identify participants and track their locations within the smart room. The identity and location of participants can be displayed within different devices, such as PDAs and desktop/laptop computers.
- Keep track of the meeting progress based on an agenda as seen in Fig. 18.3(a), which is known a priori, and retrieved from a globally accessible knowledge base.
- Provide information about the participants such as biography, contact information, affiliation etc. as in Fig. 18.3(b)
- Record the meeting, lecture, or presentation based on the best camera selection mechanism [2], a snapshot of which is shown in Fig. 18.1. The recording is tagged with metadata obtained by the situation model of the service, to allow selective retrieval and postprocessing of the recording, as in Fig. 18.3(c).
- Provide a summary of events that were missed during a person's absence, by keeping track of the identity of the person and logging the time of exit and reentry. A snapshot of this feature is shown in Fig. 18.3(d).
- Provide context-aware assistance by displaying relevant information from past meetings (e.g., presentations, URLs). Information from past meetings includes video segments recorded by the recording service.
- Assist participants by automatically carrying out casual meeting tasks, such as opening presentations, e-mailing material to the participants, displaying notes/messages, notifying the presenters about the time left for their presentation using text-to-speech, booking new meetings, etc.

Fig. 18.1. The Intelligent Video Recording System of Athens Information Technology, which selects the most frontal view of a speaker, and operates autonomously or under the control of the Memory Jog service.

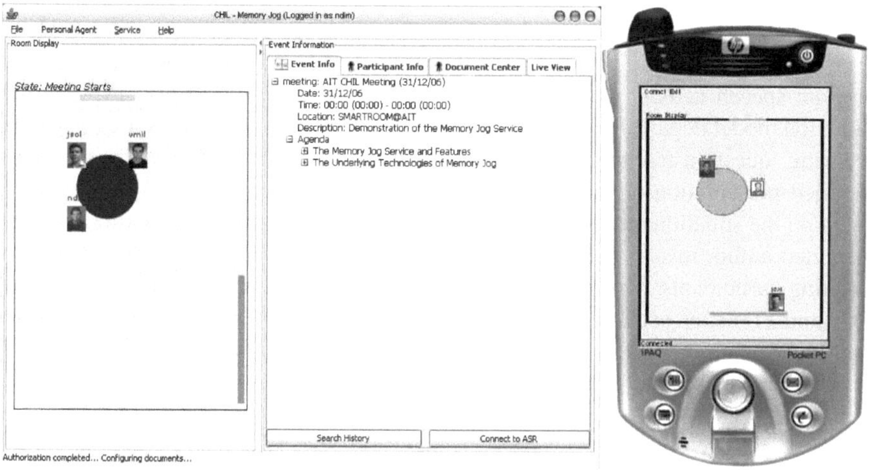

Fig. 18.2. The graphical user interface (GUI) of the Memory Jog service of AIT, as provided in the cases of a laptop and PDA.

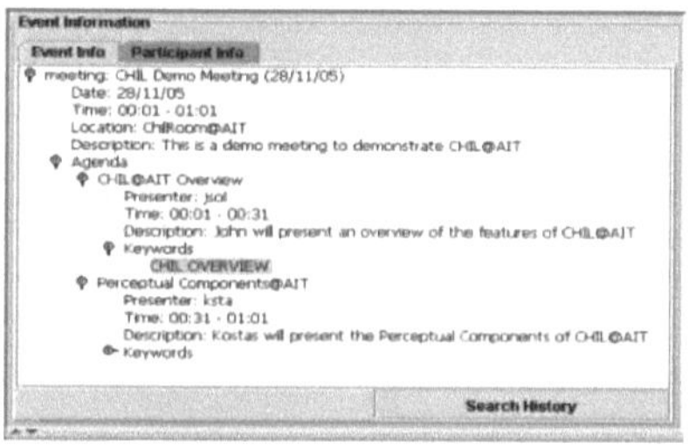

(a) Agenda monitoring in the Memory Jog service.

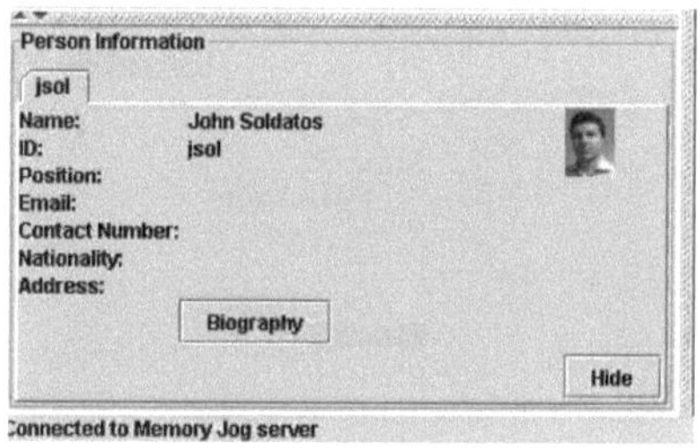

(b) Person-related information as presented in AIT's Memory Jog service.

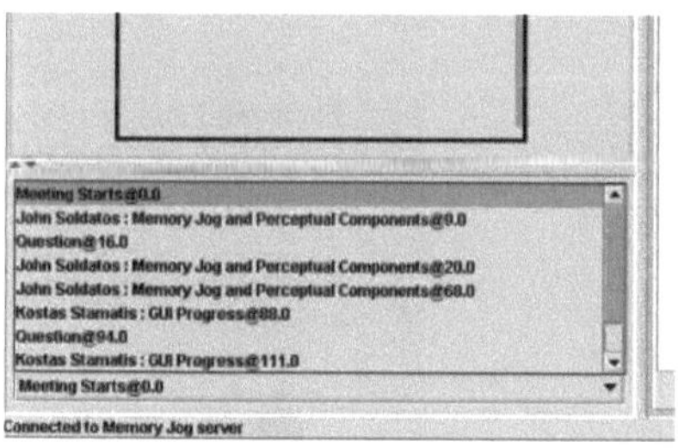

(c) Selecting the appropriate video segment from AIT's Memory Jog service. Presenter and time information as presented.

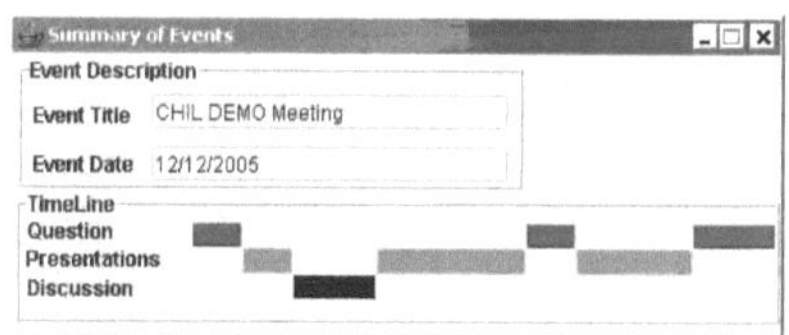

(d) The "What Happened While I Was Away?" feature of the Memory Jog service, providing an outline of the events that a person missed during his or her absence.

Fig. 18.3. Features of the AIT Memory Jog service.

The Memory Jog service exploits a variety of perceptual components including face recognition [10], acoustic localization [4], visual body tracking [11], face detection, speech activity detection [12], and other components. It also exploits the situation model depicted in Fig. 18.4 to follow higher-level situations such as agenda tracking, question tracking, meeting commencement, and meeting finish. A more detailed presentation of this model is described in Section 18.1.1, where we also mention the situation model mapping issues. The service exploits actuating services (targeted audio, text-to-speech, display services) as well, to output information to meeting participants. Moreover, information is rendered in GUIs, which run on both PCs and PDAs, as seen in Fig. 18.2. In the case of a PDA, the visualization detail and rate of communication have been adjusted to meet the limitations of the device such as screen size and issues related to batteries.

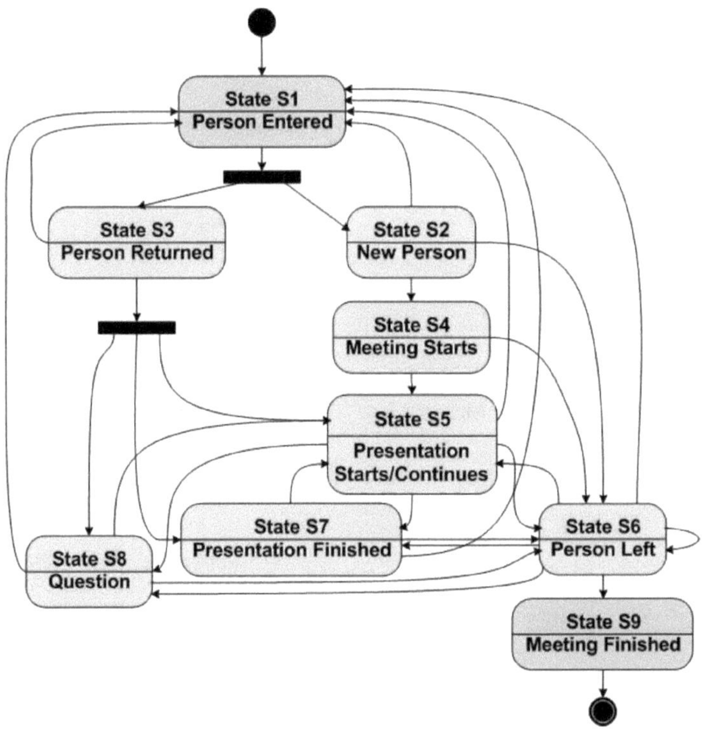

Fig. 18.4. The situation model for monitoring meetings as adopted by the AIT Memory Jog service.

The Memory Jog service was designed, implemented, and evaluated using the CHIL Reference Architecture, exploiting the whole set of sensors, actuators, and perceptual components that enrich the Memory Jog with a constant feed of contextual information. In addition to the support it provides for local participants in the smart space and mobile ones (using PDAs as seen in Fig. 18.2), support is also implemented for participants who want to interface with the smart space from remote smart spaces, which we call *satellite spaces*. These satellite spaces are equipped with a smaller set of sensors (such as a camera and a microphone) that undertake the process of face identification, speech activity detection, etc. Finally, the Memory Jog service has been integrated with two features that can be used during the event, the Skype library for Voice-over-TCP telephony and Google Calendar, features that have been integrated using the Skype and Google Calendar libraries.

The graphical user interface (GUI) is a visualization tool with capabilities exceeding the simple monitoring of the smart space. The left-hand side of the GUI, as seen in Fig. 18.2, outlines the smart space area, including the artifacts such as the tables, presentation areas, doors etc. This information is retrieved by the Memory Jog service as soon as a user successfully logs in. On the right-hand side, the service provides information about the event that is currently taking place or is about to start. Information such as the meeting title, the participants, the expected time that the meeting will start, the time that each agenda item is assigned to, and the presenter is also provided in a tree-like fashion. For each agenda item, appropriate keywords are added, which can facilitate the users of future meetings in searching previous presentations. The agenda tracking is illustrated in Fig. 18.3(a)

In the room outline (on the left-hand side), the users can select a participant in the meeting and perform a number of tasks, either if they are in the smart space, or if they participate remotely from a satellite space. These tasks can be either sending a private message (text or using the targeted audio device) or initiating a call using Skype (more helpful if the participant is not inside the smart space, as in a satellite space), as seen in Fig. 18.6, or checking the biography of the user. The GUI can also help the participants in organizing other meetings, which can be booked using the Google Calendar interface in the Google Calendar website. This calendar can be made public so that all participants can see the schedule. This is illustrated in Fig. 18.5, which shows the scheduling of another meeting and shows it appears in the Google Calendar website of the group. Moreover, the users, either local or from satellite spaces, can initiate a request to see videos of past events that were stored using the Intelligent Video Recording feature of Memory Jog. This sequence allows the users to select the video of their choice, based on a set of metadata gathered during the time of the recording: The date and time of the presentation, meeting title, agenda item title, presenter, and meeting state. The meeting state refers to whether this recording is about a question or the actual presentation and is extracted by the situation modeling component of the CHIL framework.

When a user wants to participate from a satellite space in the meeting, he can initiate a request using the GUI, which forwards the request to his *PersonalAgent*. His *PersonalAgent* undertakes the required connection action as soon as the user provides the IP of the "listening end" of the smart space. This coupling is made by two special agent-members of the CHIL framework the *ExtendedServerAgent* and the *ExtendedAgent*. These two agents actually interpret ontology messages to XML-formated ones and initiate an exchange. The remote users, after registering to the Memory Jog service, are given the same rights as the local ones and can use the Memory Jog service as if they were in the smart space.

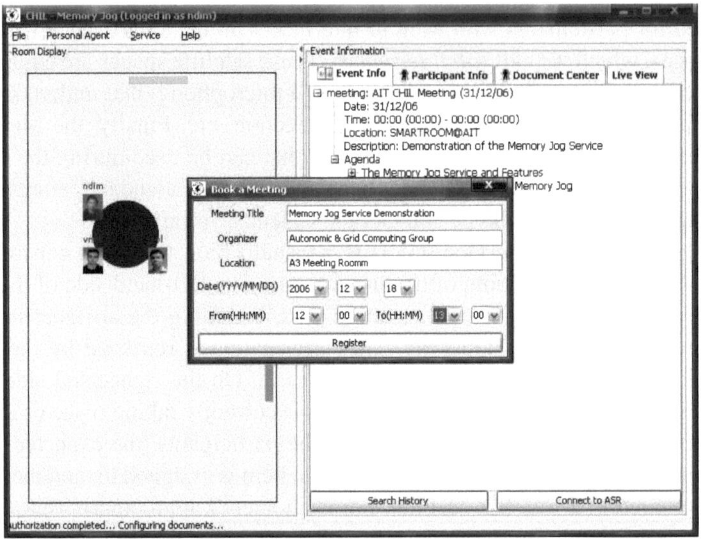

(a) Scheduling a new event using the Google Calendar library from the AIT Memory Jog. Entries such as time, organizer, date, and location can be specified using the dialog window.

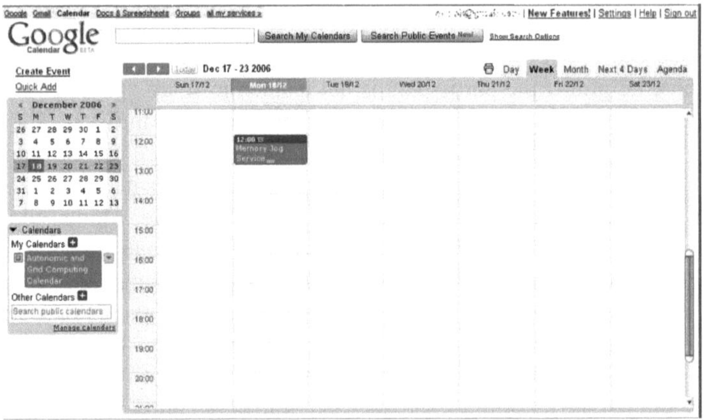

(b) The online Google Calendar of the Autonomic & Grid Computing Group of AIT. The scheduled meetings are announced in this publicly available calendar.

Fig. 18.5. Scheduling a new event using the Google Calendar library from the AIT Memory Jog. The calendar is publicly available for viewing.

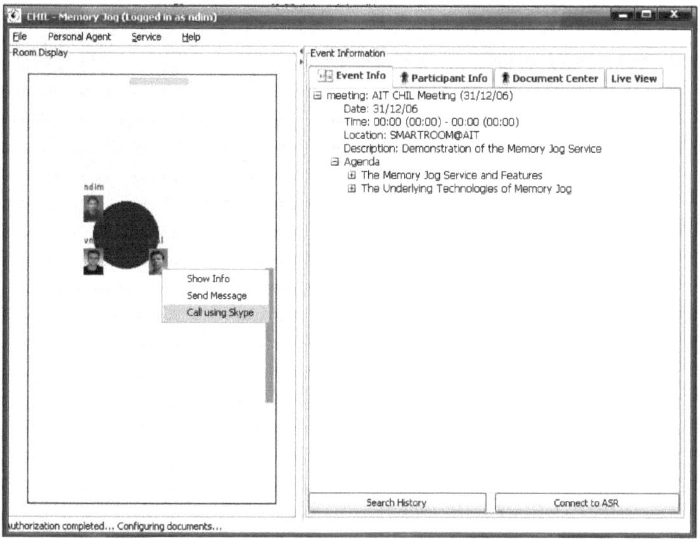

Fig. 18.6. Calling a participant using the Skype library

Each user-participant is modeled by his *PersonalAgent*, a corresponding agent that undertakes the tedious tasks of interfacing with the rest of the CHIL framework. All internal interactions in the Memory Jog service, are filtered by the "heart" of it, the *MemoryJogAgent*, who is actually responsible for orchestrating the communication requests from within the framework. At all times, it is aware of the current participants in the smart space, including information about their locations, their IDs, their biography, etc., the meeting agenda that is accompanied by keywords for easier searching, the persistent storage areas (such as databases, knowledge bases, etc.), the presence and status of all sensors and actuators in the smart space, the status of all perceptual components that provide contextual input to the situation modeling agent, the key agents of the CHIL framework (such as the situation modeling agent, the agent manager, the autonomic agent manager), etc. This information is bundled in a data model which contains all this information and is accessible to the Memory Jog. This data model is updated either by user request, or implicitly if needed.

Despite the fact that the Memory Jog service is a service by itself, the CHIL framework enables the integration of numerous subservices into a larger one. In our case, AIT´s Memory Jog service subsumes two other services that operate in parallel. For example, the Intelligent Video Recording service is controlled by the *DirectorAgent*, which is managed by the Memory Jog service. The Memory Jog service notifies the *DirectorAgent* about the changes in the current state (meeting starts, question, etc.) including the required information such as the agenda item or the presenter, and the *DirectorAgent* in turn adjusts the sensor streams, which are directed to the Intelligent Video Recording component and stores the metadata to the database. The Memory Jog service can control additional services such as the Surveillance Service, a service responsible for monitoring the indoor environment about unauthorized people accessing restricted areas.

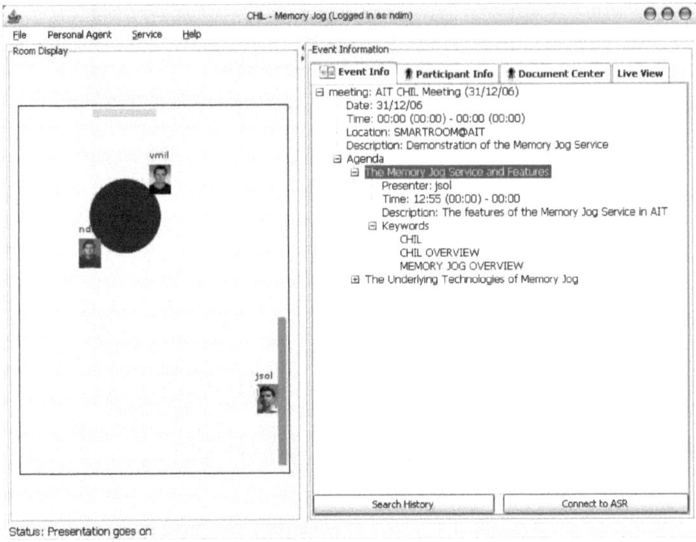

Fig. 18.7. Tracking the meeting agenda using the AIT Memory Jog. This is an example of identifying the start of a presentation.

In addition to functionality features that enrich the Memory Jog service, a plethora of nonfunctional features operate transparently and ensure the robust operation of the service, without interrupting the participants. Such features are the autonomic handling of the service in terms of the service itself as well as its also regarding the sensor data streaming [5], the interfacing with the CHIL Ontology which provides the CHIL "world model" comprising information about sensors, perceptual components, artifacts, room, people [9], as well as a resource management system for the infrastructure [14].

18.1.1 AIT Memory Jog Situation Model for Meeting Support

As an example of a situation modeling application, we present our apporach in the AIT Memory Jog service. The situation model, depicted in Fig. 18.4, shows a series of individual states, all of which are interconnected to form a directed graph. The state *NIL* is considered the initial state of the situation machine; it will signify the state traversal throughout the event evolution. This state transition is made possible only when a set of rules that couple the model during its design is satisfied. These rules can be organized in the form of a truth table. Table 18.2, a truth table that accompanies the AIT Memory Jog service situation model.

An integral part of the Memory Jog service is the situation modeling component (implemented as an agent of the CHIL agent framework). This component is fuses the underlying perceptual component outputs (e.g., body tracker, face ID, audio source localization etc.) and attempts to identify which of the situation model states (as explicitly defined in the model of Fig. 18.4), best matches the current state. This process is crucial for the optimal operation of the service, as the identification of the current state will trigger the application logic that will be propagated throughout the CHIL framework. This model is implemented as an XML file in which the transitions and rules are placed. In Table 18.1, we outline a snapshot of this file, which shows the requirements for going to state *Question*; A person needs to be near the presentation area but not speaking, and at least one person in the table area needs to speak (asking the question).

```
...
<State>
  <Uid>S5</Uid>
  <Name>Question</Name>
  <PreviousStates>
    <PreviousState>S4</PreviousState>
    ...
  </PreviousStates>
  <NearToArtifact>
    <Room Relation='>' Quantity='0'>0</Room>
    <Table Relation='>=' Quantity='0'>2</Table>
    <Whiteboard Relation='==' Quantity='0'>1</Whiteboard>
  </NearToArtifact>
  <DoingNearToArtifact>
    <Talking>
      <Table Relation='==' Quantity='0'>1</Table>
      <Whiteboard Relation='==' Quantity='0'>0</Whiteboard>
    </Talking>
  </DoingNearToArtifact>
</State>
...
```

Table 18.1. Sample situation model XML file.

In Table 18.2, we illustrate the state-transition truth table. The left column shows the possible state transitions as defined by the model. The right column specifies the combinations of the perceptual components that trigger the state transition. *PeopleCount* pertains to the number of people for each case (e.g., Room, Table, Whiteboard), *SpeakerLocation* refers to the speaker's location (e.g., Whiteboard or Table area), and *FaceID* is related to the identity of the person who enters the room. Moving from one state to the other requires that all arguments in the right column are satisfied. For example:

Situation Transition	Combinations of Perceptual Components Outputs
NIL $\Rightarrow$ S1	PeopleCount(Room) = PeopleCount(Room) + 1
S3, S4, S6, S7, S8 $\Rightarrow$ S1	PeopleCount(Room) = PeopleCount(Room) + 1
S1 $\Rightarrow$ S2	FaceID(NewPerson) = NewPersonID
S1 $\Rightarrow$ S3	FaceID(NewPerson) = OldPersonID
S2 $\Rightarrow$ S4	PeopleCount(Room) = N **AND** PeopleCount(Table) = N
S3, S4, S6, S7, S8 $\Rightarrow$ S5	PeopleCount(Room) $\geq N$ **AND** PeopleCount(Table) ≥ 1 **AND** PeopleCount(Whiteboard)=1 **AND** SpeakerLocation=Whiteboard
S2, S3, S4, S5, S6, S7, S8 $\Rightarrow$ S6	PeoplCount(Room) = PeopleCount(Room) - 1
S3, S5, S6 $\Rightarrow$ S7	PeopleCount(Room) $\geq N$ **AND** PeopleCount(Table) $\geq N$ **AND** PeopleCount(Whiteboard) = 0 **AND** SpeakerLocation = Table
S3, S5, S6 $\Rightarrow$ S8	PeopleCount(Room) $\geq N$ **AND** PeopleCount(Table) $\geq N-1$ **AND** PeopleCount(Whiteboard)=1 **AND** SpeakerLocation = Table
S6 $\Rightarrow$ S9	PeopleCount(Room) = 0

N: The expected number of people participating in the meeting.

Table, Whiteboard: Key-areas of the smart space (e.g. the table area and the presentation area).

NewPerson: The FaceID perceptual component operates on the person who has entered the room. The NewPerson tracking ID is provided by the Body Tracker perceptual component.

OldPersonID, NewPersonID: Each person in the smart space is assigned an ID. In this case, the *OldPersonID* implies that the person who has returned has an ID of someone who was previously in the smart space, whereas the *NewPersonID* indicates that this person has a new ID.

Table 18.2. Situation model transition truth table for the model depicted in Figure 18.4.

(S1 $\Rightarrow$ S3 : FaceID(NewPerson) = OldPersonID): This describes how a newcomer in the smart space is considered as someone who had left and returned. The FaceID perceptual component determined that the ID of this person (the *NewPerson*) belongs to a person who had left. This context enables the triggering of some predefined steps that (in this case) will initiate the visualization of a summary of agenda items this person missed during his or her absence (a feature that we conveniently call "*What happened while I was away?*").

(S3, S5 $\Rightarrow$ S7 : PeopleCount(Table) = N, SpeakerLocation = Table): This establishes the requirements for the transition to the state *Presentation Finished*. As can be seen from the rules, this transition is possible if the number of people listed in the table equals those participating in the meeting and the speaker is located in the table area, given that the previous state was *Presentation continues*.

18.1.2 Focus Group Evaluation

As a first attempt to evaluate the Memory Jog service, we organized a series of focus groups to which we demonstrated the service, along with a brief description of the technologies, avoiding, however, details such as implementation techniques, assumptions, etc. This demonstration served two purposes; on the one hand, to demonstrate the service as a whole focusing on the key areas it targets; and, on the other to allow the participants to become familiar with Memory Jog's technologies, especially with their limitations, so that we could get a better view about whether they believe such services can be useful. This demonstration followed a thorough discussion among the AIT CHIL team and the participants during which the participants were asked questions regarding the nature of the meetings they attend, the type of information they usually need, problems that arise during the meeting, etc. Following the discussion, the participants had the chance to experience the service itself at the AIT Smart Laboratory, of the Autonomic and Grid Computing Group. On average, each focus group dedicated 1.5 hours to the evaluation, including discussions and presentation.

A diverse group of people participated in the evaluation; students, faculty, engineers, administration, members, etc. The main outcome was that both faculty and students were very interested in seeing this service being used in conferences and lectures, and not so much in cases of meetings. where the scenario is more static. Furthermore, the features individually were very appealing to the users, especially the Targeted Audio device and the Intelligent Meeting Recording [2] with turn and content annotation.

With respect to the external connectivity abilities that the Memory Jog has (the mobile and remote users as in the satellite spaces case), the comments of the evaluation groups were that, especially in the case of meetings, these features assist significantly the possible future expansion of such applications. This belief is shown in the last two entries of the following table, which summarizes the overall approval rating. It shows that the participants of the evaluation focus groups were not so impressed by the fact that remote smart spaces can be as useful as the mobile devices. This is because they were not very familiar with such complicated environments, whereas close to 75% of the participants had used or owned a portable device, such as a PDA. The ability to "participate" remotely in a meeting using such devices was appreciated by many of the participants, especially people working in managerial positions.

In Table 18.3, we summarize the results of these evaluations. The approval rating reflects the overall approval of the users based on how useful the corresponding feature is. Finally, one very important comment, which was mentioned by groups working in management positions, is that this service *introduces a novel interaction procedure and not just a supporting environment.*

The outcome of this evaluation process proves that people who are involved in meetings, lectures, and presentations consider such new types of services useful as long as user distraction is avoided. The majority of the participants appreciated the

Feature Description	Approval Rating
Intelligent Recording	100.0%
Agenda Tracking	56.0%
What Happened While I Was Away	72.0%
Participants' Biography	68.0%
Searching Past Events	84.0%
Ability to connect using PDAs	95.0%
Ability to connect remote Smart Space	43.0%

Table 18.3. Memory Jog focus group evaluation summary.

service, expressing, however, their concern about the possible market penetration of such sophisticated services.

18.1.3 User Evaluation

In addition to the focus group evaluations that we conducted, a series of actual user evaluations was organized that aimed at evaluating how user-friendly, intrusive, and useful this service is in real-case scenarios. These evaluations took place in the group's Smart Room from the October, 19, 2006, until November, 10, 2006. In the scope of this process, we compiled a detailed scenario that covered as much of the Memory Jog features as possible. This scenario comprised a series of presentations about a specific event (a murder case); these presentations these were made by two different people demonstrating the evidence and clues on which the participants should base their decision about who the murderer was.

The evaluation process included

- filling-out a questionnaire that aimed at getting feedback regarding the participants' knowledge of solving mysteries, (e.g., watching TV shows, reading novels etc.),
- a brief introduction to the Memory Jog service and limitations of the underlying components,
- a 10-15 minute discussion regarding state-of-the-art technologies in meeting-, lecture-, and presentation-support systems
- the murder-case scenario, which the participants had to solve with and without the Memory Jog service, and
- a 20-30 minute final interview during which the participants elaborated on their experience with Memory Jog as a tool for improving collaborative work[1].

In all cases, the evaluation process was completed by a questionnaire the participants had to fill out. This questionnaire focused in their experience solving the case with and without the Memory Jog. In total, 42 participants were involved in this

[1] A similar approach was followed but without using the Memory Jog service, so that we could have comparative results. In this case, the participants solved the murder-case by discussion, rather than using the tools provided by the Memory Jog service. The discussion concerned how the participants could be assisted in such a collaborative problem.

process, 18 of whom participated using the Memory Jog service and 24 without the service.

With this process we wanted to explore the behavior of the service in terms of its "non-functional features" which we have already defined. These features are summarized below:

Usability The system should have good usability
Intuitiveness Users should be able to use the main functionality of the Memory Jog with limited training.
Intrusiveness The service should be nonintrusive.
Usefulness The service should be useful.

In order to further facilitate the evaluation process, in the scope of the actual service evaluation, we provided with the option to the participants of filling in their first questionnaire regarding their experiences with mystery novels and films, online, using the Autonomic and Grid Computing group's Wiki-based website. As seen in Fig. 18.8(a), participants could access their account in the Wiki and fill in the forms. Similarly, we also provided the biographies of the presenters online, in the same system (18.8(b)). The reason for choosing this option is to get the participants to use their computers as soon as possible, instead of having them answer the questions on paper. On the other hand, the scenario without the Memory Jog required the participants write their answers on a form.

The results are shown in Table 18.4, and illustrate the participants' experiences in using the service and its features. The interview was organized in such a way that the participants were acting as the "jury", and a member of the AIT team acted as the police officer, who listened to the arguments of the participants about who they considered to be the murderer. During this interview, we noted how the participants followed the clues in determining the possible suspects.

The main differences between the participants who used the Memory Jog and those who did not were as follows:

1. The participants using the Memory Jog followed a more structured approach in determining the suspects. During the non-Memory Jog process, the participants were mostly making assumptions rather than reviewing in depth to each suspect.
2. The participants in the Memory Jog case took much less time to narrow down the suspects and select a murderer. In the non-Memory Jog case, the participants almost never decided on a single suspect alone.
3. The arguments provided in the Memory Jog case seemed more complete and accurate, whereas in the non-Memory Jog case, they were slightly vague and general.
4. On the other hand, during the Memory Jog case (especially in the beginning), the participants were collaborating less than in the non-Memory Jog case. This is because they had the GUI in front of them, which gave the impression that they should work alone. However, this phenomenon was experienced only in the first minutes of the evaluation process.

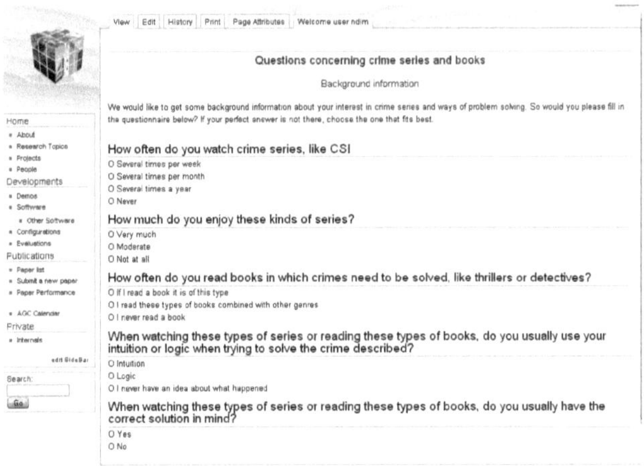

(a) Filling the pre-questionnaire online, using the Autonomic & Grid Computing group's Wiki-based website.

(b) Accessing the presenters' biographies from the Wiki-based website of the Autonomic & Grid Computing group of AIT

Fig. 18.8. Facilitating the evaluation process, using the Wiki-based website of the Autonomic and Grid Computing group at AIT.

18.2 The UPC Memory Jog Service

The UPC Memory Jog service has evolved to serve three specific application scenarios since the start of the CHIL project. Each successive application scenario has progressively added new functionalities to the service while keeping the functionalities developed in the previous versions:

- In the first version, the UPC Memory Jog addressed a functionality called *"What was said while I was away?"*, providing a short report to a latecomer about the

Usability	The average score on the ease-of-use scale of the questionnaire was 5.73 on a 7-point scale. From group interviews: One of the groups indicated that they would like to see the GUI a bit more organized.
Intuitiveness	The participants found that the service is simple to use. However, an initial training should take place, not exceeding 5-10 minutes.
Intrusiveness	The average score on a 7-item intrusiveness scale was: 4.88 (on a 7-point scale where 7 means not intrusive at all). From group interviews: Some participants found the GUI very intrusive, as they kept looking at it. Other groups did not specifically mention anything about the intrusiveness.
Usefulness	The mean score on the usefulness scale of the questionnaire was 5.6 on a 7-point scale. From group interviews: People saw the recording as useful for some types of meetings, because fewer notes need to be taken and things can be checked later on. Other functionalities were considered less useful in the test situation, but they are seen as more useful for other types of meetings were people, e.g., do not know each other and the groups are larger.

Table 18.4. Memory Jog user evaluation summary.

development of a small meeting. Smart-Room monitoring and text-based summarization of the presentation slides until the arrival of the latecomer were implemented to provide accurate up-to-the-minute information on a proactive basis.

- A second version of the UPC Memory Jog was developed in 2005, addressing teamwork in a learning scenario, that of a lab assignment with one teacher and a group of students. The Memory Jog service assisted both the teacher and the students, the teacher assisted Memory Jog by providing support and giving explanations to the students, monitoring the development of the lab session, and calling the teacher if students had questions. The students could ask questions through a question and answer interface on subjects related to the lab session the Memory Jog service gave "hints" if lab assignment was progressing slowly.
  In this case, monitoring of the smart room and the students' progress was again essential. Perceptual components such as Speech Detection, Acoustic Localization, Person Tracking, Face Detection and Identification, and Object Analysis (laptop) were running in parallel during live demonstrations. Several service components were available: student and teacher interaction interfaces, question and answering engine, politeness loop in spoken room messages, highlight recorder, and highlight player.
- The third version of the UPC Memory Jog was developed in 2006 and presented in 2007. In this case, the scenario addressed was a group of journalists discussing the current events of today with the purpose of selecting the most appropriate headlines for the front page of the next day's newspaper. In addition, a latecomer

contributed an additional piece of news, and a field journalist brought yet another headline in a live audiovisual connection to the smart room.

In the following, we describe the third version of the UPC Memory Jog service as it covers most of the functionalities of the previous versions. In this specific implementation of the Memory Jog at UPC, information was provided to a group of newspaper journalists gathered together in the CHIL smart room. They have 10 minutes to decide the front page of tomorrow's edition of their newspaper. In addition to the information provided to the journalists in the smart room, a field journalist and a latecomer propose an additional news story. In this case, the Memory Jog service makes available headlines and news that have been created elsewhere (information-shift). The Memory Jog service is also capable of providing background information about the relevant pieces of news (information-pull) and in some occasions may decide to jump in human-to-human communication to provide a proactive service (information-push). The design paradigm behind these three ways of providing information was to enhance human-to-human communication; i.e., the journalists are helped to freely interact with each other instead of being forced to focus on how to interact with the Memory Jog service. The two most outstanding ways for the Memory Jog to interact with the journalists are

- A Talking Head not only informs the journalists about available resources and points out events such as the arrival of a latecomer or news being contributed by remote colleagues, but also facilitates information requests from the journalists in a human-like interface based on automatic speech recognition technologies.
- A remote field journalist is enabled to easily communicate with the journalists in the smart room. A Skype-based bi-directional audio communication is supported by real-time video managed by an automatic cameraman. The video is further enhanced by text annotations that reflect the context awareness of the Memory Jog.

The task of the journalists is to decide on the two most important pieces of news to appear on the front page of tomorrow's newspaper. The news from which the journalists have to select are imported by an up-to-date RSS feed. Fig. 18.9 shows an example of the resulting front page.

18.2.1 Hardware Setup of the UPC Smart Room

The unobtrusiveness desired for a natural interaction between humans and computers sets limitations on the positioning of the sensors in the room: The acoustic technologies applied in the UPC smart room limit themselves to far-field wall-mounted microphones that allow the participants to freely move around in the room without being concerned about how and where their voices and other sounds are picked up. However, the signals that these microphones deliver show an unfavorable signal-to-noise ratio and contain a large amount of reverberation due the scarce furniture and the acoustically hard floor and walls.

The UPC Newspaper

Barcelona, January 18, 2007

Cisco to sue Apple on iPhone name

from BBC world news on technology

Cisco Systems is suing Apple for trademark infringement in a US federal court, for using the iPhone name.
Apple launched its new handheld mobile phone device under the iPhone name on Tuesday, at the hotly anticipated Macworld event in San Francisco.
Following the launch Cisco said it hoped to resolve the matter by Tuesday evening after negotiations.
Apple responded by saying the lawsuit was "silly" and that Cisco's trademark registration was "tenuous at best".
"We think Cisco's trademark lawsuit is silly," Apple spokesman Alan Hely said. "There are already several companies using the name iPhone for Voice Over Internet Protocol (VOIP) products."
"We are the first company to ever use the iPhone name for a cell phone, and if Cisco wants to challenge us on it we are very confident we will prevail."
'Revolutionary'
This issue is not about money, and it's not about the phone itself; it is about Cisco's obligation to protect its trademark in the face of Apple using it without our permission
Jon Noh, Cisco spokesperson

Cisco, which has owned the trademark since 2000, said it thought Apple would agree to a final document and public statement regarding the trademark.
"Cisco entered into negotiations with Apple in good faith after Apple repeatedly asked permission to use Cisco's iPhone name," said Mark

Munsters star De Carlo dies at 84

from BBC world news on entertainment

Actress Yvonne De Carlo, best known as vampire Lily Munster in US sitcom The Munsters, has died aged 84.
"She passed away in my arms on Monday," said the star's son, Bruce Morgan, adding that his mother's health had been in decline for several years.
De Carlo's heavily made-up matriarch presided over the raucous, cobwebbed Munsters house from 1964-1966.
Although the spoof horror comedy only ran for two years, it had a long life in syndication and two films were made.
"It meant security. It gave me a new, young audience I wouldn't have had otherwise," she said when it ended.
She was the vampire mom to millions of baby boomers
Kevin Burns, friend
"It made me hot again, which I wasn't for a while."
Born Peggy Yvonne Middleton in Vancouver, Canada, De Carlo was raised in poverty and dropped out of high school.
After winning a beauty contest, she landed bit parts in Hollywood and later graduated to leading lady status.
Film studio Paramount initially signed her, it was said, because she resembled its major star, Dorothy Lamour, and executives wanted to warn Lamour she could be replaced.
Her first starring role was in 1945's Salome - Where She Danced, where she played a ballerina who became a spy in the wild west.
The film was a box office success

Fig. 18.9. The final result of the work in the journalist scenario: This example shows the front page generated by a group of journalists on January 18, 2007.

The consumer-type video sensors were similarly chosen and mounted to yield an unobtrusive observation of the whole room. For example, the angle of the corner cameras is wide enough to cover each point in the room by at least two cameras. Consequently, the quality and details obtained from these lenses are limited. Even the zoom camera that points at the entrance of the room and a webcam on the console show close-ups of faces that are not more than 60 times 80 pixels in size.

18.2.2 Integrating Technologies in the UPC Service

The two multimodal perceptual components described in the following receive the raw audio and video streams as input data. The analysis of these data is often inspired by our knowledge about human perception in vision and hearing. Most of the technologies on which both perceptual components are based have their typical applications with well-established evaluation methods. However, the criteria that determine their usefulness can unexpectedly change when they are integrated in a multimodal system that aims at acquiring context awareness for providing services in real time. In some cases, astonishingly large error rates can be tolerable if a technology is backed up by a similar one that uses a different modality. In other cases, strict criteria of synchronization and low delay can arise as a consequence of the integration.

When humans experience a computer-driven service like the Memory Jog, another subjective bias naturally arises: Unexpected actions of the service triggered by a false-positive detection of one of the technologies turn out to be far more annoying than a service not provided due to false-negative detection.

The following sections look at each of the applied technologies and briefly describe our experience with their role in the Memory Jog service. Special focus is given to the usability of the output of their analysis rather than to implementation details or to their individual performance.

Multimodal Perceptual Component: Person and Object Tracking

Person tracking is based on an acoustic localizer and a multicamera 3D person and object tracker. The latter detects regions of interest (e.g., persons, chairs, or laptops) via foreground segmentation in each of the five cameras. A three-dimensional representation of these regions of interest is obtained by a ShapeFromSilhouette algorithm [7] that receives the binary foreground masks from all five cameras. These 3D regions of interest are consequently labeled and tracked over time. To resolve ambiguities (people crossing, someone sitting on a chair, etc.), a color histogram is acquired from each person and object. The output of the multicamera 3D person and object tracker is enriched by (1) an algorithm that is able to distinguish between an object and a person assuming an average range of physical properties of adult humans and (2) an algorithm that analyzes human body posture [3] (standing, sitting, etc.) with a standard model of the human body that is aligned to the 3D regions of interest earlier classified as "person".

The real-time multimicrophone acoustic localization and tracking system [1] is based on the cross-power spectrum phase from three T-shaped microphone arrays. It is robust to the speaker head orientation and provides one or more 3D localizations with detected acoustic activity. The output of the multimicrophone acoustic localization is enriched by an acoustic event classifier [16] that is based on a combination of ASR features and acoustic features. Typical events such as door opening/closing, phone ringing, chair moving, speech, cough, laugh, etc. can be detected.

The combination of video- and audio-based tracking systems allows the system to gain a basic understanding of what happens in the smart room:

- A person of interest (e.g., the latecomer) can be tracked in the room. This location is used to direct the talking head and an automatic cameraman (cf. Section 18.2.2) to his current position.
- The position of all participants can be used to guesstimate changes of the state of the session, e.g., among the states "people enter", "meeting starts" or "coffee break".
- The position of a sudden acoustic event can be determined. The automatic cameraman has been configured to capture these events by choosing the camera that is positioned farthest from the location of the acoustic event.
- In the current implementation of the context awareness, the detection of a latecomer is based on a multitude of criteria among which the first two depend on the person and object tracking: increase in the number of peoples, appearance of a new object close to the door, detection of the acoustic signal of a knock on the door, a doorslam or steps close to the door,

Person identification is based on Face ID [17] and Speaker ID technologies [8]. The Face ID algorithm is applied to either faces that are captured close to the door or images that are captured by the webcam mounted on top of the monitor at the console. In a preprocessing step, a face detector is applied on those parts of the image that have previously been classified as two-dimensional regions of interest in a binary foreground mask in the preprocessing stage of the person and object tracker. Multiple face-like regions collected at different time instants are then analyzed to select a frontal view of the face for further processing by the Face Identification technology. Face ID matches these frontal views against faces stored in a database. If no frontal view is available, the algorithm is capable of basing the face ID on side and profile views, although the identification is less reliable.

The second technology, Speaker ID, provides real-time information about the identity of an active speaker. The SpeakerID algorithm is based on the comparison of Gaussian mixture models. Apart from the poor signal-to-noise ratio and the reverberation in the signals obtained from the far-field microphones, this technology is challenged by the necessity to separate the microphone signal into segments uttered by a single speaker. In an unscripted scenario, this would require the detection of speech activity, the detection of a shift of speakers, and the detection of the number of persons speaking simultaneously. In order to circumvent this segmentation problem, the ID of the active speaker is only determined during the usage of the dialog system (explained next), because in this situation the signal naturally stems from a single speaker talking in a quiet background.

Instead of simply listing all recognized detections of the multimodal Person Identification system, the ID output is assigned to the corresponding person in the person tracker. This allows for accumulating the IDs obtained for a person from both Speaker ID and Face ID in the course of the session. The Person Identification technology also provides valuable feedback to the 3D person tracker about whether it is still tracking the same person.

Since Person Identification was also used to allow the talking head to address the session participants with their name, reducing false-positive IDs was emphasized during debugging and error minimization. For the same purpose, both audio- and video-based ID technologies have incorporated a modelclass for unknown IDs.

Dialog System

The Dialog System allows human-like verbal interaction with the Memory Jog system. It is based on two components: commercially available 2D animation of a talking head (PeoplePutty by Haptek [6], shown in Fig. 18.10) and an ASR based dialog system that utilizes the Cambridge University Engineering Department's HTK [18] recognizer.

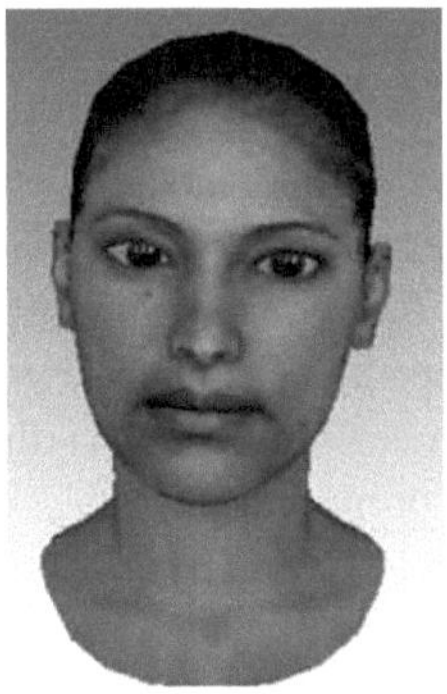

Fig. 18.10. Face of the PeoplePutty talking head: The parameters of the talking head software allow its voice, emotions and look to be adapted as appropriate for the Dialog system according to the context.

The speech synthesis part of the talking head was enhanced by a politeness delay unit that acquired a speech activity flag and obliged the speech synthesis engine not to interrupt a human-to-human conversation. The visual representation of the talking head was projected on one of the walls of the smart room next to the graphical user interfaces (GUI) used by the journalists to publish the front page (cf. Fig. 18.9 and 18.12). The talking head also serves as the voice of the Memory Jog, e.g., giving indications about how to use the GUIs, commenting on acoustic events, welcoming the participants or a latecomer, pointing out when the participants run out of time, congratulating upon a successful contribution, saying good-bye, etc. The KTH-based speech recognition system was trained for two different grammars: the trigger sentences and the commands. The possible phrases of the two grammars are listed in Fig. 18.11.

Trigger sentences			
	[PLEASE]	CHIL ROOM	[PLEASE]
	[PLEASE]	CHIL SERVICE	[PLEASE]
	[PLEASE]	CHIL SYSTEM	[PLEASE]
	[PLEASE]	CHIL ASSISTANT	[PLEASE]
Commands			
		NOTHING	
		NOTHING NOTHING	
	[PLEASE]	SHOW ME THE FRONT PAGE OF OUR COMPETITOR	[PLEASE]
	[PLEASE]	SHOW ME THE FRONT PAGE OF OUR COMPETITORS	[PLEASE]
	[PLEASE]	SHOW ME YESTERDAYS FRONT PAGE	[PLEASE]
		WHO IS OUR FIELD JOURNALIST	[PLEASE]
	[PLEASE]	SHOW ME YESTERDAYS BUSINESS NEWS	[PLEASE]
	[PLEASE]	SHOW ME YESTERDAYS ENTERTAINMENT NEWS	[PLEASE]
	[PLEASE]	SHOW ME YESTERDAYS POLITICS NEWS	[PLEASE]
	[PLEASE]	SHOW ME YESTERDAYS SCIENCE NEWS	[PLEASE]
	[PLEASE]	SHOW ME YESTERDAYS TECHNOLOGY NEWS	[PLEASE]

Fig. 18.11. The trigger sentences and the commands understood by the dialog system. Most of the sentences can be accompanied by an optional "please".

When not active, the dialog system is listening in the human-to-human conversation to detect one of the trigger sentences. If a trigger sentence is detected with a high enough confidence level, the talking head utters *"Yes, please?"* as positive feedback to inform the person in the room who raised the question. When the following command is understood, the corresponding action is initiated. If the dialog system wrongly detects a trigger sentence, the user can reset the dialog system by uttering a simple *"nothing"*.

Personalized Question-Answering System

The journalists in the smart room have an advanced question-answering system [15] at hand that allows them to ask questions related to the news from previous weeks. In comparison to a Google-based search engine, this system is capable of giving the exact answer for factual questions instead of merely listing promising text-snippets.

To personalize the output of the question-answering system, the result of the Face ID obtained from the webcam at the console is utilized. For this purpose, the field of expertise of each journalist has been preconfigured. For example, a journalist working on business news would receive a different answer to the question "Who is meeting in New York?" than a journalist responsible for entertainment news. In order to allow a journalist to access the news from all fields, the automatic selection of specific areas from which the question-answering aystem gains its knowledge can be manually overridden in a small GUI.

Decision GUI

With the graphical user interface in Fig. 18.12, the journalists decide on the front page. They can edit the automatically downloaded news text, preview the resulting front page, and initiate the final publishing stage (cf. Fig. 18.9).

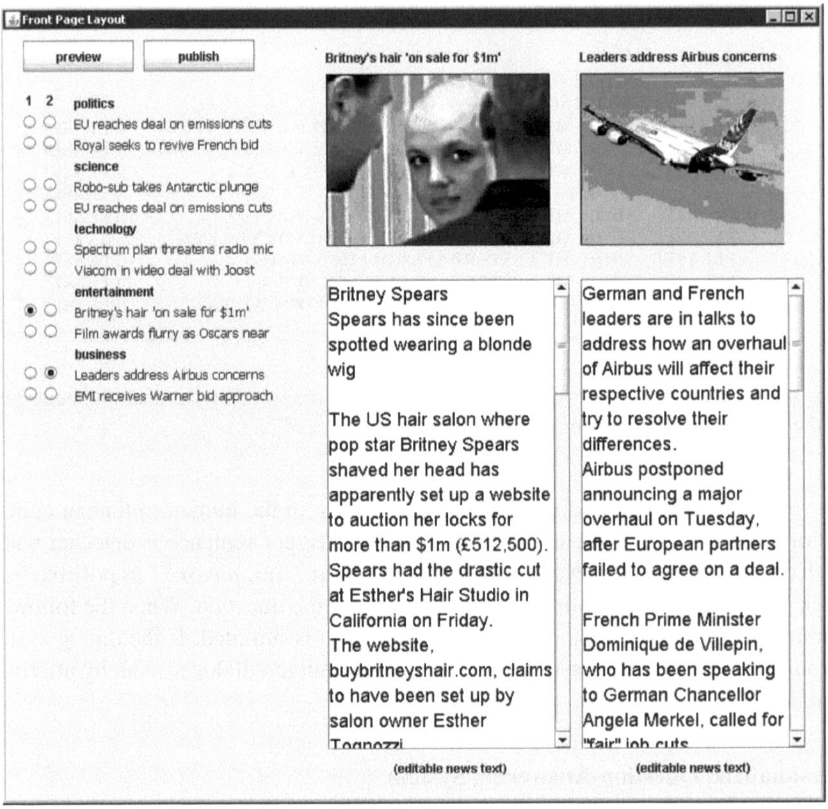

Fig. 18.12. Screenshot of the Decision graphical user interface: this GUI is the main tool of the field journalist. If shows and allows editing of two selected top news. Clicking on the related image directs the journalists to the Internet page with the source of the information. The GUI is also projected on one of the walls of the smart room to allow the journalists to move freely in the room while discussing the news.

RSS Feed of BBC News

A Bash script is executed every five minutes on one of the Linux servers to download the latest RSS feeds from the BBC world news. The downloaded Web pages are processed with the text-based Lynx Internet browser (lynx.browser.org). The text of the news and a corresponding image are extracted with awk and made available via Samba to the computers that display the graphical user interfaces for the journalists.

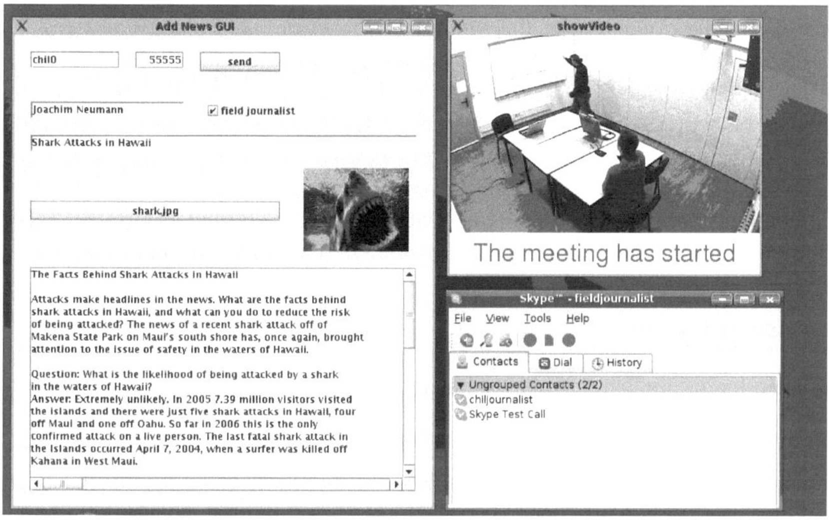

Fig. 18.13. Screenshot of the Field Journalist's laptop: in the lower right, the Skype-based bi-directional audio communication allows talking to the journalists in the room. The upper right shows the real-time video stream from one of the cameras of the meeting room. An automatic cameraman is choosing the optimal camera from five possible angles. This decision is based on the location of the last acoustic event and smoothed by a hysteresis to avoid rapid camera-changes. The real-time video streaming also displays annotations in the form of subtitles that explain the situation, e.g., "people enter", "interaction with ASR", "sound of keys", "front page published", etc or as shown here: "The meeting has started". On the left side of the screen, a graphical user interface allows the field journalist of add a piece of news (a test and an image) to the decision GUI of the journalists in the room

18.2.3 UPC Memory Jog (UPCMJ) Evaluations

All three versions of the UPC Memory Jog service were available for interested visitors. The second version was formally evaluated by specialists at TUE.

Introduction

In the second version of the Memory Jog service, the triggering mechanism for hints is primarily time-based: The system knows which issues have already been resolved, if it decides that it takes too long for participants to resolve the next issue, it will provide a hint. A more advanced triggering mechanism might relate to content analysis: The system overhears the conversation, and sees the notes on the flipchart, and thereby identifies information needs from the conversation. Such smart services may be at odds with a basic design heuristic for interactive systems, namely user control. Through their proactive nature, control over the timing of events is with the system rather than the user. Results from several investigations indeed point out that users are reluctant to accept such proactive services. This lead to the following central hypotheses for the evaluation of the Memory Jog service. (1) Proactive systems lead to faster task completion than reactive systems. (2) People will accept a service that overhears their conversation, watches their behavior, and acts on the information collected in this way. (3) People will accept a service that acts autonomously and proactively, compared to more conventional services where the user controls the actions of the system. We set up an experiment to compare a reactive (user-controlled) version in which hints are provided on demand with a proactive (system-controlled) version of the Memory Jog in which hints are triggered on the basis of content analysis.

Experimental Set-up

The experiment applied a within-subjects (or rather "within-group") design. Each group was presented with two mathematical problems to be solved by the group members using the Memory Jog service. One problem concerned the leaning tower of Pisa (participants had to compute how more the tower could lean over before it would tip over); the other concerned The Lake Baikal (participants had to compute how deep a rope spanning the lake enters the water). Participants were presented with two conditions of the Memory Jog service: In one condition, they could ask for hints (on-demand condition), whereas, in the other condition, hints were provided by the system automatically (proactive condition). The groups were asked to deliberate about how to solve the problems and to make notes on a flipchart so the system would be able to keep track of their way to arrive at a solution and provide the proper hints at the right time. The time needed to complete the task and the number of questions and hints were automatically logged. Participants' attitudes toward the system were collected via a questionnaire and group interviews.

The smart hints triggering mechanism used in the evaluation requires the system to recognize and interpret the spoken conversation and written text. To anticipate a more sophisticated implementation of the system, for the current evaluation a Wizard of Oz setting is used. In this setting, one of the experimenters follows the participants in a hidden room, where he answers the factual questions and provides the proper hints at the proper time.

Sixteen groups participated in the experiment (seven groups of three people and nine groups of two people). The 39 participants (23 males and 22 females ranging age from 17 to 33) had various educational and social backgrounds and all were native Dutch speakers.

UPCMJ Evaluation Results

Table 18.5 shows the objective differences between the two conditions.

	On-Demand	Proactive
Time to complete	17:11 min	15:11 min
Number of hints	2.9	2.5
Number of questions	8.0	6.0
Correct solutions	81%	81%

Table 18.5. Objective differences between conditions.

These results indicate that objectively there are no significant differences between the system-control condition and the user-control condition. There appears to be a tendency, however, for the system-control condition problemsolving to be more efficient and for students to require fewer hints and factual questions.

The questionnaire and interview data are based on all 39 participants. All scales used in the questionnaire have sufficient reliability (i.e., Cronbach's alpha exceeds 0.7). The questionnaire results are summarized in Table 18.6.

Dimension	Score
Usefulness	5.4
Ease of use	5.7
Enjoyment	5.1
Control	4.2
Intrusiveness	5.6
Privacy	5.4
Trust	5.0
Attitude	5.7
Intention to use	5.5

Table 18.6. Average questionnaire scores.

In order to get an impression of the participants' intention to use the system, after the second task participants were asked whether they would use the system for their studies if possible. Also, they were asked whether they preferred the on-demand condition or the pro-active condition on several aspects. For an overview of the results, see Table 18.7.

Almost all students indicated that the service was pleasant to use and practical. Most students mentioned they would use the service for their studies if a dedicated implementation could be provided. They also recognized the usefulness for the system in secondary education (for instance, in physics or mathematics class). One of the merits mentioned was the fact that the hints provided by the system guide the student in the right direction and prevent him or her from struggling too long.

	On-Demand	Proactive	No Preference
Problem solving	23	38	38
Ease of use	38	18	44
Enjoyment	36	33	31
Future use	54	33	13

Table 18.7. User preferences (in %).

For both hinting strategies, several advantages were mentioned. On-demand hinting enables users to get a hint whenever they think they need one. On the other, hand proactive hinting may force users to wait for a hint, so that students are more challenged to do some more thinking themselves, before the system comes to their aid. Also, the proactive control condition is very useful when students do not have a clue how to solve the problem. If students do not know how to tackle the problem. they do not have to admit this by asking for a hint. This prevents them from looking for a solution too long without making progress.

Discussion and Conclusion on the UPCMJ Evaluation

The results of our evaluation show a tendency toward shorter problem-solving times with the proactive version of the service. This observation can partly be explained by the fact that in the proactive condition, participants needed fewer factual questions and hints to complete the task than in the reactive condition. This provides some support for our first hypothesis, stating that proactive systems result in higher productivity (although the differences were not significant). It should be kept in mind, though, that productivity in a learning context should not be measured only by time but also by more intrinsic measures, and this awaits further research.

User attitudes toward the service were in general positive with regard to privacy and trust, indicating that users accept a service that watches and analyzes their behavior to infer information needs. This confirms our second hypothesis. General attitude towards use, usefulness, and intention to use were rated positively by the students as well, providing support for our third hypothesis. These findings indicate that people are willing to accept CHIL-like intelligent services that take the initiative to provide help.

A striking finding is that even though the proactive version of the system appears to enhance students' problem-solving ability, a majority of students had a preference for the on-demand version for future use. This is further corroborated by the relatively low score for perceived control.

We conclude that a proactive hinting strategy has potential benefits but that its implementation requires a very careful design in order to deal with user concerns about locus of control.

18.2.4 Conclusions on the UPC Memory Jog

Our experience with the integration of multiple technologies into two multimodal perceptual components and an integrated Memory Jog service that interacts with humans in our smart room was very positive. The technology developers were challenged by the computational demands of a real-time implementation of their technology, signals from unobtrusive sensors, and noise from a real-world scenario. These constraints tested the technologies at their limits of performance, processing delay, and robustness. Still, the system has been successfully tested with real users visiting the smart room at UPC. A demonstration of the Memory Jog service is available at the UPC smart room.

Additionally, the implementation of the multimodal perceptual components and the integrated service sparked numerous ideas of how to combine technologies in a new way to introduce a higher level of robustness in real-world applications. Subjects that have experienced the Memory Jog service agree that the service was helpful and some added ideas to our wishlist of future services.

References

1. A. Abad, C. Segura, D. Macho, J. Hernando, and C. Nadeu. Audio person tracking in a smart-room environment. In *Interspeech - ICSLP*, pages 2590–2593, Pittsburgh, PA, Sept. 2006.
2. S. Azodolmolky, N. Dimakis, V. Mylonakis, G. Souretis, J. Soldatos, A. Pnevmatikakis, and L. Polymenakos. Middleware for in-door ambient intelligence: The PolyOmaton system. In *4th IFIP TC-6 Networking Conference, 2nd Next Generation Networking Workshop(NGNM)*, Waterloo, Canada, May 1-6 2005.
3. C. Canton-Ferrer, J. R. Casas, and M. Pardàs. Human model and motion based 3D action recognition in multiple view scenarios (invited paper). In *14th European Signal Processing Conference, EUSIPCO*, University of Pisa, Florence, Italy, 4–9 Sept. 2006.
4. A. G. Constantinides, L. C. Polymenakos, and F. Talantzis. Estimation of direction of arrival using information theory. *IEEE Signal Processing Letters*, 12(8):561 – 564, Aug. 2005.
5. N. Dimakis, J. Soldatos, L. Polymenakos, M. Schenk, U. Pfirrmann, and A. Bürkle. Perceptive middleware and intelligent agents enhancing service autonomy in smart spaces. In *IEEE/WIC/ACM International Conference on Web Intelligence and Intelligent Agent Technology*, pages 276–283, Hong Kong, Dec. 2006.
6. Haptek's peopleputty website (`http://www.haptek.com`).
7. J. L. Landabaso and M. Pardas. Foreground regions extraction and characterization towards real-time object tracking. In *Machine Learning for Multimodal Interaction (MLMI)*, LNCS 3869, pages 241–249. Springer, 2006.
8. J. Luque, R. Morros, A. Garde, J. Anguita, M. Farrus, D. Macho, F. Marqués, C. Martínez, V. Vilaplana, and J. Hernando. Audio, video and multimodal person identification in a smart room. In *Multimodal Technologies for Perception of Humans. First International Evaluation Workshop on Classification of Events, Activities and Relationships, CLEAR 2006*, LNCS 4122, pages 258–269, Southampton, UK, Apr. 6-7 2006. Springer-Verlag.

9. A. Paar, J. Reuter, J. Soldatos, K. Stamatis, and L. Polymenakos. A Formally Specified Ontology Management API as a Registry for Ubiquitous Computing Systems. In *3rd IFIP Conference in Artificial Intelligence Applications and Innovations*, volume 204/2006 of *IFIP International Federation for Information Processing*, pages 137–147. Springer Boston, Jun. 2006.
10. A. Pnevmatikakis and L. Polymenakos. An automatic face detection and recognition system for video streams. In *2nd Joint Workshop on Multi-Modal Interaction and Related Machine Learning Algorithms*, Jul. 11-13 2005.
11. A. Pnevmatikakis and L. Polymenakos. Kalman tracking with target feedback on adaptive background learning. In *Joint Workshop on Multimodal Interaction and Related Machine Learning Algorithms*, LNCS 4299, pages 114–122, 2005.
12. E. Rentzeperis, A. Stergiou, C. Boukis, G. Souretis, A. Pnevmatikakis, and L. Polymenakos. An adaptive speech activity detector based on signal energy and LDA. In *3rd Joint Workshop on Multi-Modal Interaction and Related Machine Learning Algorithms*, May 1-3 2006.
13. J. Soldatos, N. Dimakis, K. Stamatis, and L. Polymenakos. A breadboard architecture for pervasive context-aware services in smart spaces: middleware components and prototype applications. *Personal and Ubiquitous Computing Journal*, 11(2):193–212, Mar. 2007.
14. J. Soldatos, K. Stamatis, S. Azodolmolky, I. Pandis, and L. Polymenakos. Semantic Web technologies for ubiquitous computing resource management in smart spaces. *International Journal of Web Engineering and Technology*, 3(4):353 – 373, 2007.
15. M. Surdeanu, D. Dominguez-Sal, and P. R. Comas. Design and performance analysis of a factoid question answering system for spontaneous speech transcriptions. In *Interspeech - ICSLP*, Pittsburgh, PA, Sept. 2006.
16. A. Temko, R. Malkin, C. Zieger, D. Macho, C. Nadeu, and M. Omologo. Evaluation of acoustic event detection and classification systems. In *Multimodal Technologies for Perception of Humans. First International Evaluation Workshop on Classification of Events, Activities and Relationships CLEAR 2006*, LNCS 4122, pages 311–322. Springer-Verlag, Southampton, UK, Apr. 6-7 2006.
17. V. Vilaplana, C. Martinez, J. Cruz, and F. Marques. Face recognition using groups of images in smart room scenarios. In *International Conference on Image Processing (ICIP'2006)*, Atlanta GA, 2006.
18. S. J. Young. The HTK hidden Markov model toolkit: Design and philosophy. Technical Report TR 152, Cambridge University Engineering Dept, Speech Group, 1993.

19

The Connector Service: Representing Availability for Mobile Communication

Maria Danninger[1], Erica Robles[2], Abhay Sukumaran[2], Clifford Nass[2]

[1] Universität Karlsruhe (TH), Interactive Systems Labs, Fakultät für Informatik, Karlsruhe, Germany

[2] Stanford University, Department of Communication, Stanford, CA, USA

This chapter will discuss the Connector service. The goal was to build a mobile communication system that could help people better manage the risk of untimely interruptions in an always-on world, where we can reach each other anytime and anyplace. In an iterative process of design, we have developed technologies that intelligently and nondisruptively integrate into human communication processes.

Because interpersonal communication does not typically occur in environments comparable to laboratories, instantiations of the Connector service were explored and evaluated with real users engaged in real-life activities. Almost 100 students at Stanford University and dozens of employees from a large software company in Silicon Valley used the system while engaged in their normal daily activities.

19.1 The Always-On World: Benefits and Burdens of Mobile Communication

By making communication possible from anywhere at any time, mobile technologies overcome traditional limitations on communication. With no need for co-location with conversational partners, or for disengaging from one's current activity, everyday social dynamics have been dramatically transformed. This shift fosters new opportunities for building and maintaining social relationships, communicating efficiently in the workplace, and for being generally unfettered by constraints of place. However, mobile communication technologies simultaneously undermine their value by introducing the constant risk of interruptions. Poorly timed calls disrupt social norms [14, 16], break mental concentration, and make continued task engagement difficult. Effectively managing the tension between opportunities for productive conversation and risks of interruption constitutes the central challenge in our research.

Prior research on "interruptibility" approaches the problem of alleviating cognitive costs associated with untimely disruptions by developing systems that actively manage the trade-off between interruptions and information [1]. By sensing information about a context such as the home or the office, these systems computationally process the relationship between the user's state and their context. They are then capable of preempting communication attempts, ensuring that users need not be interrupted to make an explicit determination about their current availability.

Alternately, 'awareness' research construes the problem of managing interruptions by framing the availability for communication in primarily social or situational terms. Technologies informed by this approach seek to augment mediated communication channels with useful contextual features that are naturally available in face-to-face interactions. These indicators, such as chat status messages, allow participants, rather than the system, to make more socially appropriate communicative choices [9].

Mobile contexts involve specific challenges that call for approaches to system design positioned between these two bodies of work. The wide variety of potential circumstances for mobile phone use radically increases the difficulty of correctly interpreting the context. Furthermore, the phone remains a limited platform for sensing, making it very technologically challenging to create a true "interruptibility"-aware mobile device. Thus, social aspects of phone use, like conventions around talking while in public places, become critical to managing the communication. Mobile phones often inadvertently encourage interruptions. This is because often ubiquitous and always-on, they provide no cues about an appropriate time for conversation. Consequently, senders navigate a socially sterile environment each time they place a call. They make simplifying assumptions, often violating social norms in the process. Receivers, assumed to be available anytime they answer their phone, can often be heard answering calls in public places with the self-contradictory statement, "I can't talk right now".

Our design approach reflects CHIL's guiding principles and technologies. Recognizing the role of social factors as well as the difficulty of building socially suave algorithms, we focused on crafting representations of availability for use by humans. Starting from the design and implementation of a system that transmitted simple calendar-based availability, we demonstrated that socially appropriate behavior could be facilitated by a mediating mobile communication system. Working within an established design space, we built and tested 'Connector' and 'Virtual Assistant' prototypes that incorporated situational information in addition to planned information derived from calendars. In parallel, to fully articulate and relate the different conceptions of availability, we conducted basic social science research that confirmed our initial insights and suggested future design directions. Finally, we expanded the design space, through the 'Touch-Talk' and 'One-Way Phone' prototypes, which allowed the user to turn an interruption into an opportunity for lightweight communication without social disruption.

19.2 The Connector: Representing the Receiver's Plans to the Sender

We began by implementing a relatively simple system capable of representing the receiver's availability for mobile communication. A network-based agent, the 'Connector', accessed the receiver's personal calendars to determine whether they were busy at the time of a call (see [7] for implementation details). It then warned callers when there was a chance that their communication attempt might be perceived as an untimely interruption. Thus, the Connector system was designed to reduce inappropriate phone calls while increasing the likelihood of the sender successfully reaching the receiver.

We chose to base the system's inferences on calendar information because calendars represent people's prospective judgments, or plans, about their availability. Calendars are already in common usage as a technology through which people attempt to explicitly indicate times when they are busy or free, and thus require no special effort or learning on the user's part. Determining whether these prospective representations have enough resolution for a system to infer availability based on them was a guiding research motivation. If this were the case, then providing even rudimentary situational cues might guide conversational partners toward more appropriate behavior. Potentially, this might recreate the useful social information rendered obsolete by the decontextualized nature of contemporary mobile systems.

To interact with our prototype system, callers would dial a special number assigned to each Connector user. The Connector agent, represented by a synthetic voice, would then provide the caller with information:

> Connector [individual]: *"Hello, this is Bob's Connector agent. Bob is currently busy. He will be available next time at 3:30 pm. Please call back later. To still connect your call, please press 1 now."*

The system offered availability cues but had no agency in the decision-making process. Control remained firmly in the hands of the sender. As such, the system gave us a way to test the usefulness of giving people availability indicators, without making any forays into actively mediating the communication. We were able to determine whether or not senders would respect this information and act accordingly.

Our design process suggested an additional feature based on the Connector's ability to know the availability of multiple potential recipients. The 'group calling' feature thus gave the sender the option to pick a different member of the group to connect with in the event that the individual he or she trying to reach was unavailable. We anticipated that this feature might be especially useful in workplace settings, where reaching anyone that filled a general organizational role might suffice.

To use this feature, callers dialed a number assigned to a group of individuals. They then heard a Connector agent offering them a series of choices about how to proceed:

> Connector [group]: *"Thanks for calling the reading group. The following persons are currently available: Anna, Peter, Jane. To call Anna, please press 1; for Peter, press 2; for Jane, press 3."*

We tested the system through field studies designed to determine if the Connector could improve mobile communications by reducing the number of inappropriate interruptions.

Evaluating the Connector

Ninety college-aged mobile phone users participated in two field experiments over the course of two consecutive weeks. In both experiments, participants worked together (in randomly assigned groups of 11 or 12 members) to accomplish tasks via their mobile phones. We based our assessment of the Connector on team performance as well as individual-level attitudes.

The eight teams were randomly assigned to four experimental conditions. Condition I operated as the experimental control, with team members in this condition calling each other in the standard fashion. Condition II offered callers a 'one-to-one Connector' capable of communicating availability information for a team member whom they called. Condition III participants were given access to a 'One-to-N Connector' capable of transmitting availability information about all members of the group. Finally, teams in condition IV received access to both kinds of Connectors.

Hypothesizing that different types of tasks would benefit from different Connector features, we designed two experiments that used these conditions. The first required collaborative problem solving, and the second required coordinating members of the group. This strategy was designed to tease apart the relative benefits of the Connector for teams employing different social strategies as they proceeded toward their goal.

Experiment 1, collaboration through the Connector, tasked teams with solving a 'Mystery Person' puzzle (see Fig. 19.1). We chose this task because it required mutual interdependence among team members to arrive at the correct solution. Each team member was given a set of clues (such as facial features) about the identity of a mystery character that, when exchanged with others, would help the team collectively identify the person.

Experiment 2, coordinating through the Connector, tasked teams with arranging a face-to-face meeting at a well-known campus landmark, assembling as many team members as possible (see Fig. 19.1). Teams took group photos as evidence of successful coordination. This task was chosen for its external validity, since group coordination tends to be a frequent use for mobile phones.

At the end of each experiment, participants completed questionnaires, rating their experiences on a series of 10-point Likert scales.

A total of 710 calls was made across all conditions in both experiments, with participants spending a total of 15.5 hours on the phone. Using chi-square and analysis of variance (ANOVA) statistical techniques, we analyzed the data for team performance as well as individual participant attitudes of frustration, sense of team cohesion, and interest levels in each task (for a detailed description of the results, see [5]).

Across both studies, the Connector significantly affected team performance [1]. Connector-mediated calls had a higher rate of successful connection than those in the control condition ($\chi^2 = 3,78$, $df = 1$, $p = 0.05$). In Experiment 1, teams with both Connector features were the only ones that managed to correctly identify their

[1] All results reported are statistically significant at $p < 0.05$

Fig. 19.1. Collaboration and coordination tasks: Mystery person puzzle (left) and group meeting at the Gates of Hell (right).

mystery person.[2] In Experiment 2, the one-to-one Connector teams had the best performance, assembling at least seven members for their group picture.

Individual participants with access to Connector features during Experiment 1 reported significantly greater interest in the task and lower frustration than their counterparts in the control condition. Moreover, individuals using the one-to-one Connector in Experiment 2 reported the greatest sense of team cohesion, even controlling for the actual performance outcome of the group [3].

These results confirmed that even a simple representation of availability in this case based upon prospective information, or calendaring is useful at the time of a call. By representing receiver state in individual instances, such a system can facilitate performance improvements and affect subjective experiences across a series of communications. Moreover, allowing senders to be the judge of whether a particular request constitutes a justifiable interruption is perfectly reasonable. Human social judgments can be trusted to take into account information from a "machine in the middle" [19]) and make the appropriate decision. Finally, different underlying social dynamics influence the way these judgments play out, as shown by the relative success of different Connector versions in the two experiments.

Building on our Connector research, we then decided to refine the representation of availability by augmenting prospective information with situational data. By offering a more accurate understanding of availability, we believed we might better facilitate people's decision making about when to interrupt.

---

[2] the probability of this outcome is $0.04(= (1/8)(1/7) + (1/8)(1/7) = 2/56)$ which is less than 0.05.

[3] Cohesion was not significantly affected by the success covariate, $p = 0.65$.

19.3 Situated Aspects of Availability

In the next stage of our research program, we performed a variety of studies aimed at understanding and designing for a more situated understanding of people's availability. Pursuing parallel lines of investigation that used distinct research strategies, our goal was to develop techniques for folding situational information into representations of availability. The first line of studies focused on the capacity for a sensor-based, context-aware system to accurately determine the receiver's state. To determine the usefulness of this approach, we developed a "Virtual Assistant" prototype for use in an office environment. Based on a suite of CHIL technologies, it gave us the ability to investigate a Connector system within a relatively stable context. After initial investigations, we then ported our approach to mobile contexts. The second line of research employs social science techniques to answer basic questions about the relationship between a user's situation at the time of a call and her availability for conversation. Finally, we discuss how learnings from both strands of research motivate a reconceptualization of interruptibility and availability, and in turn inform the development of novel technological solutions.

19.3.1 Virtual Assistant: Sensing Office Context

We began our study of situational approaches to representing availability by taking advantage of the context-aware CHIL technologies already deployed at the Karlsruhe research site. Cameras installed inside offices could be used to detect any activity in special regions of interest like the desk, the door, or visitor seating (see Fig. 19.2). Audio and visual data served as input to a rule-based system that used confidence measures to determine the current state: office empty vs. occupied, occupant alone vs. in a meeting, and whether or not the occupant was currently on the phone [5].[4]

The area immediately outside the office was equipped with a small webcam to detect visitors and transmit information to a face recognition system capable of identifying each member of the research lab (about 15 people), and distinguishing between familiar people and strangers. [5]

Leveraging this infrastructure, we created a "Virtual Assistant" capable of communicating a rich representation of the office occupant's availability to potential visitors and callers, based on the observed situation in the office.

Mediating Calls Based on Contextual Cues

The Virtual Assistant mediated all calls made to the office phone, warning senders when the current situation was likely to be inappropriate for talking on the phone. When placing a call, the sender heard the following:

[4] The correct office state was reliably detected in the single-desk office of a senior researcher in 90.6% over a period of 8 days.

[5] A standalone evaluation of the face recognition system with five known and 20 unknown persons reported 87.2% accuracy on a per-frame basis, and 99.5% accuracy when accumulating frames over time. This system was developed by Hazim K. Ekenel and Lorant Toth.

Fig. 19.2. Camera views of offices. Bright bounding boxes indicate activity in areas of interest.

Hello Jane. Thank you for calling Bob Smith. This is his virtual assistant. Bob is currently in a meeting with Eric. Your call is important. Please hang on, or press 1, to leave a message. Instead, you may now press 2 to be connected to the office phone. To schedule a call at the next available time, please press 4.

If the caller chose to leave a message, the receiver's phone did not ring. Instead, the receiver was sent an audio attachment via email, thus mitigating an unnecessary interruption while still transmitting relevant information. If the sender opted to schedule a call, then the Virtual Assistant automatically initiated a call-back when the receiver next became available, as determined by the situation in the office:

Hello Jane, this is an automatic call-back from Bob's virtual assistant. Bob is now available. Please hold while your call is being connected, or you may hang up the phone.

In order to maintain a reasonable level of privacy for the receiver, we programmed the system to moderate the levels of disclosure about the receiver's current situation based on the relationship between the office occupant and the caller (who was identified by caller ID).

Mediating Office Visits

The research team placed a screen, speakers, and webcam outside the office door (see Fig. 19.3), enabling the Connector to communicate with visitors while capturing identity information via facial recognition. As with phone-based interactions, the relationship between the visitor and the occupant determined different levels of information disclosure about the circumstances within the office. Finally, the system maintained a record of all visitors so that if the office occupant had been out for some time, he could see a list of missed connections upon his return.

We hypothesized that this purely sensor-based approach would also create some positive impact on managing interruptions, much like the calendar-based one had effected. The Virtual Assistant was field-tested with a three-week deployment at Karlsruhe.

Fig. 19.3. A visitor consults the Virtual Assistant.

Evaluating the Virtual Assistant

The Virtual Assistant was installed in the office of a busy senior researcher. By comparing interruptions during times when the system was activated and deactivated, we assessed just how well a model of availability informed by totally situational cues might operate, from the perspective of the receiver.

In order to capture the receiver's perceptions of each interruption, we created a Web-based diary application (see Fig. 19.4). The system captured the day's activities in an interactive timeline that provided detailed information about each communication attempt, including memory aids like snapshots of the office or of visitors. The receiver logged in at the end of each day, annotated each interruption[6], and noted how available he had been at the time[7].

Over the course of three weeks, 150 attempts were made to contact the researcher in his office. Of these, 76 occurred via phone and 74 took place in person. Half (49.4%) of all contact attempts led to a conversation. We found no significant differences between connection success rates in the experimental versus control conditions. There were also no significant differences in success rates between phone and in-person attempts.

We then analyzed the diary ratings for each social interaction to determine whether or not the Virtual Assistant affected the number of inappropriate interactions experienced by the office occupant. The Virtual Assistant did, in fact, decrease the proportion of interruptions rated as inappropriate by the user [8]. Where the Virtual Assistant mediated interactions, there were no interruptions in the "not appropriate"

[6] Interruptions were rated on a 4-point scale: "very appropriate", "quite appropriate", "somewhat appropriate" and "not appropriate"

[7] Availability was rated on a 4-point scale: "not available", " somewhat available", "quite available" and "very available"

[8] without virtual secretary ($M = 2,3, SD = 1.1$), vs. with virtual secretary ($M = 1,5, SD = .6$), $t = -3.93, df = 65, p < .001$)

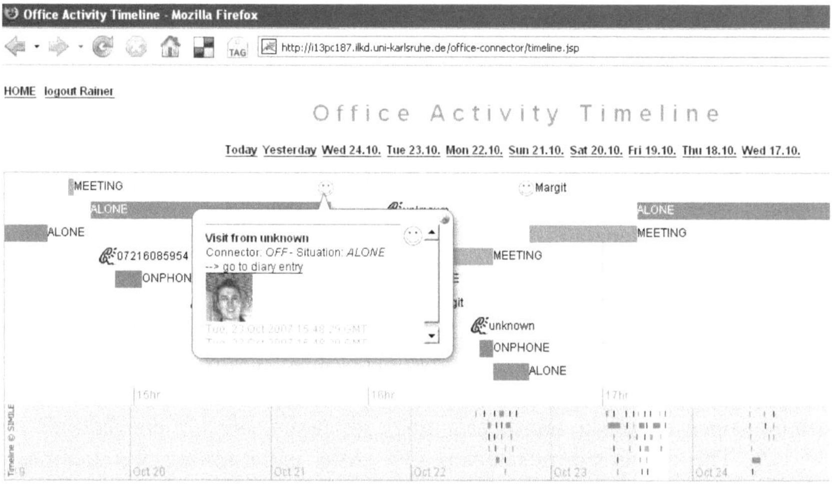

Fig. 19.4. The Office Activity Diary was shown as interactive time line widget.

category, and only 6% of interruptions fell into the "somewhat appropriate" category, as compared to 19% "not appropriate" and 15% "somewhat appropriate" interruptions in the control condition.

Given the reduction in the inappropriateness of interruptions, we might expect that the office occupant would indicate greater availability for communication during those times. However, analysis indicated that it was quite difficult to self-report availability with much precision. Most of the time, the office occupant indicated that he was "somewhat available", rather than being in a state of "very available", or "not at all available". This result may derive from a general expectation that people should be available while in the workplace.

Finally, we conducted user observations and informal interviews with both visitors and the office occupant. All parties felt that the Virtual Assistant interface was relatively easy to interact with; most people seemed to find the system understandable despite the lack of any prior experience with it. Thus, we could be relatively confident that evaluations of system efficacy were not particularly colored by any usability breakdowns.

While our study confirmed the usefulness of representing the availability of receivers based on their current context, the technological implementation of our Virtual Assistant was tied to a static context, the office. This strategy benefited from a preexisting sensor-based infrastructure, a limited set of possible receiver states, and a greater degree of common ground [3] between senders and receivers about the social expectations of the office space. Ultimately, our design goal for this line of research was to extend this context-aware strategy into the mobile domain.

19.3.2 Sensing Mobile Context

The most useful place to deploy an intelligent mediator is within a mobile context, where the situational features are far more diverse and the meanings assigned to interruptions may differ much more widely, thus creating the greatest risk of disruptions. However, without the luxury of a heavily instrumented smart-room environment, building a mobile sensing platform presents a remarkable challenge. Our research team explored two separate strategies - classification and learning - to determine how a mobile phone might infer contextual information.

The first strategy captured acoustic signals in the user's immediate environment via a wearable microphone. Based on five-second audio samples, the system categorized the environment as one of nine types of settings: airport, bus, gallery (loosely any large indoor space not meant for travel), park (any outdoor nonurban setting), plaza (any open-air, urban, nonstreet setting), restaurant, street, train, and train platform [15]. The system's error rate was 15%. After initial testing, we attempted to refine the implementation so that all audio input and processing could occur onboard the mobile device. However, current mobile phones have relatively low audio quality and perform their own background noise cancellation, significantly limiting our ability to differentiate among environments. We revised our set of environmental categories, instead distinguishing among only three settings: inside, outside, and vehicle, at an error rate of 16% [15].

We then pursued a second strategy, foregoing the construction of categorical models for mobile contexts, for a learned activity model that used acoustic evidence to estimate whether or not the user was interruptible. In other words, rather than trying to accurately classify a context, we learned what kinds of soundscapes were associated with greater and lesser degrees of availability. This approach presented the additional advantage of graceful negotiation of user's privacy concerns; the system did not need to understand where a user was, only that her availability was correlated to the audio signal.

We evaluated this approach using both high-quality and mobile phone audio signals. The total misclassification rate for high-quality audio in a single-user study was 6.5%. In the case of mobile phone audio signals collected via a two-user study, the misclassification rate was 16% for the first user and 27% for the second. For additional information about these technologies and approaches to interruptibility modeling, see [15]. Like the first strategy, this implementation requires computationally expensive real-time audio signal processing. Thus, porting the entire system to a mobile phone is still a technological feat slightly out of reach.

Nevertheless, we remained committed to "smart phones" as the appropriate form factor for our designs. Despite their current context-sensing limitations, their platforms still provide a degree of flexibility key to iterative prototyping processes (see Fig. 19.5). Connector-enabled phone models have been used to control types of phone alerts and to automatically switch to silent mode in meetings, or extra-loud in noisy environments. Moreover, we enabled speaker-independent speech recognition that could be used to issue a set of simple voice commands (such as dialing or sending short predefined messages) [6]. Finally, smart phones have the advantage of supporting custom graphical user interfaces, which are better than voice for some purposes like browsing.

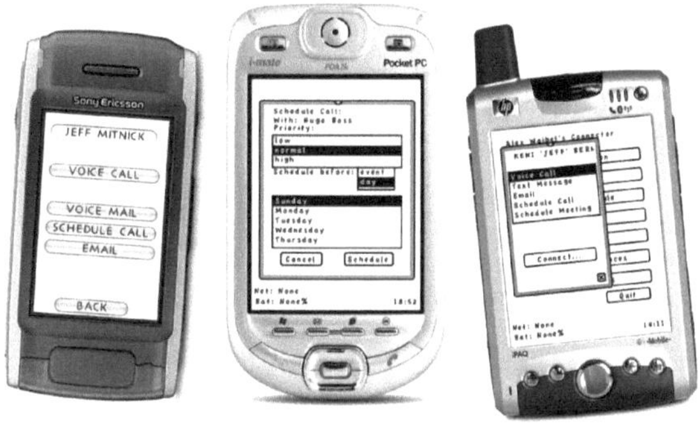

Fig. 19.5. Three generations of context-aware CHIL phones: Sony Ericsson P900, HP iPAQ h6315, and i-mate PDA2K GSM/GPRS Pocket PC.

19.3.3 Availability: Situated Dimensions of Mobile Phone Use

Parallel to our research on context-aware Connector implementations, we employed social science methods to understand the relationship between availability and situational features. By conducting a large-scale, multimethod research study of everyday mobile phone use, we were able to derive a series of contextual dimensions that correlate with people's availability for mobile communication: activity category, typicality, and social, physical and mental engagement.

Study Design

Ninety college-aged mobile users participated in a week-long study of the context of mobile phone use. As in the Connector studies, participants provided calendar information beforehand. Using an "experience sampling" methodology [13, 18], we also assessed participants' availability at various times throughout the day via their mobile phones. This enabled us to consider the relative contributions of "in-the-moment" features of context, compared to prospective estimates of availability. Finally, each participant retrospectively provided open-ended, qualitative descriptions about their context. The study attempted to capture a rich picture of availability by triangulating these descriptions (Fig. 19.6).

Experience sampling was conducted via a telephony server preprogrammed to call each participant at four random times each day. If the call was answered, the system played a prompt asking the receiver to indicate his or her current availability for a conversation by pressing a digit on his or her phone keypad. The scale used ranged from one ("not available") to nine ("very available").

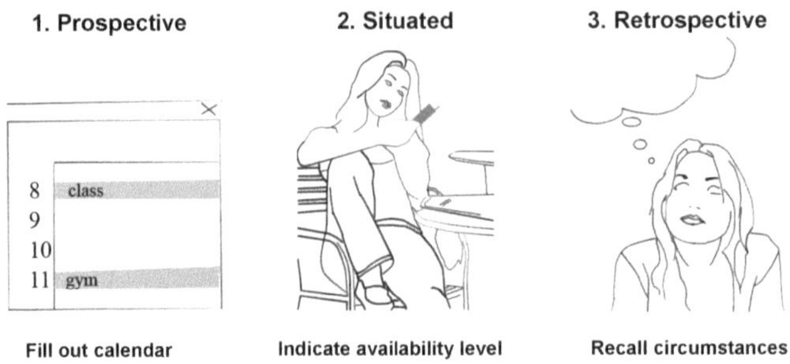

Fig. 19.6. Three kinds of availability assessments.

Participants completed an online questionnaire at the end of each day. They provided open-ended descriptions of their situation at the time of each call, and reported considerations influencing their decision to answer. Responses included accounts such as, "*I was visiting with a friend. We were talking and not too busy, but he is a very close friend,*" or, "*I was at work, but I was packing up to head home for the night. I was not busy. My boss was in the room and asked what the call was about.*". These qualitative responses constituted a corpus of data from which we could systematically code responses into meaningful categories that might help us better understand the dimensions of context that influence people's availability.

Dimensions of Mobile Context

We analyzed the open-ended descriptions via content analysis. Three independent coders categorized each response with respect to seven dimensions: activity category, typicality, physical engagement, mental engagement, social engagement, presence of others, and private vs. public context. These dimensions were derived through an iterative coding process, informed by previous literature that indicates that type of task and degree of engagement with current activity affect people's willingness to communicate [2].

For the *activity category*, we distinguished among categories such as *basic* (eating, showering, sleeping, etc.), *transportation* (driving, cycling, etc.), *required* (homework, job, etc.), *alone/personal* (leisure reading, browsing the Web, etc.), and *social* (hanging out with friends, dining out, etc.). *Typicality* was rated on a five-point scale (from "not at all typical" to "very typical") as a measure of whether the current situation was generally perceived as one in which people would normally talk on their phones. The engagement measures, *physical*, *mental*, and *social* were assessed on 5-point scales ranging from "not at all engaged" to "very engaged". *Presence of others* was a binary indicator of whether other people were present, separate from social engagement with them. Finally, situations were rated as either *private vs. public contexts* (e.g., home, dorm, car vs. sidewalk, institutional building). For all these dimensions, intercoder reliability was confirmed via Krippendorf's α scores. Analysis indicated that social engagement, presence of others, and private vs. public were highly correlated dimensions. Thus, we collapsed these into a single dimension, "social context" for use in subsequent regression models.

Predicting Availability

Our next step was to determine the relative contribution made by each of these dimensions to people's availability in-the-moment. We conducted a series of regression analyses to determine the contribution calendaring and our proposed dimensions of mobile context made to people's behaviors, in the moment (for further details, see [5]).

Not surprisingly, *typicality*, or how typical it would be according to social norms to answer a phone call in a given situation, significantly predicted whether or not participants chose to answer their phone. In other words, if the setting was one in which accepting calls was generally accepted or routine, participants were more likely to do so.

Further, activity category proved to be a significant predictor of availability. For example, people were significantly more likely to answer a call when they were alone or engaged in a personal activity as opposed to when in social situations. In other words, people seemed to generally respect social conventions, or the prioritizing of those with whom they were already engaged. When participants were involved in required activities, or in transportation-oriented tasks, they were far more likely to either report extreme unavailability or not answer their phones at all, as opposed to indicating nuanced degrees of availability. Different categories of activity might thus differ in the degree of flexibility they allow for communication via mobile phones.

Calendar entries were relatively effective predictors of availability, supporting the relationship between people's plans and their in-the-moment behaviors. However, calendar entries on their own were only weakly predictive; when coupled with dimensions like social context, or physical engagement, the predictors became twice as powerful.

Social context consistently and significantly predicted availability. People were more likely to answer calls when alone or in private situations, reinforcing the results of analysis of activity categories; people tend to reduce their availability in response to co-located engagement with other people.

Physical engagement correlated with extreme unavailability but was not particularly useful as a predictor for more nuanced treatments of availability. Unlike traditional communicative sites like an office desk or a fixed phone, physical engagement in a mobile context tends to be more challenging and hence results in a stronger influence on unavailability.

Mental engagement was not a significant predictor in our regression models. It is possible that this aspect of context might be better measured via behavioral methods like secondary-task reaction time [10]. However, it might also indicate that mental states of flow [4]; which have been the motivating basis of prior research, may not, in fact, be the chief issue for interruptibility. Purely cognitive models of availability overlook the notably social and physical aspects of mobile use.

Taken together, these results confirm the value of prospective representations of availability, like calendaring. They also demonstrate that situated dimensions, like social context, physical engagement, and different types of activities, predict certain kinds of availability. Most important, these data constitute a large body of empirical evidence that confirms social influence on mobile context: Where activities involve interacting with others, people are less willing to communicate via mobile phone; during events where accepting calls is a more typical behavior, people are likely to do so.

This evidence, while useful in constructing a rich availability representation, also points to the possibility of a different pathway for communication. If social norms and engagement with others constrain availability, then a design opportunity exists for mobile services that enable communication in situations where the user is socially engaged but mentally available. Such designs would explore the possibility of managing interruptions, not through preemption, but through making available more options for interaction that subtly circumvent restrictions of the current context. The user in such cases would be empowered to decide whether to use such a system to increase her availability without being forced to move to a more interruptible context.

19.3.4 Available but Not Interruptible: A New Design Space

Our two lines of research analyzed the relationship between the receiver's situation and availability. We first deployed a series of context-aware systems that determined the receiver's availability based on the state of his surroundings. A "Virtual Assistant", offered a representation of availability to visitors and senders based on its classification of the state of the receiver's office. Though initial testing provided promising results, success remained predicated on a heavily instrumented, static setting. Correctly identifying mobile contexts proved a far less tractable problem due to both the technological limitations of mobile phones as sensing platforms and the extreme diversity of potential circumstances.

The Virtual Assistant project illuminated a key distinction between interruptions and availability. While the receiver confirmed that unwanted interruptions were reduced, her self-reported availability did not correspondingly increase at the time of these contact attempts. Rather, the receiver generally reported being "somewhat available", perhaps a result of the social expectations of behavior within the workplace. This distinction between interruptibility and availability suggests that mobile users might be willing to communicate in many contexts were it not for the social rules that make doing so inappropriate.

The second line of research, availability, confirmed the role played by social norms and expectations in influencing the receiver's decisions to engage in a mobile communication. Situations where it might be more typical for people to receive phone calls corresponded to higher levels of availability. Engagement in social activities, or being in social contexts, made people far less likely to answer a call. Being mentally engaged, however, did not significantly affect how available people felt.

Taken together, the results indicate a design space for mobile communication systems in which the receiver is considered available, but not necessarily interruptible because doing so violates social norms or expectations currently in place. This in turn generated an insight about a novel direction for mobile systems, in which they offer senders new options to communicate that would allow receivers to accept communication quickly, privately, and quietly. We then made initial attempts in this direction.

19.4 New Communication Modalities: Implicit Availability Representations

The insight that social contexts often restrict a user's decision to engage in phone conversation opens a question about the users willingness to do so if possible without disrupting conventions or norms. For instance, text messaging has been shown to be used to work around the limitations of not having a private space to talk in the home or on public transit [12].

Prior research suggests that users sometimes regard interruptions as welcome or even important to everyday tasks. Managers in a previous study [11] claimed that they desire to retain control over their interruptions, since filtering them is an important part of their job. Thus, rather than prevent the interruption and try to deflect it to a more appropriate time or place, this approach relies on building smarter, more flexible interfaces that afford greater choice of interaction style.

By providing users with greater agency in managing untimely phone calls, we located control entirely at the user rather than at the system, with the attendant advantage of having humans make the critical decisions. The system plays a passive mediator that facilitates communications to employ the best modality or medium for each recipient.

We describe two systems, One-Way Phone and Touch-Talk, each intended to provide users with novel forms of answering calls in social situations where they are unable to talk aloud or leave the current context. Both systems were deployed in business environments to determine their potential advantages.

19.4.1 One-Way Phone: Meta-information in Restricted Channels

As an initial exploration we created "one-way" calling, allowing senders to speak and receivers to listen. Senders dialed the receiver and then pressed 1 or 2 to indicate whether they wanted to place a one-way or two-way call (see [8] for complete implementation details). The receivers' Caller ID displayed both the sender's name and the kind of call being placed (see Fig. 19.7).

Though seemingly restrictive, this kind of asymmetric channel is rich in meta-information. The sender's perceived conversational needs were now embedded into the channel itself; the medium quite literally became the message. The fact that the sender chose to place a one-way call communicates intent to make the conversation brief and informative, requiring no response or acknowledgment. We automatically muted the receiver's phone during one-way calls to reinforce a lack of social obligation to perform conversational maintenance. Research related to meta-information in chat-based contexts reveals that by providing recipients with an expectation about the degree of involvement necessitated, they can make more informed decisions about whether or not to engage [2].

The one-way service addressed those situations where a receiver, might in fact be available and willing to communicate but could not do so because he would be interrupting others in his immediate context if he started talking on the phone. For instance, a corporate user could take a one-way call in a meeting, quietly obtaining potentially time-critical information, while respecting the need of others not to be disturbed. Moreover, the people immediately engaged with the recipient were fully shielded from the sender, since no receiver-side audio entered the one-way channel, making such communication an option even in situations where the privacy of others was critical.

If a receiver's situation did not allow for conversation, she could opt to switch a two-way call to one-way mode. The receiver did not, however, have the ability to change a one-way call into a normal two-way conversation.

Fig. 19.7. One-Way Phone interface.

Evaluating the One-Way Phone

We studied one-way calling via a field test with 10 employees from a large Silicon Valley software company [8]. We chose corporate users precisely because they represent mobile users without time for full-blown conversation because of frequent meetings, conferences, and discussions. Participants were from a number of departments within the company (finance, sales, HR, research, product strategy, etc.) and had different professional roles, ranging from assistants to managers.

The users were given daily mocked-up group scheduling tasks to be accomplished using one-way calls. We hypothesized that one-way calls would be efficient for scheduling purposes since a lot of the individual actions involved, such as sending confirmations or reaching someone currently in a meeting, are rather well-suited to a one-way calling system.

It turned out that one-way calls were, in general, usable and especially advantageous for receivers, who indicated that they appreciated the ability to choose the modality most suited to their current circumstances. However, the inability to get any feedback from the receiver was too limiting. People desired synchronous confirmations during communication to build shared conversational grounding [3] and found it rude to just give instructions and leave the other party with no say in the conversation. The key advantage of phone communication – synchronicity – was suppressed if the conversation was not two-way. These insights informed the subsequent design of Touch-Talk.

19.4.2 Touch-Talk: Managing Availability Through Multimodal Conversation

Our next system allowed receivers to transmit feedback to the sender without disrupting their current situation by talking. "Touch-Talk" mapped pre-programmed messages to digits on a phone keypad. When a button was pressed, the corresponding prompt was synthesized by a text-to-speech engine and played to the sender. By pressing these buttons while listening to the other person talk, receivers could literally make their phones speak on their behalf (see [8] for implementation details).

The default set of prompts was selected via an iterative process of user testing. Ultimately, we settled on a series of basic utterances like, "Yes", "No", "Okays", and "More information". These responses support a basic level of conversational grounding that carries little content but assures the sender that the receiver has understood and oriented appropriately to the topic of conversation. Furthermore, we designed the system to obey turn-taking protocols for human dialog, playing distinctive audio cues to signal mode switches and to preface prompts. This technique was based on prior research demonstrating that people have similar social responses during conversations with automated voice agents as with humans [17]. Finally, concise responses delivered via synthetic voice mitigated any sender expectation that a full, natural conversation could occur.

Since Touch-Talk was implemented entirely on the network, users could keep using their normal phones; the only change was dialing a special number to reach someone on the Touch-Talk service. Receivers could personalize their prompts by logging into a Web-based interface. The interface showed current settings and allowed receivers to share their custom messages with others. The relative popularity of different prompts was conveyed by prompt "tag clouds" (see Fig. 19.8), encouraging the social construction of conventions about the appropriate usage of this system. Finally, receivers were able to switch to a two-way conversation at any point.

We then conducted a final field study to assess the effects of Touch-Talk in a business setting.

Touch-Talk for Corporate Communication

We deployed the Touch-Talk system in a large Silicon Valley software company as part of routine business communications over a one month period. Fifteen participants used the system on a regular basis, enabling our research team to make initial

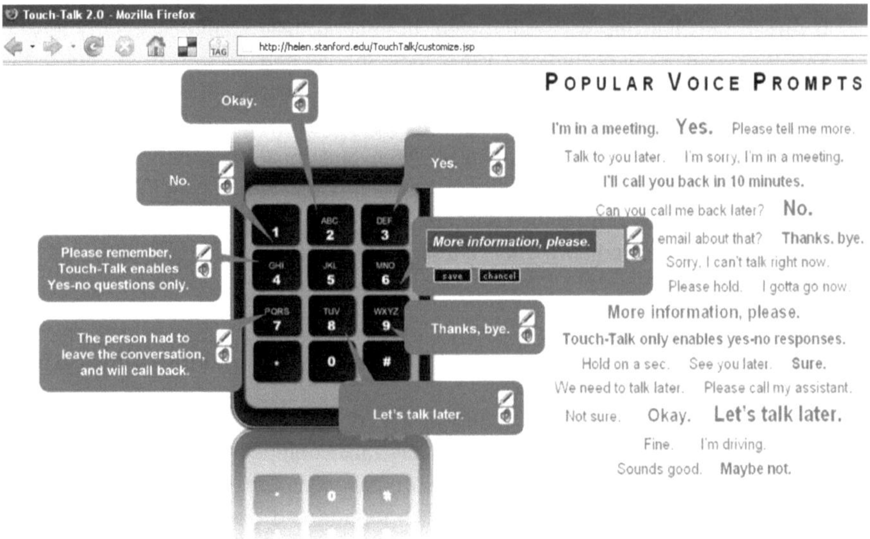

Fig. 19.8. Touch-Talk web interface. On the right are popular voice prompts from other people.

determinations about the potential for this system to expand the reach of mobile communications by enabling conversational partners to circumvent socially conventional restrictions on communication (see [8] for complete study details).

Analysis of usage patterns indicated that of all calls made to Touch-Talk numbers, 34% were actually carried out in Touch-Talk mode. The distribution of key presses indicated that the most common prompts selected were "Yes" and "No". Surprisingly, "Okay", a phrase used liberally in conventional calls for maintaining the flow of a conversation, was rarely employed. This result hints at a potential direction for future research about how conversational grounding operates in such unusual, asymmetric communication configurations.

An examination of call endings revealed that "thanks, bye" was used at the end of 41% of all Touch-Talk conversations. This statistic suggests that conversational partners effectively reach some sort of mutual conclusion. Twenty-four % of calls closed with "Let's talk later", or "The person had to leave the conversation and will call back later". These endings indicated that the issues discussed would require a follow-up conversation, perhaps in a more traditional fashion. Finally, 18% of calls switched to a normal two-way discussion at some point during the course of the conversation.

In addition to an analysis of use patterns, we administered biweekly questionnaires and conducted interviews with all participants at the end of the study. Users were generally positive about their experience and wanted to continue using Touch-Talk after the study ended. Participants reported using the service in meetings, in public spaces, and during transit. Touch-Talk was received enthusiastically as a way to let callers know that they were busy, and to ease negotiations about a future communication opportunity. Finally, one nonnative English speaker reported interest in the system because it helped him communicate more clearly with native speakers.

Finally, we detected a relationship between social hierarchies and the appropriateness of utilizing different communication modalities. Users were generally comfortable conducting conversations in Touch-Talk mode but tended to avoid it during interactions with authority figures, managers, or spouses. This is a rather generative result, pointing toward a series of possible studies about the relationship among social structures, availability, and modality.

19.5 Conclusions

Reflecting the broader ideology of the CHIL paradigm, the work presented in this chapter was motivated by a guiding interest in designing technology that gracefully integrates into human communication processes. To that end, our particular target has been mobile communication systems that intelligently manage the risks for untimely interruptions engendered by their capacity to enable complete connectivity anytime and anyplace.

Informed by research on interruptibility and awareness, we developed the Connector, a network-based system that provided information about a receiver's availability at the time a sender placed his call. The Connector based its judgments of availability on prospective, self-reported plans, represented by personal calendars. This initial proof-of-concept also included a group feature, enabling senders to learn about multiple potential recipients at the same time.

After field studies confirmed the value of the Connector, we then focused on alternate strategies for deriving a representation of availability. Through a series of studies on contextual aspects of availability, we found that social aspects of people's situations are consistently significant predictors of their willingness to answer mobile calls. Studies of situational cues carried out in a smart, context-aware office confirmed that communicating the receiver's current state helps senders avoid interrupting him or her at inappropriate times. In a multimethod study of everyday mobile phone use, we found that representations of availability that combined calendaring with information about social context were twice as powerful as those utilizing only calendar information.

We then turned toward a third area of exploration, designing novel interaction techniques that embedded meta-information about the sender's communicative intent into particular varieties of channels. The One-Way Phone and Touch-Talk systems explored the possibility that receivers might be willing to communicate in socially sensitive contexts if only they were able to do so in a more appropriate manner.

This work generated a number of design insights we believe are broadly applicable within the space of mobile communication, context-awareness, and interruptibility:

Sender behaviors are shaped by availability information. By judiciously exposing meaningful availability representations based on either prospective data like calendaring or situational information (either sensed or self-reported or a blend of both), a communication system creates social expectations that guide user behavior in predictable and beneficial ways. Our studies consistently show fewer social violations when such representations are made available. Senders will, in fact, obey suggested conventions perhaps because doing so offers the built-in incentive of increased probability for achieving a successful connection. A successful context-aware system need not necessarily make decisions about interruptibility on the user's behalf; merely representing the receiver's state can suffice.

On extending sensor-based approaches to mobile contexts. We conducted a series of studies aimed at representing the receiver's state of availability via a sensor-based, context-aware system. Aligned with the dominant approach in prior work on interruptibility, we began by leveraging existing CHIL infrastructure to create a smart office capable of inferring availability based on a classification of the state of a receiver's office. Though this technique proved rather reliable, porting the technique to phones met with resistances we believe to be rather inherent to mobility. With less ability to classify contexts, given wide ranges of possibilities, we shifted our technique, instead learning, based on audio samples, the ways environments sounded when receivers were more or less available. Rather than identifying contexts, our system aimed to act appropriately with respect to receiver behaviors driven by those contexts.

Methodologies for mobile studies. Mobile use is notoriously difficult to study because of the lack of control inherent when situations vary widely. The availability study demonstrates a relatively lightweight means for collecting data on mobile use for large sample sizes. Using a telephony server, we were able to "experience-sample" 90 users, probing participants for information via their familiar mobile device on a daily basis. The scalability of this approach guarantees the ability to gather sufficiently large data sets for confident generalization. Complementing this data with calendaring information and open-ended diary entries, we could triangulate results across multiple data collection techniques. Diary entries complement behavioral data by engaging people's experiences through their own perceptions and ontologies. Performing a content analysis bridges from a qualitative to quantitative approach, opening the resultant data to numerical representations that can be verified through statistical analysis and used in computational models of availability.

Interruption is a social construction. Availability does not always bear a straightforward relationship to interruptibility. The Virtual Assistant and availability studies both indicated the consistent importance of social aspects in determining receiver behaviors. Interruptibility, especially in mobile environments, may have to do more with people not wanting to be disruptive or inappropriate than with mental engagement. There exist circumstances where even though available, people are deprived of potentially important opportunities for information exchange because they are busy preserving the social fabric. We responded to this insight via One-Way Phone and Touch-Talk, in which we embedded meta-information in the channel so that conversation was possible in previously untenable situations. Senders and receivers would negotiate in implicit ways about the capacity for conversation within their current contexts.

Given the design space collectively articulated by these findings, we aim in future work to implement an intelligent, context-aware Connector on a single device. Ideally this device would construct a representation of availability, and then gracefully communicate this information to human conversational partners. The system would be capable of suggesting the best modality to communicate, creating options in real time for voice-to-text, voice-to-voice, text-to-voice, and text-to-text options for all mobile conversations. Moreover, it might artfully foreground or background its agency by either actively mediating to lessen the burden of the user, or fading into the background and offering up flexible tools when context did not determine availability but constrained it. Such a service would indeed be the ultimate connector, and a worthy addition to the CHIL suite of technologies that seek to empower human communication.

References

1. P. Adamczyk and B. Bailey. If not now, when?: the effects of interruption at different moments within task execution. *Proceedings of the SIGCHI conference on Human factors in computing systems*, pages 271–278, 2004.
2. D. Avrahami and S. Hudson. QnA: augmenting an instant messaging client to balance user responsiveness and performance. *Proceedings of the 2004 ACM conference on Computer supported cooperative work*, pages 515–518, 2004.
3. H. Clark. *Using Language*. Cambridge University Press, 1996.
4. M. Csikszentmihalyi. *Flow: The Psychology of Optimal Experience*. Harper Perennial, New York, Mar. 1991.
5. M. Danninger. *Intelligently Connecting People – Facilitating Socially Appropriate Communication in Mobile and Office Environments*. PhD thesis, Universität Karlsruhe (TH), 2008.
6. M. Danninger, G. Flaherty, K. Bernardin, H. K. Ekenel, T. Köhler, R. Malkin, R. Stiefelhagen, and A. Waibel. The connector: facilitating context-aware communication. In *ICMI '05: Proceedings of the 7th international conference on Multimodal interfaces*, pages 69–75, New York, NY, 2005. ACM Press.
7. M. Danninger, E. Robles, L. Takayama, Q. Wang, T. Kluge, R. Stiefelhagen, and C. Nass. The connector service - predicting availability in mobile contexts. In *MLMI*, LNCS 4299, pages 129–141. Springer, 2006.
8. M. Danninger, L. Takayama, Q. Wang, C. Schultz, J. Beringer, P. Hofmann, F. James, and C. Nass. Can you talk or only touch-talk? a voip-based phone feature for quick, quiet, and private communication. In *ICMI '07: Proceedings of the 9th international conference on Multimodal interfaces*, New York, NY, 2007. ACM Press.
9. T. Erickson and W. A. Kellogg. Social translucence: an approach to designing systems that support social processes. *ACM Transactions on Computer-Human Interaction*, 7(1):59–83, 2000.
10. J. Fogarty, S. E. Hudson, C. G. Atkeson, D. Avrahami, J. Forlizzi, S. Kiesler, J. C. Lee, and J. Yang. Predicting human interruptibility with sensors. *ACM Transactions on Computer-Human Interaction*, 12(1):119–146, 2005.

11. J. M. Hudson, J. Christensen, W. A. Kellogg, and T. Erickson. "i'd be overwhelmed, but it's just one more thing to do": availability and interruption in research management. In *CHI '02: Proceedings of the SIGCHI conference on Human factors in computing systems*, pages 97–104, New York, NY, 2002. ACM Press.
12. M. Ito. Mobile Phones, Japanese Youth, and the Re-Placement of Social Contact. *Annual Meeting of the Society for Social Studies of Science*, 2001.
13. R. Kubey, R. Larson, and M. Csikszentmihalyi. Experience sampling method applications to communication research questions. *Journal of communication*, 46(2):99–120, 1996.
14. S. Love and M. Perry. Dealing with mobile conversations in public places: some implications for the design of socially intrusive technologies. In *CHI '04: CHI '04 extended abstracts on Human factors in computing systems*, pages 1195–1198, New York, NY, 2004. ACM Press.
15. R. Malkin. *Machine listening for context-aware computing*. PhD thesis, Carnegie Mellon University, 2006.
16. A. Monk, J. Carroll, S. Parker, and M. Blythe. Why are mobile phones annoying? *Behavioral Information Technology*, 23(1):33–41, 2004.
17. C. Nass and S. Brave. *Wired for Speech: How Voice Activates and Advances the Human-Computer Relationship.* The MIT Press, 2005.
18. L. Palen, M. Salzman, and E. Youngs. Going wireless: behavior & practice of new mobile phone users. *Proceedings of the 2000 ACM conference on Computer supported cooperative work*, pages 201–210, 2000.
19. N. Wiener. *The Human Use of Human Beings: Cybernetics and Society.* Da Capo Press, 1988.

20

Relational Cockpit

Janienke Sturm and Jacques Terken

Technische Universiteit Eindhoven, Eindhoven, The Netherlands

Recently, the observation that socially inappropriate behavior during meetings may result in suboptimal group performance inspired researchers to develop systems that monitor and give feedback on social dynamics [3, 6, 12, 13]. These systems capture observable properties of the meeting participants, such as speaking time, posture, and gestures, analyze the interaction of people, and give feedback by offering visualizations of the social data. In [3], for instance, a wide range of vocal features, aspects of body language, and physiological signals is measured to calculate a behavior-based index of group interest, which is then shown to the participants on either a private or a public display. In [6], feedback is provided about the speaking time of different participants, visualized through a histogram presented on a public display. Evaluations showed that real-time feedback on speaking activity can result in more equal participation of all meeting members. In the framework of the CHIL project, these findings and observations lead us to believe that automatic feedback on audiovisual behavior of meeting participants may help to improve the social dynamics of the meeting and increase the satisfaction of the group members with the discussion process. We designed a CHIL service (the *Relational Cockpit*, or RC) that generates unobtrusive feedback to participants about the social dynamics during the meeting, presented in realtime on the basis of captured audiovisual cues. Our goal is to make the members aware of their behavior, and in this way influence the group's social dynamics.

We define social dynamics as the way verbal and nonverbal communicative signals of the participants in a meeting regulate the flow of a conversation (who has the floor) [1, 14]. The three most relevant determinants of the flow of conversation are the following.

- *Plain speaking time.* Since interrupting the speaker is bound to social conventions, within certain limits the current speaker determines how long he or she will speak. Speaking means having the opportunity to control the flow of conversation and influence the other participants.
- *Speaker eye gaze, as an indication of the focus of attention.* The current speaker controls the flow of conversation by having the privilege of selecting the next speaker. This may be done through verbal means, such as when the speaker

names another participant and asks for his opinion, but often it is done in a more subtle way, by nonverbal means such as eye gaze [9, 8, 2, 17]. In addition, when addressing all participants, the speaker should take care to look at all participants in due time in order to avoid giving the impression that he or she is neglecting particular participants.

- *Listener eye gaze*. The participant who is speaking is being looked at by the other participants, indicating that he or she is the focus of attention [15, 18]. When the speaker is speaking for a long time, other participants may lose interest, which is signaled by their gazing elsewhere.

The RC is intended for meetings with a protocol that invests participants with equal rights and responsibilities to contribute to the meeting, as, for instance, in a case where a committee needs to make a joint decision and every participant has information relevant to the decision, or the members of a team need to reach an agreement about a further course of action. In such collaborative meetings, everyone should be able to contribute to the meeting, regardless of the quality of the individual contributions and their impact on the final decision. Obviously, in other types of meetings, for example, instructive meetings or presentations, balancing the participation may be less favorable.

We formulated the following hypotheses concerning the influence of feedback on social dynamics in small, collaborative meetings:

1. Speaking time will be distributed more equally in sessions with feedback than in sessions without feedback. Concretely, participants who underparticipate without feedback will participate more in the presence of feedbac,k and participants who overparticipate without feedback will participate less when feedback is presented.
2. Speakers' visual attention will be distributed more equally among listeners when feedback is present than without feedback.
3. Visual attention from listeners for the speaker will be higher in sessions with feedback.
4. Participants' satisfaction about group communication and performance will be higher in the presence of feedback.

20.1 Prototype

20.1.1 Visualization

Functional and nonfunctional requirements for the RC service were formulated on the basis of literature, focus groups, unpublished ethnographical studies of meetings conducted at Technische Universiteit Eindhoven (TUE) and general usability considerations. The service should not influence the users' performance on their primary task; therefore, the feedback should be easy to understand at a glance. The visualization should be presented on a shared display and should provide the following information:

- cumulative speaking time of each participant since the beginning of the meeting;
- duration of the current turn;
- cumulative visual attention from the speaker for the listeners;
- cumulative and current visual attention from the listeners for the speaker.

With respect to the visualization of the speaker's attention for the listeners, we have chosen to display the attention each listener receives summed over all speakers. The rationale for this choice is that in order to receive attention from the speaker, a listener should actively participate in the conversation. Therefore, displaying the amount of speaker attention a given listener receives may encourage him to become more active.

The visualization is cast as a peripheral display and projected in the center of a table, as shown in Fig. 20.1 for a meeting with four participants.

Fig. 20.1. Visualization of social dynamics during a meeting.

Fig. 20.2. Visualization of current and cumulative speaking activity and visual attention.

The visualization contains the following components (see Fig. 20.2): (1) The middle circle (coded S, for Speaking activity) represents the participant's cumulative speaking time since the beginning of the meeting. For the current speaker, this circle is surrounded by a lighter-colored ring, the size of which represents the duration of the ongoing turn. (2) The leftmost circle (coded AS, for Attention from Speaker) indicates how much visual attention the participant – as a listener – has received since the beginning of the meeting from the other participants while they were speaking (added up across all other participants). (3) The right-hand circle (coded AL for

Attention from Listeners) represents how much attention a participant has received while speaking from the other participants since the beginning of the meeting. For the current speaker, this circle is surrounded by a lighter-colored ring representing how much visual attention she currently receives from the other participants. The information is updated dynamically in real time. The different circles are distinguished by different colors (the codes are not shown in the actual visualization). In order to facilitate users' understanding of the meaning of the different circles, a short description of its meaning is displayed underneath each circle.

20.1.2 Technology

The visualization is generated on the basis of combined audio (speech) and visual (focus of attention) cues, captured in real time during the meeting. Our first prototype was evaluated using a Wizard of Oz approach [11], which means that the behavior of meeting participants (speaking activity and eye gaze/head orientation) was monitored by human observers rather than automatically by technological components. Promising effects of the service on the social dynamics were obtained, but post hoc analyses of the data showed that the reliability of the monitoring task was below acceptable standards, in particular for eye gaze/head orientation. It was therefore decided to build a new prototype service in which the required perceptual technologies are implemented.

In the new prototype, speaking time for individual participants is determined by means of a close-talking microphone. From the audio input, the system can extract voice onset and offset patterns for each individual participant, so that the system can determine who is speaking when and for how long (this is known as speaker diarization). Recent developments toward speaker diarization apply multimicrophone arrays distributed across the room, but this technology is not yet robust enough. Instead, we applied a solution by which each participant is equipped with a close-talking microphone, and the microphones are connected to a multichannel audio controller. The output of the controller is sent to a server that continuously detects if participants are speaking or silent.

The visual focus of attention of meeting participants is now estimated on the basis of head orientation. Technology for monitoring eye gaze has greatly progressed over the last few years, and unobtrusive solutions are available, for instance, for determining where someone is looking on the screen of a desktop computer. However, there are no easily available solutions for monitoring the eye gaze of multiple participants in a meeting who may move around more or less freely. Fortunately, given certain situational constraints, eye gaze may be estimated from head orientation. The constraints have to do with the number of participants and the physical arrangement, and decrease the likelihood that a participant may switch visual attention between other participants only by moving the eyes. In such situations, head orientation can be considered a reliable indicator of gaze direction, as was shown in [15]: Focus-of-attention estimation can get an average accuracy of 88.7% in a meeting scenario with four participants. To detect head orientation, participants wear headbands with two pieces of reflective tape, which are tracked by infrared sensitive cameras mounted to the ceiling of the meeting room. The two pieces of tape enable the cameras to pick up two separate coordinates for each headband, which are sent to a server. On the basis of these two coordinates, the server estimates the angle of the headband relative to the perpendicular axis (looking straight ahead) in a two-dimensional horizontal plane (i.e., the orientation of the headband), which is the basis for determining the visual focus of attention of the person wearing the headband (see Fig. 20.3).

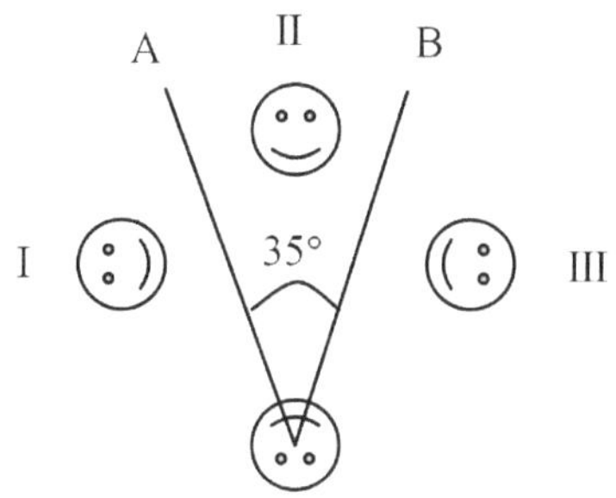

Fig. 20.3. Schematic diagram showing the relationship between the measured orientation of the headband and the participant's visual focus of attention.

If the orientation of the headband is between lines A and B, the focus of attention is on participant II; when it is to the left of line A, the focus of attention is on participant I; and if it is to the right of line B, the focus of attention is on participant III. The optimal angle between lines A and B was determined empirically during several pilot tests.

Finally, combined audio and visual data are sent to a server that controls the visualization shown on the meeting table.

Although we consider the technological equipment used (headbands and close-talking microphones) suitable for evaluating our concept in a laboratory setting, it is obvious that for real meetings, this setup is far too intrusive. Less intrusive technology that is suitable for real meetings has been developed within CHIL and other projects, for example, speaker localization on the basis of microphone arrays and camera-based head pose estimation. However, at the time of our evaluations, the performance of this technology was not yet sufficiently accurate.

20.2 Evaluation

20.2.1 Experimental Setup

The evaluation of the Relational Cockpit applied a within-subjects (or, rather, "within-groups") design. Each group participated in two discussion sessions in which the members had to reach an agreement on a particular topic. Two different discussion topics were provided. Subjects were presented with two conditions: In one condition, feedback about speaking time and visual attention was provided in the form of the visualization shown in Fig. 20.2 (the Feedback condition), whereas in the other condition, no feedback was provided (the NoFeedback condition). To avoid order effects, the order of Feedback and NoFeedback conditions was balanced across groups. The same was done for the two discussion topics.

20.2.2 Experimental Task

We used two adjusted hidden-profile decision tasks, adapted from those used by DiMicco [6]. The goal for both tasks is for a group to reach consensus about the optimal rank-ordering of a set of alternatives (e.g., three locations for a new supermarket). Each of the group members has to defend a different position, representing a particular set of beliefs and values (a profile). For instance, one group member would emphasize financial incentives, whereas another member would emphasize environmental issues. For more information about these adapted hidden-profile tasks, see [11] or [16].

20.2.3 Procedure

Each group participated in two conditions, the Feedback condition and the NoFeedback condition. In both conditions, the group first had a five-minute warm-up discussion about a topic of their choice. This short discussion served for the group members to get used to each other and to the environment, and to familiarize with the feedback. The warm-up discussion was included in the NoFeedback condition as well, to keep both conditions as similar as possible.

After the warm-up discussion, participants had 10 minutes to individually study their profile and the alternatives, to make a preliminary rank-ordering, and to memorize their arguments. (During the discussion, all paperwork was taken away, to avoid a situation where participants looked at their papers rather than at each other). During the next 20 minutes, the participants discussed the three alternatives and tried to reach an agreement.

After each discussion, participants were asked to fill in a questionnaire. In the NoFeedback condition, the questionnaire addressed group-related issues, whereas in the Feedback condition, the questionnaire addressed both group and service-related issues. A group interview followed the questionnaire; addressing questions about how the discussion went in the NoFeedback condition, extended with questions concerning the visualization in the Feedback condition.

In order to get an impression of the participants' intention to use the system, a fake third task was introduced and the participants were asked to indicate individually whether or not they would prefer to use the system for this third and final task. After this, participants were told that the third task did not exist and that the experiment had finished.

20.2.4 Evaluation Metrics

Measures for speaking time, speaker's attention, and attention from listeners were obtained from log files of the speech activity and head orientation trackers. For speaking time, each participant's speaking time is expressed as the percentage of time that the participant had been speaking of the total speaking time for that session. In addition, we calculated to what extent the amount of speaking time for the participants is equally distributed. We use the Gini coefficient as a measure of equality. Equation (20.1) defines the Gini coefficient for groups of four participants:

$$\text{Gini} = 2/3 \sum_i |participation_i - 25\%|, \tag{20.1}$$

The Gini coefficient sums, over all group members, the deviations of each person from equal participation (25% for a group of four), normalized by the maximum possible value of this deviation [19, 5]. Its values range from 0 for very high equality to 1 for low equality. For speaker's attention, we used the Gini coefficient to calculate to what extent the speaker's attention is distributed equally over the other three participants (the listeners) during the whole meeting. For attention from listeners, for each individual speaker we calculated the average number of listeners (i.e., participants for whom the speaker was the focus of attention) throughout the meeting. The average number of listeners is expressed as a percentage of the maximum number of listeners.

Subjective judgments about participants' attitudes toward the system and toward the group were collected by means of Likert scale-type questionnaires and group interviews. The group-satisfaction questionnaire (83 items) combined existing questionnaires about team member satisfaction, task cohesion, and perceived viability (capability of the group to continue working as a team in the future). The service-related questionnaire (28 items) combined existing questionnaires addressing issues of control, privacy, ease of use, usefulness, intrusiveness, enjoyment, trust, attitude, and intent to use. More information about the questionnaires can be found in [7]. The group interviews addressed several specific topics in more detail, such as the positive and negative aspects of the system, the influence of the visualization on the discussion, and the perceived reliability of the information that is shown.

20.2.5 Participants

A total of 21 groups participated in the experiment. Nineteen groups had four people and two groups had three people, for a total of 82 participants. Twenty-four participants with various educational and social backgrounds were recruited from a database listing volunteers for experiments. The other 58 participants were students of the faculty of Industrial Design at TUE. All participants were native Dutch speakers and were paid a small fee for participation. In some of the groups, some of the members already knew each other. One group was an existing student team.

20.3 Results

20.3.1 Reliability Analysis

To assess the reliability of the speech activity and head orientation trackers in an objective way, the automatic loggings of the trackers were compared with manual annotations. Three meeting fragments of two minutes were randomly selected. For these fragments, an expert coder manually annotated speech activity as well as head orientation for each participant, using the ANVIL video annotation tool [10]. The resulting annotation is referred to as the reference annotation.

Speech diarization can be considered a segmentation task, i.e., detecting when a person speaks. Detection of visual focus of attention, on the other hand, is a combination of a segmentation task and a classification task: Besides deciding when a participant changes his gaze direction, where the participants's gaze is directed should be detected. Different evaluation measures were used for the segmentation and classification tasks. For segmentation (identifying the onset and offset of speech activity and changes in gaze direction) we used segmentation accuracy. Segmentation accuracy is defined as 100 - SbER (segment boundary error rate), where SbER is the sum of segment boundary insertions, deletions, and misplacements divided by the total number of segment boundaries according to the manual segmentation (the reference). To calculate the SbER, we set a level of tolerance, indicating the time window within which the segment boundaries can still be considered to match the boundaries in the reference annotation. The tolerance window was set to 1 second. Segmentation accuracy was 57.3% for speech and 40.7% for head orientation. For speech segmentation, 8.1% segment boundary deletions, 13.7% insertions, and 21.0% misplacements were found. For head orientation, 41.6% segment boundary deletions, 3.1% insertions, and 14.6% misplacements were found. This means that our algorithms are not very precise at detecting the timing of transitions from one state to another. However, further comparison of the automatic and manual segmentations showed that the total distribution of speaking time for individual speakers was preserved well enough to be used in our system and for the purpose of initial user studies. The same applies with respect to head orientation. The reliability of the classifications of gaze direction was measured using Cohen's kappa. Kappa measures pairwise agreement among a set of coders making category judgments, while correcting for chance agreement [4]. Kappa has a range from 0 to 1, with large values indicating better reliability. Kappa was calculated using those segments for which there was agreement on both segment boundaries (37.5% of the segment boundaries for head orientation) and amounted to a value of 0.81, which is considered excellent, meaning that the head orientation tracker could properly detect the direction of the gaze.

20.3.2 Social Dynamics

Speaking Time

Due to technical problems in some sessions, we had complete speech and head orientation data from only 15 groups (13 groups of four and two groups of three). As expected, the average speaking time across all participants was 25% in groups of four participants and 33% in groups of three participants. The total speaking time of individual participants was relatively well correlated between the NoFeedback and Feedback conditions. The Pearson correlation coefficient was $r = 0.44$ ($N = 58$, $p = 0.001$). There were no statistically significant changes in the speaking time of individual participants between the two [$t(57) = 0.036$, $p = 0.971$]. Speaking time was divided fairly equally over the participants in both conditions: We found Gini coefficients of 0.14 in the NoFeedback condition and 0.11 in the Feedback condition. The difference in equality between the two conditions was not statistically significant [$t(14) = 0.942$, $p = 0.362$].

In order to test the hypothesis that especially participants who speak less than average (the underparticipators) or more than average (the overparticipators) will adapt their behavior as a result of the feedback, we categorized the participants into three categories. This was done only for the groups with four participants. Those whose total speaking time was more than one standard deviation below average in the NoFeedback condition are categorized as underparticipators (seven speakers, 13.5%); those who speak more than one standard deviation above average were categorized as overparticipators (seven speakers, 13.5%); the rest was categorized as middle participators (38 speakers, 73%). The results are shown in Table 20.1. T tests were performed on the speaking time data of each category. We found that the underparticipators significantly increased their speaking time in the Feedback condition as compared to the NoFeedback condition [$t(6) = -3.302$, $p = 0.016$]. Moreover, overparticipators significantly decreased their speaking time with feedback [$t(6) = 2.318$, $p = 0.060$]. No significant difference between the two conditions was found for the middle participators [$t(37) = 0.745$, $p = 0.461$. The results thus indicate that participants who are at the extremes of the speaking-time range tend to change their behavior so as to become less extreme.

	N	Average Speaking Time	
		NoFeedback	Feedback
All (groups of four)	52	25.0%	25.0%
All (groups of three)	6	33.0%	33.0%
Under-participators	7	12.4%	22.7%
Middle participators	38	25.3%	24.6%
Over-participators	7	36.4%	29.2%

Table 20.1. Speaking time results.

This finding could be explained simply in terms of a regression toward the mean, the phenomenon that measures that have extreme values at one point in time are likely to be less extreme when measured on a different occasion, for statistical reasons. However, closer inspection of the results renders this explanation unlikely. The distribution of participants over different percentage bins turned out to be narrower in the Feedback condition than in the NoFeedback condition, with participants being centered more closely around the mean. Under an explanation in terms of a regression to the mean, the shape of the distribution should remain approximately the same. Furthermore, related research has also indicated that people tend to change their behavior on the basis of visual feedback, while an explanation in terms of a regression to the mean was ruled out [5]. Therefore, we consider it safe to assume that regression to the mean is not a conclusive explanation for our findings.

Attention from Speaker

The distribution of the speakers' attention throughout the meeting was rather unequal in both conditions (Gini coefficients are 0.54 in the NoFeedback condition and 0.55 in the Feedback condition). Unequal distribution of speakers' visual attention was to be expected, because the default position of the head is straight ahead. Since many people do not turn their head as much as they move their eyes, it is likely that there is a bias toward the participant sitting opposite the speaker. Closer inspection of the data confirmed that for most speakers (73% in the NoFeedback condition and 83% in the Feedback condition), the participant seated opposite was the main visual focus of attention. The difference between the Gini coefficients in the two conditions is not statistically significant [$t(57) = -0.686, p = 0.495$], indicating that feedback about the way speakers divided their attention across listeners did not incite speakers to divide their attention more equally.

Attention from Listeners

The average attention level (i.e., the average percentage of listeners looking at the speaker) is 41% in the NoFeedback condition and 42% in the Feedback condition. The difference in attention level between the two conditions is not statistically significant [$t(57) = -1.246, p = 0.218$], indicating that listeners did not pay more visual attention to the speaker as a result of the feedback.

20.3.3 Questionnaire and Interview Results

The questionnaire and interview data are based on all 82 participants. The questionnaire data concerning group satisfaction showed only minor differences between the Feedback and NoFeedback conditions (Table 20.2). Participants' attitudes toward the system were moderately positive: The average scores on different subscales were between 4 and 5 on a seven-point scale (Table 20.3). Lower scores were obtained for usefulness (average 3.5) and control (average 3.8). Fifty-one percent of the participants indicated that they preferred to use the system for a third task (which actually did not exist), for various reasons, such as, "The system shows interesting information about my behavior", and "It is fun to use the system". Thirty-eight percent of the participants preferred not to use the system again. The most prominent reason for not wanting to use the system again is the distraction from the meeting task that it causes.

Category	Score	
	NoFeedback	Feedback
Perceived viability	5.1	5.0
Task cohesion	5.1	5.0
Team member satisfaction	5.6	5.6

Table 20.2. Average scores for group-related dimensions in both conditions.

Category	Score
Trust	4.3
Usefulness	3.5
Privacy	4.5
Intrusiveness	4.5
Ease of use	4.5
Enjoyment	4.4
Control	3.9
Attitude	4.4
Intent to use	4.0

Table 20.3. Average scores for service-related dimensions.

During the interviews, several participants indicated that the meaning of the circles was not immediately clear to them. They would need more extensive training with the system in order to fully grasp the meaning of the circles and be able to use the information during the discussion before the system can be really useful (which is in line with the relatively low questionnaire scores for usefulness). For most participants, the speaking-time circle was the most intuitive one, and therefore this information was most used. Several participants mentioned that the circles enabled them to better divide their attention, while other participants found that the circles introduced some kind of competition. Some participants indicated that measuring head orientation was not the most reliable way to measure attention, because it captures only visual attention and they may pay attention to the speaker even when they are not looking at her. Most people, however, found that the circles adequately reflected speaking activity and focus of attention during the meeting.

20.4 Conclusion and Lessons Learned

The Relational Cockpit provides real-time feedback in small, collaborative group meetings on the social dynamics of the meeting. A dynamic visualization of speaking time and gaze behavior is offered to meeting participants through a peripheral display. Analyses of the reliability of the speech and head orientation trackers showed that the prototype is able to detect speech activity and visual focus of attention with sufficient reliability. The service was evaluated in a within-subjects evaluation with 82 participants. The study confirmed our hypothesis that the visualization affected the social dynamics of the meeting. A significant effect of the visualization of speaking time was found for under- and overparticipators who, as a result of the feedback, changed their speaking behavior to become less extreme. For visual focus of attention, the effects that were found were in the right direction but failed to reach statistical significance. It also turned out that the visualization did not influence the participants' satisfaction with their team. Questionnaire and interview data showed that participants were moderately positive about the system, although several participants had concerns about the fact that the system distracted them from the discussion. Half of the participants indicated that they would like to use the system in a future occasion. Here we describe the most important lessons that we learned from this study.

The results of our evaluation indicate that speaking time is more affected by the feedback than visual attention. A possible explanation might be sought in the controllability of the behavior. Although both speaking activity and visual attention can be consciously controlled, intuitively it appears much easier to control speaking activity than visual attention. Noticing that one has already been speaking for a long time, one may simply decide to stop speaking and give someone else a turn. Similarly, speakers who are not very active may decide to become more active when the evidence of their under- or overparticipation is clearly shown on the table. Visual attention, on the other hand, seems to be less under conscious control and to be ruled more by events in the environment and by entrained habits, such as paying attention to all members of the audience instead of looking at papers or at a single member of the audience. Thus, it may well be that speaking time can be more easily changed on the basis of feedback than visual attention.

Our results may also be accounted for in terms of the concrete properties of the visualizations. After an explanation by the experiment leader, participants had only five minutes to get to know the system by using it in a warm-up discussion. Although this was enough for participants to understand the concept of the visualization, it may have been insufficient to really understand the meaning and impact of the information shown. Several participants indicated that they did not really use the visualization, because thinking about what to do with the information would distract them too much from the actual discussion going on. In particular, the circles representing visual attention ("attention from listeners" and "attention from speaker") appeared to be difficult to interpret. It may take more than one meeting to be able to understand the meaning of the circles at a glance and do something with the information (i.e., to change one's behavior) without being distracted too much. Future research should be aimed at developing visualizations that are easier to interpret and therefore less distracting and less intrusive than the current visualization.

However, just signaling information related to the social dynamics through a peripheral display with the aim of improving nonoptimal social dynamics may not be sufficient, as participants have to focus on the primary task (the meeting) and may neglect the peripheral information. For that reason, it would be interesting and useful to explore opportunities for actual intervention. For instance, if the distributions of speaking time and visual attention fall outside certain standards of acceptability, the system might draw the attention to the visualizations and encourage the participants to discuss the displayed information.

Finally, it may be noted that the majority of the groups taking part in our experiment consisted of people who did not know each other, so they did not have a history together and they would not be in any meeting together afterwards either. In such a situation, people are often rather polite, friendly, and lenient (almost all participants indicated that they found the other group members kind and not irritating). This may have influenced the results since, in such a situation, the discussion often goes well in terms of social dynamics. Moreover, if were be problems concerning the social dynamics – for example, if one person took the lead and disregarded some of the other participants – people may not have had the drive to change the situation, because it was just this one time that they had to deal with it. The situation is completely different when people have to meet with the same group every week. Therefore, providing feedback about social dynamics may be more useful for groups that have just started and will continue to work together for some time, or for existing groups experiencing problems.

References

1. M. Argyle. *Social Interaction*. Methuen, London, 1969.
2. I. Bakx, K. van Turnhout, and J. Terken. Facial orientation during multi party interaction with information kiosks. In *Interact*, 2003.
3. S. Basu. Towards measuring human interactions in conversational settings. In *IEEE International Workshop on Cues in Communication (CUES)*, 2001.
4. J. Cohen. A coefficient of agreement for nominal scales. *Educational and Psychological Measurement*, pages 37–46, 1960.
5. J. DiMicco, J. Hollenbach, and A. Pandolfo. The impact of increased awareness while face-to-face. *Human-Computer Interaction*, 22(1), 2007.
6. J. DiMicco, A. Pandolfo, and W. Bender. Influencing group participation with a shared display. In *CSCW04: Proceedings of the ACM Conference on Computer Supported Co-operative Work*, pages 614–623, 2004.
7. I. Graziola, A. Eyck, and J. Sturm. Modelling intention to use services supporting human-human communication and collaboration. Technical report, CHIL Project, 2007.
8. A. Kalma. Gazing in triads: A powerful signal in floor apportionment. *British Journal of Social Psychology*, 31:21–39, 1992.
9. A. Kendon. Some functions of gaze direction in social interaction. *Acta Psychologica*, 25:22–63, 1967.
10. M. Kipp. Anvil – a generic annotation tool for multimodal dialogue. In *Eurospeech01: Proceedings of 7th European Conference on Speech Communication and Technology*, pages 1367–1370, 2001.
11. O. Kulyk, C. Wang, and J. Terken. Real-time feedback based on nonverbal behaviour to enhance social dynamics in small group meetings. In *MLMI'05: Proceedings of Joint Workshop on Multimodal Interaction and Related Machine Learning Algorithms*, LNCS 3869, pages 150–161, 2006.
12. A. Madan, R. Caneel, and A. Pentland. Groupmedia: Distributed multimodal interfaces. In *ICMI04:Proceedings of the Sixth International Conference on Multimodal Interfaces*, 2004.

13. R. Rienks, D. Zhang, D. Gatica-Perez, and W. Post. Detection and application of influence rankings in small group meetings. In *ICMI06: Proceedings of the Eighth International Conference on Multimodal Interfaces*, 2006.
14. H. Sacks, E. Schegloff, and G. Jefferson. A simplest systematics for the organisation of turn-taking for conversation. *Language. Journal of the Linguistic Society of America*, 50:696–735, 1974.
15. R. Stiefelhagen and J. Zhu. Head-orientation and gaze direction in meetings. In *CHI'02: Proceedings of Human Factors in Computing Systems*, pages 858–859, 2002.
16. J. Sturm, O. H.-V. Herwijnen, A. Eyck, and J. Terken. Influencing social dynamics through a peripheral display. In *ICMI07: Proceedings of the International Conference on Multimodal Interaction*, 2007.
17. Y. Takemae, K. Otsuka, and N. Mukawa. Video cut editing rule based on participants' gaze in multiparty conversation. In *11th ACM International Conference on Multimedia*, pages 303–306, 2003.
18. R. Vertegaal, R. Slagter, G. van der Veer, and A. Nijholt. Eye gaze patterns in conversations: There is more to conversational agents than meets the eye. In *CHI'01: Human Factors in Computing Systems*, pages 301–308, 2001.
19. S. Weisband, S. Schneider, and T. Conolly. Computer-mediated communication and social information: Status salience and status awareness. *Academy of Management Journal*, 38:1124–1151, 1995.

21

Automatic Relational Reporting to Support Group Dynamics

Fabio Pianesi, Massimo Zancanaro, Alessandro Cappelletti, Bruno Lepri, Elena Not

Foundation Bruno Kessler, irst, Trento, Italy

The complexity of social dynamics occurring in small group interactions often hinders the performance of teams. The availability of rich multimodal information about what is going on during the meeting makes it possible to explore the possibility of providing various kinds of support to dysfunctional teams, from facilitation to training sessions addressing both the individuals and the group as a whole. A necessary step in this direction is that of automatically capturing and understanding group dynamics. In order to improve the performance of meetings, external interventions by experts such as facilitators and trainers are commonly employed. Facilitators participate in the meetings as external elements of the group, and their role is to help participants maintain a fair and focused behavior as well as to direct and set the pace of the discussion. Training experiences aim at increasing the relational skills of individual participants by providing offline (with respect to meetings) guidance – or coaching – so that the team eventually will be able to overcome or cope with its dysfunctionalities.

In this chapter, we present a multimodal system, called "the relational report (RR)", that monitors groups and generates individual reports about the participants' behavior. The system observes the meeting as a coach would, and not as a recorder. This means that the system does not keep a verbatim record of what people said and/or did during the meeting. The generated reports are not minutes; they do not address content, but present a more qualitative, meta-level interpretation of what happened in the social dynamics of the group. The reports are delivered privately to each participant after the meeting, and their purpose is that of informing participants about their behavior rather than evaluating it. Hence, the system acts as a coach for the individual group participants.

21.1 Background and Related Work

In discussing the role of collaboration for teachers and, in particular, peer coaching, Andersen [1] suggests that coaching sessions provide a scheduled opportunity to think reflectively and that the coaching process allows the externalization of both through contents and processes that are normally internal, making them available to examination. By bringing a different perspective to the relationship, the coach can see circumstances and possibilities that the coachee cannot. According to Boud and colleagues [5], there are three stages in the reflective process: (1) the return to experience ("What happened?"); (2) attending to feelings ("How did I feel?", "Why did I (re)act this way?"); and (3) the reevaluation of the experience ("What does it mean?"). In the present work, we mainly focused on the first stage. Our Relation Report generator is focus on the first stage. Our relational report generator is focused on the first stage and is meant to support the user to better engage in the other two.

In the field of CSCW, the focus is often on distributed meetings, and the social relationships among meeting participants have been recognized as a fundamental aspect of the meetings' efficacy since the seminal work of Tang [22]. Many different attempts have been made to bring the social dynamics to a "visible" level. For example, Dourish and Bly [11] investigated the effects on groups of providing information about the distributed meeting context without using a full video-conferencing system. They designed a system, called Portholes, consisting of a simple chat-based system augmented with a shared database of regularly updated visual information available at all sites. Their findings suggest that across-distance awareness can lead to more effective communication, and improved interactions, and can contribute to a shared sense of community. Another example in this respect is the work of Erikson and colleagues [12], which proposed the idea of "social translucence", that is, graphical widgets that signal cues that are socially salient. The claim is that such a functionality – by supporting mutual awareness and accountability – makes it easier for people to carry on coherent discussions; observe and imitate others' actions, create, notice, and conform to social conventions; and engage in other forms of collective interaction. In our work, we deal with face-to-face communication; therefore, awareness and visibility of the context are not problems; the impact of participants' perception of their own activity on the others could play an important role, though. An example of a work closer to ours in this respect is DiMicco and colleagues [9], which investigates the effects of providing team members with feedback about their own speaking activity during a face-to-face meeting. Our approach, though similar in spirit, is different, especially because we address a larger set of basic information (beyond speech activity) to bear on the automatic understanding of relational behavior. Other work closer to our approach is Maloney-Krichmar and Preece's [17] research on the dynamics of an online group community. They used a coding scheme similar to the one we will discuss later, and inspired by the same source as ours, and investigated interrater agreement by considering agreement rate (proportions). The schema was basically meant to serve analytical and theoretical building purposes, while ours was devised to serve the automatic annotation of meetings so that functionalities such as the relational report can be built. Our work has deep roots in the field of multimodality. Most of the current research in this area aims at providing easy access to computerized services for the group to efficiently accomplish its tasks. For example, most of the services provided in the CHIL project aim at offering better ways of connecting people (the Connector service; see Chapter 19) and supporting human memory (the Memory Jog service; see Chapter 18).

21.2 The Survival Task Experiment

In order to assess the acceptability of an automatically generated relational report, a Wizard of Oz [23] experiment was designed. Eleven groups of four people engaged in a structured 30-minute discussion. All the relational reports were produced by a human coach, but half of the participants were told that an automatic system produced them, and the other half were told the truth. The experiment addressed the same four dimensions informally examined in the focus groups: (1) the perceived usefulness of the RR; (2) its reliability (whether people think that an automatic system can reliably provide a report on such a delicate matter as individual behavior in group situations); (3) its intrusiveness (the perceived degree of intrusiveness of a service that monitors group and individual behavior to provide reports on their relational behavior); and (4) its acceptability (what affects the acceptance of the report by addressees?). The participants (40% males and 60% females) involved in the study were all clerks from FBK-irst administrative services. In all cases, they knew each other, and had often been involved in common group activities in the past. The average age was 35 years. All the groups were mixed gender.

In order to ensure more engagement in group discussions, we employed a consensus decision-making scenario where each participant is asked to express her or his opinion and the group is encouraged to discuss each individual proposal by weighing and evaluating their quality. Consensus is usually enforced by establishing that any participant's proposal becomes part of the common sorted list only if she managed to convince the others of the validity of her proposal. In our case, an element of competition was also added by awarding a prize to the individual who proposed the greatest number of correct and consensually accepted items. We used the Survival Task, which is frequently used in experimental and social psychology to elicit decision-making processes in small groups. Originally designed by the National Aeronautics and Space Administration (NASA) to train astronauts before the first moon landing, the Survival Task proved to be a good indicator of group decision-making processes [14]. The exercise consists of promoting group discussion by asking participants to reach a consensus on how to survive in a disaster scenario, like a moon landing or a plane crashing in Canada. The group has to rank a number (usually 15) of items according to their importance for crew members to survive.

The groups were videotaped using four fixed omnidirectional cameras, close-talking microphones, and seven T-shaped microphone arrays, each consisting of four omnidirectional microphones. There was no attempt to hide the recording devices since one of the purposes of the experiment was to evaluate the acceptability of being recorded. A few days after their session, participants received an individual report created by a social psychologist, describing their behavior in terms of the functional roles played during the meeting. In writing the reports, the psychologist considered only nonverbal aspects of participants' behavior, such as posture and tone of voice, and not aspects related to content. Each subject was convened individually so that he could not discuss the content of his report with the other participants; during that session, they were also asked to fill out a questionnaire designed to investigate the four dimensions mentioned above. Half of the participants were told that their report was automatically produced by an intelligent system able to monitor the group's behavior, while the other half (the control group) were told that the report was written by a psychologist. The selection was randomized and balanced with respect to gender. The attitude toward the report was tested by a seven-item questionnaire aimed at assessing the perceived usefulness, reliability, perceived degree of intrusiveness, and acceptability of the report. A semantic differential targeting the appropriateness, completeness, and clarity of the report was also used (the semantic differential was part of the six-scale questionnaire proposed by [13] with a Cronbach alpha of 0.9482). The answers to the questionnaire were analyzed by means of a multivariate ANOVA ($p = 0.05$), applied to the data from 41 questionnaires (three subjects did not fill theirs out properly). The factor was the source of the report: "human" for the control condition, and "system" for the experimental one. Generally, there were no statistically significant differences among the responses to the questionnaire in the two groups. Regarding the subscales of the semantic differential, they were also analyzed by means of a multivariate ANOVA with $p = 0.05$. The only difference we found concerned the appropriateness subscale [$F(1,39) = 4.883, p < 0.05$], where the "system" group rated the appropriateness of their report higher than the "human" one (estimated means and standard errors: $M_{expert} = 28.38, SE_{expert} = 2.03; M_{system} = 34.82, SE_{system} = 1.98$. To the end, this study did not reveal any significant difference between the two groups concerning usefulness, reliability, degree of intrusiveness, acceptability, completeness, and clarity of the report.

As far as these results are concerned, there is no substantive evidence that an automatically produced report about one's own relational behavior in meetings would be accepted any differently than one produced by a human expert. Though needing confirmation by further study, this is an encouraging result, for it supports the idea that meeting participants could indeed consider automatically produced reports to improve their own relational skills.

21.3 The Functional Role-Coding Scheme

The goal of presenting individual profiles to participants suggested that we consider those approaches to social dynamics that focus on the roles members play inside the group, as opposed to approaches that define roles according to the social expectations associated with a given position (as in [16]). These kinds of roles – called *functional roles* [21] – are defined in terms of the behavior enacted in a particular context and exploit information about what actually happened in the course of the interaction, while reducing the necessity for knowledge about the group's structure, history, position in the organization, etc. Benne and Sheats [4] provided a list of functional roles for working groups, and divided them into three classes: task-oriented, maintenance-oriented, and individual-oriented roles. The first two types of roles are directed toward the group's needs: Task-oriented roles provide facilitation and coordination in view of task accomplishment, while maintenance roles contribute to social structure and interpersonal relations in order to reduce tensions and maintain smooth group functioning. The third type of roles, the "individual roles", focuses on the individual and her goals and needs rather than the group's. During the interaction, each person can enact more than one role. Drawing on Benne and Sheats, Bales [3] proposed the Interaction Process Analysis (IPA), a framework to study small groups by classifying individual behavior in a two-dimensional role space consisting of a Task Area and a Socioemotional area. The roles pertaining to the latter stem from activities that support or weaken interpersonal relationships. For example, complimenting another person is a positive socioemotional behavior in that it increases group cohesion and mutual trust; insulting another participant, on the other hand, can undermine social relationships. The other six categories pertain to task-oriented activities, that is, behavioral manifestations relating to the management and solution of the problem(s) the group is addressing. Giving and asking for information, opinions, and suggestions related to the problem at hand are examples of task-oriented activities. Building on Benne and Sheats' functional roles and on Bales' two-dimensional approach, and drawing on observations performed on a set of face-to-face meetings, a coding scheme was produced – the Functional Role-Coding Scheme (FRCS) – consisting of five labels for the Task Area and six labels for the Socioemotional one. The Task Area consists of roles relating to the facilitation and coordination of the tasks in which the group is involved, as well as to the technical skills of the members as they are deployed in the course of the meeting. It includes the *Orienteer*, who orients the group by introducing the agenda and defining goals and procedures. He or she keeps the group focused and on track and summarizes the most important parts of the discussion and the group's decisions. The *Giver* provides factual information and answers to questions. The *Seeker* requests information. The *Recorder* manages the available resources for the group. The *Follower* merely listens and does not actively participate to the interaction. The Socioemotional area concerns the relationships among group members and roles oriented toward the functioning of the group as a group. It includes the *Attacker*, who deflates the status of others or expresses disapproval; the *Gatekeeper*, who acts as the moderator within the group; the *Protagonist*, who takes the floor; the *Supporter*, who shows a cooperative attitude; and the *Neutral*, who passively accepts the idea of others, serving as an audience in group discussion. Of course, participants may, and often do, play different roles during the meeting, but at a given time each of them plays exactly one role in the Task Area and one role in the Socioemotional one.

The reliability of the scheme was assessed on a corpus of meetings consisting of 130 minutes for the Socioemotional Area and 126 minutes for the Task Area (for details, see [20]). Two trained annotators coded five participants on the Socioemotional Area and five in the Task Area. Cohen's kappa was used to assess inter-annotator agreement. The agreement on the roles of the Task Area is good, at 0.71 ($N = 758, SE = 0.02, p < 0.0001$), with a confidence interval of $0.65 - 0.75 (= 0.05)$. The agreement on the roles of the Socioemotional Area is less high, at 0.6 ($N = 783, SE = 0.023, p < 0.0001$), with a confidence interval of $0.56 - 0.65 (= 0.05)$.

21.4 The Survival Task Corpus

The audio and video recording collected as part of the Survival Task experiment described above were automatically processed to extract voice activity and fidgeting information for each participant. The extraction of the voice activity was performed with a voice activity detector (VAD), which uses the time energy of the signal [7]. The audio track of the close-talking microphone of each participant was automatically segmented and then manually checked (several wrong assignments were expected due to the fact that the voice activity of a subject could often be captured by the close-talking microphone of her neighbor). The video recordings were used to extract figdeting information. Fidgeting refers to localized repetitive motions such as when the hand remains stationary while the fingers are tapping the table, or playing with glasses, etc. Fidgeting was tracked by means of skin region features; temporal motion is used as the trigger (see [8] for details). For each subject, the hand and body fidgeting were estimated and the values normalized to the person's most vigorous fidgeting during the entire recorded sequence; hence, they are person-specific.

Finally, the corpus was manually coded with the Functional Role-Coding Scheme.

21.5 Automatic Detection of Functional Roles

Role assignment has been modeled as a multiclass classification problem using the speech activity and the fidgeting features only.

In a first attempt, we used an SVM approach with a bound-constrained SV classification algorithm with a Gaussian RBF kernel (see [24]). In order to take the time into consideration, a time window of varying size, from 0 to 14 seconds, was given in two conditions: (1) the whole window is on the left-hand side of the time point to classify (which is preferable for online classification), and (2) half of the window is before and half is after the classification point (in this case, the classification for time t takes place with a delay of width/2 seconds, where width is the length of the window). The cost parameter C and the kernel parameter were estimated through the grid technique by means of cross-fold validation using a factor of 10. Although not yet good enough for a real application, the results were encouraging and demonstrated the feasibility of such a system. The performance of the classification for the Task Area roles was rather good, with a macro-precision of 0.55 and a macro-recall of 0.49 with a left window of 14 seconds; yet the differences among the classes were not negligible (the f-score ranged from 0.73 for the *Giver* role to 0.30 for the *Seeker*). The results are even a little bit worse for the Socioemotional Area roles: Although the best performances are quite high – macro-precision and macro-recall, respectively, 0.75 and 0.43 – the differences among the classes are very high (for example, the *Attacker*, since it was represented very low in the corpus, had an f-score of 0.02 while the *Protagonist* role had an f-score close to 0.8).

In a second attempt [19], features from all participants together were used to predict the role of each one. The SVM approach obtains higher recognition accuracy but suffers from two problems related to generalization capability: (1) the curse of dimensionality (if we make use of the observations/features of other speakers for our classification task, the length of the observation/feature vector grows linearly with a large multiplication constant); (2) overfitting (the Gaussian RBF kernel might have an infinite VC dimension).

We therefore turn to a different approach. In a recent work, we exploited a Bayesian approach called Influence Model [2]. Influence Model is a good technique to deal with the curse of dimensionality and overfitting. In fact, the latent structure influence modeling of interacting processes avoids the curse-of-dimensionality problem by using the "team of observers" approach. In this approach, the individual observers only look at the latent states of the other related observers, which best summarize the observations from the perspectives of the latter and thus are less likely to suffer from overfitting and lack of generalization. The performance obtained using the Influence Model for recognizing group functional roles was comparable to the interrater reliability on this corpus of data: We can generally get 75% accuracy in classifying both the Task area roles and the Socio-Emotional area roles (see [10]). One interesting observation that the Influence Model seems to be generalizable to different numbers of participants in the group, since the influence between participants was very similar for all subjects and all experiments. The ability to automatically adapt to differently sized groups without retraining would allow a great increase in the flexibility and applicability of automatic role classification technology. One important area for future work is that the current training algorithm for the Influence Model does not do well at classifying the low-frequency classes (*Orienteer/Seeker* for Task Area roles, and *Attacker/Supporter* for Socioemotional Area roles). A direction for improvement is adding more features and hierarchical training. In future works, we plan to add some novel features, beginning with vocal energy, 3D postures, and focus of attention.

21.6 From Coding Scheme to Relational Reports

Starting from the the time sequences of functional roles, a relational report can be automatically generated for each participant to highlight some major aspects of his or her behavior during the meeting. The report is built according to the task and socioemotional roles played by the subject during the interaction and also takes into account the roles played by the other participants. Each final relational report has the form of a multimedia presentation where different modalities are used to present the relevant information and provide evidence in support of it. The actual text of the report is intended to describe the behavior in an informative rather than normative way, helping the user accomplish the first step of the reflective process – namely, the return to experience (what happened during the meeting?). To improve effectiveness and emotional involvement, a virtual character is used that reads the report with emotional facial expressions appropriate to the content (e.g., a sad expression is used when something is unpleasant, such as when a serious contrast with a colleague is being recalled). When appropriate, the presentation is enriched with short audio and/or video clips from the actual meeting that exemplify the information presented.

The approach to content selection and organization we adopted is based on the assumption that in behavioral reports, the conventions on how information is presented by human experts play a major role. These "patterns of appropriate ordering"(more widely known in the NLG community as schemas [18] or the Generic Structure Potential [15] of a text) have been exploited by many NLG systems to guide the text planner in organizing the text structure. For our report generation, we have used a general-purpose schema-based text planner [6] that accesses a repository of declarative discourse schemata. Each schema contains applicability conditions that look in the time series of roles for specific patterns. Once the applicability conditions are satisfied, the text planner extracts from the body of the schema the instructions concerning what to say, what to show, and in which order, and the different media synchronization. The schemata have been derived from the analysis of the actual reports written by the social psychologist involved in the Survival Task experiment described above, and from additional expert knowledge elicited through interviews. To exemplify, one of our schemas says that if the total amount of active task roles played by the participant during the meeting (*Orienteer* + *Giver* + *Seeker* + recorder) is greater than 75%, then a text should be produced (1) expressing this active contribution (*"You have very actively contributed to the discussion, with many verbal contributions ..."*), (2) possibly indicating whether there has been dominance (*"You have maintained a highly dominant role, keeping the attention of the other partecipants, as highlighted in the chart by the bar in red."*), (3) including sentences emphasizing the amount of team guidance (*"You've maintained a pivotal role in defining how to proceed with the discussion and in summarizing the results, as highlighted in the chart by the bar in blue."*) and (4) information contribution (*"You have significantly contributed with your ideas and opinions."*). More elaborate discourse strategies involve reasoning about the behavior of various participants at a given time. For example, if the target participant often played the *Orienteer* role at the beginning and end of the meeting, with the others being silent or neutral, the report could include a statement like, "At the beginning of the meeting you have helped define the agenda and initiate the discussion, summing up the outcome of the meeting at the end". If sequences of Seeker-(recorder)-*Attacker* are observed, the report could be complemented with, "In some cases, however, you displayed critical and aggressive behavior as a response to your colleagues' contribution". The linguistic realization of sentences is currently template-based. As a preliminary solution, the wording of sentences has been based on the typical one found in reports authored by human experts; more work is needed, however, to study the impact of improper lexical choices on users. The generated report is composed as an SMIL[1] presentation.

Figure 21.1 shows a snapshot of a sample relational report. .

[1] SMIL, Synchronized Multimedia Integration Language, is the standard language for multimodal presentations developed within the W3C, http://www.w3.org/AudioVideo/.

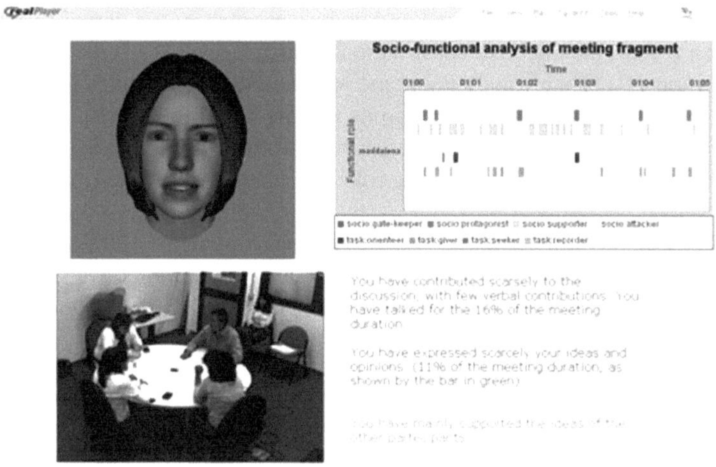

Fig. 21.1. An example of a multimedia relational report for one meeting participant.

21.7 Conclusion

The aim of this work was to contribute to an emerging new class of multimodal systems in which multimodality is used not to improve human-machine interaction, but as a core component of devices that by "observing"and staying in the loop of human-human interaction provide various types of functionalities. Besides being technically challenging, these systems raise so many user-related (intrusiveness, acceptability, privacy violation, etc.) and ethical issues that a user-centered approach to their design is virtually necessary.

Building on the availability of rich multimodal information in meeting rooms equipped with technology for audiovisual scene analysis, we have explored the prospects of a functionality inspired by coaching. It consists of a report about the social behavior of individual participants, which is generated from multimodal information, and privately delivered to them. The underlining idea is that the individual, the group(s) they are part of, and the whole organization might benefit from an increased awareness by participants of their own behavior during meetings.

In our case, we used a mix of laboratory-based testing and attitudinal study to deal with the initial phases of the user-centered design cycle.

User-related concerns were investigated by a mix of laboratory-based testing and attitudinal study. We first elicited attitudes and beliefs about the service through a focus group; then we empirically investigated the acceptance of such a system by simulating it through a WOZ experiment comparing the acceptance of automatically generated reports with that of reports produced by a professional coach. These studies provided enough evidence that the service can actually be valuable for engaging people in the first stage of the reflective process – that is, the return to experience (what happened?) – and allowed us to derive initial requirements specification.

With this much of a background and empirical support, we turned to illustrate steps taken toward the effective construction of such a service: (1) the development and validation of a reliable coding scheme to annotate group behavior; (2) the production of an annotated multimodal corpus that was used to (3) train a component for the automatic extraction of functional roles from audiovisual observations; (4) a generation component that, exploiting role information, assembles the multimedia report.

Much research is still needed before such a service might be effectively used in an ecological setting. We plan to refine the automatic classification of functional roles by exploiting more multimodal features and experimenting with different techniques. We also plan to investigate the impact of the different communicative strategies used in the relational reports (the talking head, the different graphic displays, and so on) on the reports' acceptability and effectiveness.

References

1. C. A. Andersen. Theoretical framework for examining peer collaboration in preservice teacher education. In *Proceedings of the 2000 Annual International Conference of the Association for the Education of Teachers in Science*, Akron, OH, Jan. 6-9 2000.
2. C. Asavathiratham, S. Roy, B. Lesieutre, and G. Verghese. The influence model. *IEEE Control Systems Magazine. Special Issue on Complex Systems*, 12, 2001.
3. R. F. Bales. *Personality and Interpersonal Behavior.* Rinehart and Winston, New York, 1970.
4. K. D. Benne and P. Sheats. Functional roles of group members. *Journal of Social Issues*, 4:41–49, 1948.
5. D. Boud, R. Keogh, and D. Walker, editors. *Reflection: Turning Experience into Learning*. Kogan Page, London, 1988.
6. C. Callaway, E. Not, and O. Stock. Report generation for post-visit summaries in museum environments. In O. Stock and M. Zancanaro, editors, *PEACH: Intelligent Interfaces for Museum Visits*. Springer, New York, 2007.
7. G. Carli and G. Gretter. A Start-End Point Detection Algorithm for a Real-Time Acoustic Front-End based on DSP32C VME Board. In *Proceedings of ICSPAT*, Boston, 1992.
8. P. Chippendale. Towards automatic body language annotation. In *7th IEEE International Conference on Automatic Face and Gesture Recognition, FG06*, pages 487–492, Southampton, UK, Apr. 2006.
9. J. DiMicco, A. Pandolfo, and W. Bender. Influencing group participation with a shared display. In *CSCW04: Proceedings of the ACM Conference on Computer Supported Cooperative Work*, pages 614–623, 2004.
10. W. Dong, B. Lepri, A. Cappelletti, A. Pentland, F. Pianesi, and M. Zancanaro. Using the influence model to recognize functional roles in meetings. In *In Proceedings of International Conference on Multimodal Interaction ICMI2007*, Nagoya, Japan, Nov. 2007.
11. P. Dourish and S. Bly. Portholes: Supporting awareness in a distributed work group. In *Proceedings of the ACM Conference on Human Factors in Computer Systems CHI'92.*, 1992.
12. T. Erickson, C. Halverson, W. A. Kellogg, M. Laff, and T. Wolf. Social translucence: designing social infrastructures that make collective activity visible. *Commununications of the ACM*, 45(4):40–44, 2002.

13. B. Garrison. The perceived credibility of electronic mail in newspaper newsgathering. In *In Proceedings of Communication Technology and Policy Division, Association for Educational in Journalism and Mass Communication Midwinter Conference*, Boulder, CO, Mar. 2003.
14. J. W. Hall and W. H. Watson. The effects of a normative intervention on group decision-making performance. *In Human Relations*, 23(4):299–317, 1970.
15. M. A. K. Halliday and R. Hasan. *Language, Context and Text: Aspects of Language in a Social-Semiotic Perspective*. Deakin University Press, 1985.
16. D. Katz and R. L. Kahn. *The Social Psychology of Organizations*. John Wiley, New York, 1978.
17. D. Maloney-Krichmar and J. J. Preece. The meaning of an online health community in the lives of its members: Roles, relationships and group dynamics social implications of information and communication technology. In *Proceedings of the International Symposium on Technology and Society ISTAS'02*, 2002.
18. K. R. McKeown. *Text Generation: Using Discourse Strategies and Focus Constraints to Generate Natural Language Text*. Cambridge University Press, 1985.
19. F. Pianesi, M. Zancanaro, B. Lepri, and A. Cappelletti. Multimodal annotated corpora of consensus decision making meetings. *The Journal of Language Resources and Evaluation*, 41(3-4):409–429, December 2007.
20. F. Pianesi, M. Zancanaro, E. Not, C. Leonardi, V. Falcon, and B. Lepri. Multimodal support to group dynamics. *Personal and Ubiquitous Computing*, 12(2), 2008.
21. A. Salazar. An analysis of the development and evolution of roles in the small group. *Small Group Research*, 27(4):475–503, 1996.
22. J. C. Tang. Finding from observational studies of collaborative work. *International Journal of Man-Machine Studies*, 34(2):143–160, 1991.
23. J. Wilson and D. Rosenberg. Rapid prototyping for user interface design. In M. Helander, editor, *Handbook of Human-Computer Interaction*, pages 859–875. New York, 1988.
24. M. Zancanaro, B. Lepri, and F. Pianesi. Automatic detection of group functional roles in face to face interactions. In *Proceedings of the International Conference of Multimodal Interfaces, ICMI-06*, 2006.

Part IV

The CHIL Reference Architecture

22
Introduction

Nikolaos Dimakis, John Soldatos, Lazaros Polymenakos

Athens Information Technology, Peania, Attiki,Greece

In this chapter, we introduce the motivation behind developing the CHIL Reference Architecture, which is an architectural framework along with a set of middleware elements facilitating the integration of perceptual components, sensors, actuators, and context modeling scripts. The framework and the associated middleware elements facilitate the integration and assembly of sophisticated ubiquitous computing applications in smart spaces. Specifically, they mitigate the integration issues arising from the distributed and heterogeneous nature of pervasive, ubiquitous, and context-aware computing environments. The CHIL Reference Architecture places special emphasis on the integration of perceptual components contributed by a variety of technology providers, which has not been adequately addressed in legacy middleware architectures. Therefore, legacy architectures are briefly reviewed in this chapter in order to highlight the innovative elements of the CHIL Reference Architecture. Furthermore, this chapter outlines the following chapters that describe and evaluate all the parts of the CHIL Reference Architecture.

22.1 Motivation for the CHIL Software Architecture

The vision of the CHIL project is to provide context-aware human-centric services that will operate in the background, provide assistance to the participants in CHIL spaces, and undertake tedious tasks in an unobtrusive fashion. To achieve this, significant effort has to be put in designing efficient context extraction components so that the CHIL system can acquire an accurate perspective of the current state of the CHIL space. However, the CHIL services require a much more sophisticated modeling of the actual event rather than simple and fluctuating impressions of it. In addition, an intelligent way of managing the sensors and actuators should be defined since context is derived from processing continuous sensor streams. Finally, by nature, the CHIL spaces are highly dynamic and heterogeneous; people join in or leave, sensors fail or are restarted, user devices connect to, or disconnect from, the network, numerous storage repositories and databases are used, etc. To manage this diverse infrastructure, sophisticated techniques should be defined that can map all entities present in the CHIL system and to provide information to all other components that may require it.

From these facts, one can easily understand that in addition to highly sophisticated components at an individual level, another mechanism (or a family of mechanisms) should be developed that can handle this infrastructure. The CHIL Reference Architecture for multimodal systems lies in the background and provides the solid high-performance and robust backbone of the CHIL services. Each individual need is assigned to a specially designed and integrated layer that is docked to the individual component and provides all the necessary actions to enable the component to be plugged into the CHIL framework.

22.2 Related Work

The area of architectural frameworks supporting ubiquitous computing projects is not a recent trend. Major pervasive and ubiquitous computing projects have placed significant effort in designing such platforms, which can facilitate integration, debugging, development, management, and eventually deployment of the end services.

Examples of prominent attempts in the field of software architectures for context-aware services include the Context Toolkit [3], which advocates a design-time approach to building context-aware applications, and the Interactive Workspaces project at Stanford University, which led to the design and implementation of the Interactive Room Operating System (iROS) [6]. Similar to iROS, the Oxygen Project at MIT produced the MetaGlue system [2], which constitutes a highly robust software platform based on a multiagent schema. Context-awareness is achieved using the GOALS architecture [8], which is the evolution of the MetaGlue system.

The Smart Flow middleware [7] developed by NIST is a specialized middleware that targets the sensor data transmission. It initiates a mechanism of transporting large amounts of data from sensors such as cameras, microphones, etc. to distributed nodes throughout the network, using a flexible API. The middleware generates abstract "flows" instead of transporting raw data. Apart from the data itself, these flows include information about the data frame rate, resolution, etc. This makes it easy for the other components to know the nature of the data they are receiving. The Smart Flow Control Center is the centralized point of control that undertakes the management of the data flows.

More recent attempts are the EasyLiving system developed at Microsoft Research [1, 9], a system that has also produced an architecture that enables the coordination of the devices and facilitates the fusion of contextual information, and Carnegie Mellon's Aura project [4]. Aura targets wireless, wearable, or handheld computers, and smart spaces. Aura provides all the required software architectural models that monitor an application and guide dynamic changes to it.

One of the most recent attempts for a unified ubiquitous computing architecture is the T-Engine [5], operated and managed by the T-Engine Forum [11]. T-Engine enables the distribution of software resources, including middleware developed on

T-Kernel, its compact, real-time operating system. The platform also features standardized hardware and tamper-resistant network security and enables developers to rapidly build ubiquitous computing solutions by using off-the-shelf components. T-Engine is a sophisticated platform that coordinates components designed and built using the standard T-Engine boards and focuses on embedded systems.

22.2.1 Shortcomings and Limitations

Each of the previous architectural suggestions, though individually performing according to the requirements they were assigned to meet, lacks one key "ingredient": a fully fledged proposal on how to tackle the needs that arise in modern, highly heterogeneous, multimodal, context-aware systems. Each proposal does not try to touch upon all the required layers that are needed, instead it focuses on specialized areas in service delivery, context mapping, data streaming, etc. Moreover, the majority of the current architectures, e.g., T-Engine and Easy Living, are highly application-oriented and require significant changes, both in source code and in service design, to cover the needs of other problems.

22.3 Benefits of the CHIL Reference Architecture

The CHIL Reference Architecture for multimodal perceptual systems suggests an architecture that covers all of the individual stages of information evolution: from raw sensor data to situation recognition and to end-service delivery. It provides techniques for handling the sensor infrastructure, subscribing to the sensor stream, formulating and transmitting the contextual information, as well as techniques on how to "interpret" the elementary context to sophisticated situation recognition, driving the service logic.

Moreover, it is designed as a breadboard for service development [10], which enables the service designer to easily plug new services into the framework, connect it to the rest of the framework, subscribe to receiving contextual data, make use of the rest of the infrastructure (file systems, databases, etc.), and trigger actuating logic. Finally, the CHIL Reference Architecture is equipped with autonomous mechanisms in both the sensor control and the service layer. These mechanisms have been incorporated in the core architecture implementation and have been tested and evaluated on numerous occasions.

22.4 The CHIL Architecture and the Demands of Perceptual Systems

The CHIL Architecture aims to provide a system-wide architecture for perceptual systems. It places significant emphasis on separating the layers that play a vital role in the perceptive aspect of the system such as the context acquisition layer (named the "perceptual components layer" in the CHIL Architecture), and the situation modeling layer. Each of these layers contributes significantly to the intelligence of the CHIL system: On the one hand, the perceptual components generate continuous streams of elementary context, such as the location of people, the identities of the participants in the CHIL space, the location of the speaker, etc.; on the other hand, the situation modeling layer undertakes the task of fusing these series of elementary contexts and formulating complex situations, which more accurately describe the current status of the CHIL space, such as "meeting started", "question from Bob", etc.

22.5 Overview of Part IV Chapters

In the following chapters, we describe in detail all the building blocks that make up the CHIL Reference Architecture for multimodal perceptual systems. As mentioned, the CHIL Reference Architecture, tackles the needs of these systems by introducing specialized "layers" for individual roles. We briefly introduce these layers in this section.

22.5.1 Sensor Data Streaming

In a typical CHIL smart space, it all begins from the sensor layer, as this layer is responsible for generating and forwarding the raw, uncompressed data to the next layer of processing, the context acquisition layer. Since the sensors are key in the context acquisition process, the way that they are handled –as well as the way their sensitive data are managed – plays a significant role in the performance of the end-user services. For the purpose of sensor control, the ChilFlow middleware layer undertakes all important tasks that are needed so that each sensor can provide, on demand, a constant flow of data. It models the sensor data as flows that are accompanied by sensor-specific information such as frame rate, resolution, compression (if any), etc. Each sensor driver, when locking a sensor device, initiates a flow transmission. Instead of waiting for a single component to connect and transmit the content, it forwards the sensor data (modeled as a flow) to any component that has subscribed to this flow using the ChilFlow middleware. ChilFlow is described in detail in Chapter 24.

22.5.2 Context Extraction and Formulation

The CHIL Architecture aimed from the early stages of the project to devise a series of rules that could enable each CHIL perceptual component, regardless of developing platform, operating system, implementation language, etc., to be easily exchanged with other CHIL perceptual components. Despite the fact that this may appear to be a simple task, one should consider the range of these components that are researched and developed in more than five of the Consortium's partners, all of which have their own preferences in designing, implementing, and packaging such components. Moreover, one should also consider that CHIL has in its key targets the development of context-aware services to support the end users, and not just having complicated algorithms operating alone. Furthermore, the fact that each software developing platform (as is the case for all Consortium partners) handles the sensor layer in a not-too-similar way introduces significant problems that would be difficult to tackle, such as component interchange or service portability (as they depend on perceptual components). Finally, the new extensions to meet these demands for the perceptual components should not introduce processing overhead, as these components operate on highly sensitive data, and their performance should remain real-time. In Chapter 25, we outline our approach in creating a CHIL-wide perceptual component that meets all of the above needs. We outline the needs for the CHIL compliance and present a set of APIs that facilitated this component exchange capability, which is a key aspect of the CHIL Architecture.

22.5.3 Situation Recognition and Situation Modeling

Human-centric, context-aware services depend on sophisticated context that reflects the current situation of the smart spaces. Simple information such as "person at (x, y)", or "sound at (x, y)" is of little value in the scope of providing useful services to the end users. In CHIL, the situation recognition and modeling provides sophisticated, high-level context to the CHIL services, which apply their service logic based on these contextual streams. Examples of high -level contextual representations are concepts such as "meeting is starting", "Bob is at the presentation area", "question from Alice", etc. To be able to model such events, significant effort has been put in the CHIL Architecture to place this modeling process at the very heart of the architecture. The situation modeling layer, described in Chapter 26, outlines the approaches that were followed (Bayesian networks, probabilistic models, etc) to model events that take place in the CHIL spaces such as lectures and meetings.

22.5.4 The CHIL World Model

In environments as heterogeneous and dynamic as the CHIL spaces, the presence of a central, globally accessible, intelligent "library" whose task is to maintain information about everything in the smart space is of vital importance. This world model should be controlled by intelligent algorithms able to answer sophisticated queries that could be needed at any time. A preferred option is to be able to support ontological reasoning so that it could, by using inference rules, make implicit knowledge explicit. The CHIL world model is managed by the CHIL ontology and the knowledge base server, a powerful combination of a highly detailed ontology model with a robust, accurate, and lightweight tool, presented in detail in Chapter 27.

22.5.5 Context-Aware Services

The CHIL context-aware services are designed and implemented using the CHIL agent framework, a powerful extension of the JADE framework for multiagent systems design. The CHIL services fully exploit the context generated by the perceptual components tier as well as the more sophisticated context of the situation modeling layer. The context is formulated in events that are broadcast in the CHIL framework. Following the reception of these events, the application logic is applied based on the nature of the current status in the CHIL space as described in the current state that the situation modeling component has determined. Furthermore, by design, the CHIL Architecture enables the numerous services to be plugged into the same framework, further boosting the usability of the resources. In Chapter 28, we detail the capability of this framework.

References

1. B. Brumitt, J. Krumm, B. Meyers, and S. Shafer. Ubiquitous computing and the role of geometry. *IEEE Personal Communications*, 7(5):41–43, 2000.
2. M. Coen, B. Phillips, N. Warshawsky, L. Weisman, S. Peters, and P. Finin. Meeting the computational needs of intelligent environments: The Metaglue System. In *First International Workshop on Managing Interactions in Smart Environments*, pages 201–212, Dec. 1999.
3. A. K. Dey, G. D. Abowd, and D. Salber. A conceptual framework and a toolkit for supporting the rapid prototyping of context-aware applications. *Human Computer Interaction*, 16(2-4):97–166, 2001.
4. D. Garlan, D. Siewiorek, A. Smailagic, and P. Steenkiste. Project Aura: Towards distraction-free pervasive computing. *IEEE Pervasive Computing*, 21(2):22–31, Apr.-Jun. 2002.
5. J. Krikke. T-Engine: Japan's ubiquitous computing architecture is ready for prime time. *IEEE Pervasive Computing*, 4(2):4–9, 2005.
6. S. R. Ponnekanti, B. Johanson, E. Kiciman, and A. Fox. Portability, extensibility and robustness in iROS. In *IEEE International Conference on Pervasive Computing and Communications*, pages 11–19, 2003.
7. L. Rosenthal and V. M. Stanford. NIST smart space: Pervasive computing initiative. In *WETICE '00: Proceedings of the 9th IEEE International Workshops on Enabling Technologies*, pages 6–11, Washington, DC, USA, 2000. IEEE Computer Society.
8. U. Saif, H. Pham, J. M. Paluska, J. Waterman, C. Terman, and S. Ward. A case for goal-oriented programming semantics. In *Fifth Annual Conference on Ubiquitous Computing (UbiComp '03)*, pages 74–83, 2003.
9. S. Shafer, J. Krumm, B. Brumitt, B. Mayers, M. Czerwinski, and D. Robbins. The new EasyLiving project at Microsoft Research. In *Proceedings of the 1998 DARPA/NIST Smart Spaces Workshop*, pages 127–130, 1998.
10. J. Soldatos, N. Dimakis, K. Stamatis, and L. Polymenakos. A breadboard architecture for pervasive context-aware services in smart spaces: middleware components and prototype applications. *Personal and Ubiquitous Computing Journal*, 11(2):193–212, Mar. 2007.
11. T-Engine. The T-Engine Forum. `http://www.t-engine.org/index.html`.

23

The CHIL Reference Model Architecture for Multimodal Perceptual Systems

Gerhard Sutschet

Fraunhofer Institute IITB, Karlsruhe, Germany

Due to the scale of the CHIL project, with its large number of partners contributing to the CHIL system with a diversity of technical components such as services and perceptual components, as well as their complexity, a flexible architecture that facilitates integration of components at different levels is essential. A layered architecture model was found to best meet these requirements and to allow a structured method for interfacing with sensors, integrating technology components, processing sensorial input, and composing services as collections of basic services.

Furthermore, the architecture described here supports the flexible exchange of components and the replacement of unavailable components with simulators through well-defined interfaces.

23.1 The CHIL Layer Model

In order to realize proactive and intelligent services, both context-delivering and collaborating components have to be integrated. In CHIL, context is delivered by perceptual components and learning modules. Perceptual components continuously track human activities, using all perception modalities available, and build static and dynamic models of the scene. Learning modules within the agents model the concepts and relations within the ontology. Collaboration is enabled by a set of intelligent software agents communicating on a semantic level with each other.

In order to facilitate integration of these various software components, which are situated on different functional levels (e.g., audio and video streaming functionality, perceptual components, sensor control software, situation modeling software components, software components implementing the services, user interfaces, etc.), and to reduce the complexity of this integration work, a layered architecture was developed (see Fig. 23.1).

The global architecture described here is mapped to the following layered model (see Fig. 23.1). It consists of eight horizontal layers, the ontology as a vertical layer, accessible by all components and another vertical layer, "CHIL utilities" which provides global timing and other basic services relevant to all layers.

While the lower layers deal with the management and interpretation of continuous streams of video and audio signals in terms of event detection and situation adaptation, the upper layers of the architecture enable reasoning and management of a variety of services and user-interfacing devices.

The upper layers of the CHIL architecture ("User front-end" and "Service and control") are implemented via software agents. Jade (Java Agent DEvelopment Framework) was chosen as the agent platform.

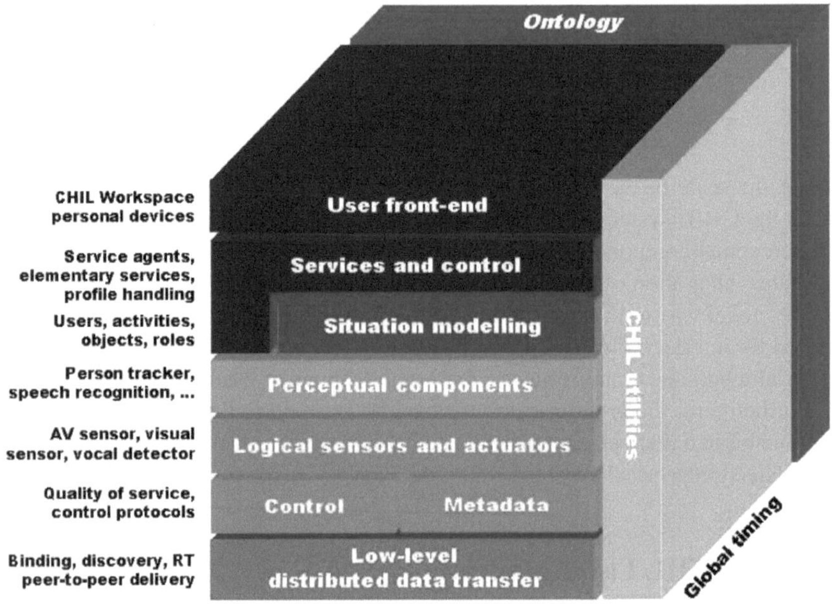

Fig. 23.1. CHIL layer model.

23.2 Brief Layer Description

23.2.1 User Front-End Layer

The user front-end layer implements all user-related components such as the personal agents, device agents, and the user profile of a CHIL user and the components responsible for the direct interaction with the user. The personal agent acts as the user's personal assistant, taking care of his demands. It interacts with its master through personal devices (e.g., notebook, PDA, smartphone) that are represented by corresponding device agents. The device agents are the controlling interfaces to the personal devices.

The personal agent also provides and controls access to its master's profile and preferences, thus ensuring user data privacy. The user profile stores personal data and service-related preferences of a user.

Another user interface issue that does not directly concern the CHIL layer model is the implementation of spoken interfaces. As they are user interfaces, they have been implemented on the user interface layer.

23.2.2 Service and Control Layer

The service and control layer contains the components that implement the different services provided by CHIL as well as service-to-service information exchange and user-service communication. The services are implemented by software agents. The service and control layer comprises both the service agents and their management. The interaction with other agents within this layer and the user front-end layer uses the communication mechanisms of the agent platform, while communication with the other layers uses internal mechanisms.

Services are reusable components that constitute elementary services and the CHIL nonobtrusive services such as the Memory Jog (see Chapter 18), the Relational Cockpit (Chapter 20), or the Connector (Chapter 19) services, which are composed of suitable elementary services.

Further details on this layer and how the agent framework can be used to build scalable services will be discussed in Chapter 28.

23.2.3 Situation Modeling Layer

The situation modeling layer is where the situation context received from audio and video sensors is processed and modeled. The context information acquired by the components at this layer helps services to respond better to varying user activities and environmental changes. For example, the situating modeling layer answers questions such as, "Is there a meeting going on in the smart room?" "Who is the person speaking at the whiteboard?" "Has this person been in the room before?"

This layer is also a collection of abstractions representing the environmental context in which the user interacts with the application. Ideally, it should act as a database that maintains an up-to-date state of objects (people, artifacts, situations) and their relationships. The situation model itself acts as a kind of directed inference engine that will watch for the occurrence of certain situations in the current environment. The searched-for situations are dictated by the needs of active services, such as the detection of the interruptibility level of a particular person in a room for whom a service has to monitor incoming calls.

In the CHIL Reference Model Architecture, the situation model takes its information from the perceptual components and produces a set of situation events that are provided to services. There was no mandatory particular implementation of such a model, and therefore we have provided a framework for handling situation models called SitCom (Situation Composer). The situation modeling layer is described in further detail in Chapter 26.

23.2.4 Perceptual Components Layer

Perceptual components are sophisticated software components that operate on a sensor output (or on a group of sensors' outputs), process it, and extract information related to people's actions. Such information may be the people's locations, IDs, hand gestures, pose recognition, etc. (see also Part II of this book). Perceptual components developed in the scope of the CHIL project are listed in the CHIL Technology Catalog available at the CHIL website, `http://chil.server.de`.

All perceptual components implemented in the scope of the CHIL project are designed based on the CHIL Compliant Perceptual Component guidelines, which specify how these perceptual components operate, "advertise" themselves, subscribe to receiving a specific sensor data stream, and forward their extracted context to the higher layer of the CHIL Reference Architecture. Figure 23.2 illustrates the structuring principles of the CHIL Compliant Perceptual Component.

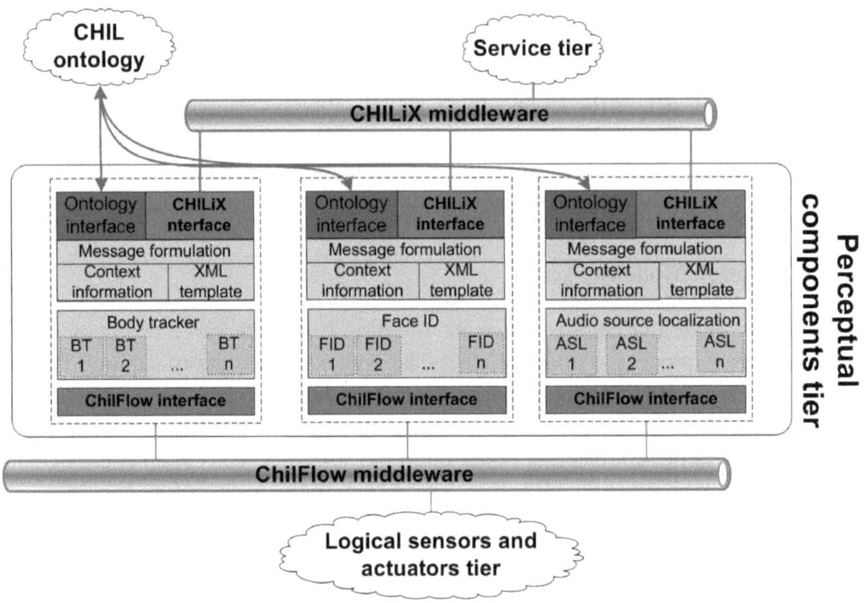

Fig. 23.2. The CHIL perceptual components tier.

The CHIL Compliant Perceptual Component is an extension of the perceptual components. It incorporates all the interfaces for communicating with the sensors, the CHIL ontology (using the knowledge base server), and service tier. These interfaces are the ChilFlow interface for gluing this tier with the logical sensors tier, and the *CHiLiX* middleware, which acts as the communication channel between the perceptual components and the situation modeling layer of the CHIL architecture. *CHiLiX* facilitates the transmission of the contextual information to the next tier by following an XML schema that encapsulates the gathered information (context) and transmits the message in either a synchronous or asynchronous manner. This capability of *CHiLiX* makes it ideal for the perceptual components tier as it is expected that all components belonging in this tier operate asynchronously; thus, a mechanism that ensures robust communication between the tiers is of high importance.

The benefits of using this expanded option in the CHIL architectural framework is that by complying with this set of rules, component interchange is possible since the flexibility of the CHIL Reference Model Architecture caters for all aspects of the communication among components. Further details of this layer will be presented in Chapter 25.

23.2.5 Logical Sensors/Actuators Layer

Sensors and actuators are key in the design and implementation of multimodal perceptual services. They act as the "eyes and ears" of the architecture and provide a continuous flow of output data to the processing components, which extract pertinent information by applying sophisticated algorithms able to extract elementary context. This layer comprises several abstractions that wrap the sensor control and transmission components for each of the sensors in the smart space. Figure 23.3 illustrates how a logical sensor can be formed for sensors (both acoustic and visual) and actuating devices such as projectors, displays, speakers, and targeted audio devices.

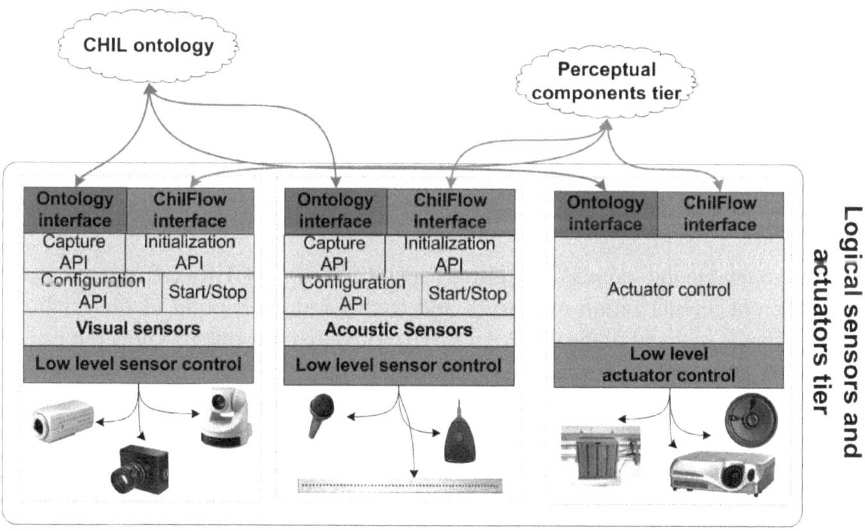

Fig. 23.3. The CHIL logical sensors and actuators tier.

Each sensor is controlled by a specified sensor controller, which provides the low-level commands to the sensor. These components produce a continuous flow of information that is transmitted using the ChilFlow middleware to any consuming component belonging in the CHIL system. This communication is achieved by the incorporation of the ChilFlow Interface, which the logical sensor implements. This interface "glues" this tier with the following one, the perceptual components tier, which is the context acquisition tier of the multimodal perceptual system and is described in the following chapter. Finally, each logical sensor is able to communicate with the framework's knowledge base, where it can register and "advertise" itself to the rest of the framework.

23.2.6 Control/Metadata Layer

Control and metadata provide mechanisms for data annotation, synchronous and asynchronous system control, data flow synchronization, effective storing and searching of multimedia content, and metadata generated by data sources.

23.2.7 Low-Level Distributed Data Transfer

The low-level distributed data transfer layer of the CHIL Architecture model is responsible for transferring high-volume and/or high-frequency data from sensors to perceptual components or between perceptual components. The NIST Smart Flow System was deployed for bootstrapping the work on this layer, but now it is implemented by the ChilFlow data transfer middleware, which is further described in Chapter 24. These systems are heavily used by developers of perceptual components to distribute their components over several networked computers to exploit more computational power. To free developers from handling networking issues, ChilFlow offers an easy-to-master yet powerful object-oriented programming interface, which provides type-safe network-transparent communication methods between perceptual components.

23.2.8 Ontology

In order to enable the intended cognitive capabilities of the CHIL software environment, the conceptualization of entities and formalized description of the relations among them is realized in the ontology. CHIL software components know the meaning of the data they are operating on and expose their functionality according to a common classification scheme. The ontology consists of two major parts:

- the CHIL domain ontology for the conceptualization of the CHIL world,
- the CHIL communication ontology, used for interagent communication.

The ontology layer in CHIL consists of the CHIL ontology and the CHIL knowledge base server. The CHIL ontology provides a high-level description of the CHIL domain of discourse that can be used to build intelligent applications by leveraging the expressiveness of the Web Ontology Language OWL. The CHIL knowledge base server was developed to make it particularly easy to manage Web Ontology Language OWL knowledge bases. Both are described in detail in Chapter 27.

24

Low-Level Distributed Data Transfer Layer: The ChilFlow Middleware

Gábor Szeder

Institute for Program Structures and Data Organization (IPD), University of Karlsruhe, Karlsruhe, Germany

To continuously observe, recognize, and interpret human activity and human-human interaction, computers must process a large amount of data captured by numerous sensors. Handling all the data acquired by different kinds of sensors, such as cameras, microphones, and microphone arrays, is rather computationally intensive: The processing power of several computers is required to perform this task in real time. Therefore, perceptual components must run distributed over multiple networked computers.

However, distributing perceptual components over multiple computers presents numerous difficulties. Figure 24.1 shows a sample arrangement of perceptual components that together realize an audiovisual speech recognition system. The sensors deployed in this scenario can be found on the left side of the picture. The sensors are a 16-channel and a 64-channel microphone array for source localization and far-field speech recognition, respectively, and a pan-tilt-zoom camera (PTZ) for lip reading. These sensors are connected to different computers and the data from the sensors are received by capture components running on these computers. Capture components and perceptual components are represented by squares. Communication channels are illustrated with thick lines with an arrowhead showing the direction of the data flow. The sample component arrangement in Fig. 24.1 shows some of the main characteristics of the network of collaborating perceptual components. Data acquired by a sensor might be processed by more than one component at the same time; for example, both the "face tracker" and "lip reader" components are processing the images captured by the "PTZ camera capture" component. Similarly, the output of a perceptual component is in some cases required by multiple components, like the output of the "audio source localizer" and "beamformer" components. Moreover, each component might have more than one input or output, like the "lip reader", "beamformer", and "audiovisual speech recognizer" components.

During the development of a system like the one in Fig. 24.1, the configuration of the system changes frequently. Such a system is mostly developed offline; i.e., previously recorded data are used for testing, fine-tuning, and experiments instead of real-time captured sensor data. The system in Fig. 24.1, for example, could be easily turned into an offline system by replacing the three capture components with

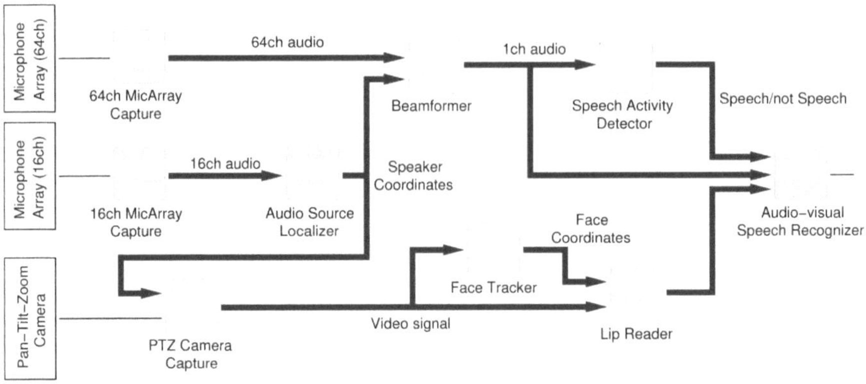

Fig. 24.1. Example arrangement of perceptual components realizing audiovisual speech recognition.

components replaying previously recorded data from hard disks. On the other hand, sometimes it is preferable for a developer to work with an individual perceptual component in isolation, independently from other components. In such cases, all inputs of that component are replayed, even those that are the output of other perceptual components. This implies that perceptual components receive input from different components in different configurations: sometimes from real sensors or perceptual components, but sometimes from replaying or simulated components. Moreover, adjacent perceptual components might run on the same or different computers, which in turn might vary with changing configurations.

Developers of perceptual components want to stay focused on enhancing processing algorithms instead of managing the connections between components and the caveats of networking. Therefore, such a communication middleware is required that is inherently network-transparent and copes with the complexity of connecting and managing a large number of perceptual components in different configurations. To facilitate the development of perceptual components even at the early stages of the CHIL project, we deployed the NIST Smart Data Flow System [1], because it solved several important issues mentioned above. However, as this system could not satisfy all of our requirements, we started to develop ChilFlow [3], our own communication middleware.

24.1 Flows

ChilFlow offers network-transparent, typed, one-to-many communication channels called *flows* for data exchange between perceptual components. Flows provide a higher abstraction level over ordinary network connections (e.g., TCP sockets) that fits the communication characteristics of components and simplifies the work of the components' developers significantly.

- *Network-transparent*: To hide the ever-changing addresses (hostname, IP address, port) of source and consumer perceptual components, flows are identified by an abstract name. This name is specified at flow creation time by its source component. Flow names must be unique in the system. Consumer components can subscribe to a particular flow by requesting it based on its name. In turn, ChilFlow will transparently connect the consumer to the source of the requested flow. This way developers of consumer components do not need to worry about from which source their component will receive input, where this source component runs, and how to connect to it.
- *Typed*: To prevent misinterpretation of the transferred data, flows are typed at two levels. First, there is a distinction between various types of flows each having specific parameters. For example, the parameters of an audio flow are the number of channels, frequency, and number of frames in a buffer, while a video flow has horizontal and vertical resolution and color channels (e.g., RGB, monochrome). Second, the type of the data is distinguished, too. This data type determines the bit depth of audio and the color depth of video flows.
  The flow's type, parameters, and data type must be defined together with the flow's name at the flow's creation by its source component. When a consumer component requests a particular flow by its name, it also has to specify the flow's type and parameters and the data type it wants to receive. If the flow with the given name has a different type or parameters or data type, then the consumer will not receive that flow.
- *One-to-many*: It is quite common that data from a single source must be processed by more than one independent consumer component at the same time. For example, video from a camera might be processed by a body tracker, a face tracker, a face recognizer, and a lip reader component as well. Flows support one-to-many communication transparently. The programmers of data source components do not need to take explicit actions to handle the multiple and possibly changing consumer components.

To adapt to the practice of perceptual component developers, data are transferred through flows in abstract units developers usually work with. Such a unit is a whole frame from a camera in case of a video flow, or a speaker's 3D coordinates with confidence value. These units are called *buffers*. Apart from the actual data, buffers also encapsulate some basic metadata (e.g., timestamp).

Consumer components have a buffer history for every received flow, where a couple of recently received buffers are stored. A newly arriving buffer is automatically stored in the history, replacing the oldest one. Buffers are then pulled from the buffer history in turn for processing. This buffer history can help compensate temporary slowdowns of the consumer component, as it prevents data loss to a certain degree. Furthermore, the buffer history is particularly useful by such a component arrangement as the "PTZ camera capture", "face tracker", and "lip reader" triple in Fig. 24.1. In this case, the output of "PTZ camera capture" is processed by the other two components. However, to process an image in a buffer received on this flow, the "lip reader" component also needs the coordinates of a face within that image. These coordinates are extracted from the same image by the “face tracker” component, and outputted through buffers of a vector flow. Since the “face tracker” obviously needs some time to process its input, the buffers containing the resulting coordinates will arrive with a short delay at “lip reader”. If the “lip reader” component stores some of the recently arrived video buffers in the buffer history, it will be able to find the matching video and vector buffers.

Consumer perceptual components are enabled to select only a subset of the data sent by the source component. For example, in case of a multichannel audio flow, the consumer can specify which particular audio channels it wants to receive. In case of a video flow, consumers can select a subregion of the images sent through the flow. In order to save bandwidth, the source component sends only the selected subset of the data through the network. However, handling selections is still completely transparent to the source component. There are no explicit actions needed at all on the source side to handle the different and changing selections of the different consumer components.

24.2 ChilFlow’s Architecture

ChilFlow consists of three different components: daemons, controlling applications, and a client library.

Daemons are primarily responsible for organizing the distributed system and for coordinating connections between perceptual components. There is a daemon process running on each computer of the system connected to other daemons via TCP sockets building an overlay network. Daemons share information about all perceptual components (e.g., their network address) and flows (e.g., their name, parameters, source component, etc.) present in the system. When a consumer component requests a flow, daemons use this information to check the flow’s parameters and to look up the address of the flow’s source component. Daemons do not participate in data exchange through flows, and if a daemon fails, it will not affect connections between source and consumer components. Daemons are also responsible for fulfilling requests of control applications by starting or stopping daemons and perceptual components. Furthermore, daemons notify control applications about various events in the system, such as creation of a flow, exiting component, etc., and forward the standard output and error of perceptual components to control applications.

A *controlling application* serves visualization and management purposes. It offers a graphical interface for convenient starting, stopping, and monitoring daemons and processing components on different computers in the distributed system. It shows a map of the running components and the flows between them similar to Fig. 24.1. Controlling applications also include plug-ins for visualizing the actual data transferred through flows. The system can be controlled by multiple controlling applications run by multiple users at the same time, a significant asset when multiple developers want to use the same sensor setup.

The *client library* is used by the perceptual components to communicate with other components of the system and to send and receive data through flows. To hide networking from the developers and avoid unfavorable interferences between processing data and networking, the library has a multithreaded design. A background thread handles incoming messages, such as accepting new consumer components, registering their selections, handling control messages from daemons, and receiving data on input flows, while the main thread processes the received data. The background thread is hidden from the developer and does not interfere with the main thread, except the critical sections necessary for accessing shared data structures, such as the buffer history. This approach decouples the processing in the consumer component from the source component(s). Even if the main thread of a consumer component if blocked or long in processing a buffer, the background thread still handles incoming messages and buffers and avoids the blocking of other components of the system.

24.3 Programming Interface

ChilFlow's API for flows is based on the push-pull model for sending and receiving buffers. For illustrating the basic functions of this API, Fig. 24.2 and 24.3 show the source code of a simple example source component with a video flow output and a consumer component with a video flow input, respectively. These two components build a processing pipeline of two stages: The source component captures data from a video camera or reads previously recorded data, while the consumer processes the data.

```
#include <chilflow.h>
using namespace ChilFlow;
extern void fill_buffer(
    VideoOutputFlow<unsigned char>::Buffer & buffer);
int main(int argc, char * argv[])
{
    unsigned int width = 640, height = 400;
    Main * cf = Main::create(argc, argv, "sample source component");
    VideoOutputFlow<unsigned char> * flow
        = cf->create_video_flow<unsigned char>("sample video flow",
            VIDEO_RGB, width, height);
    while (cf->is_running()) {
        VideoOutputFlow<unsigned char>::Buffer * buffer
            = flow->get_buffer();
        fill_buffer(*buffer);
        buffer->send();
    }
    flow->close();
    cf->finalize();
    return 0;
}
```

Fig. 24.2. An example source component with a video flow output.

ChilFlow's classes are defined in the chilflow.h header file, where everything resides in the ChilFlow namespace (lines 1 and 2). The initialization of the client library is done in line 8 by calling Main::create(). This single call encapsulates the initialization of the networking back end, creation of the TCP socket for future flow data transfer and establishing connection with the local daemon through its socket. The last string argument of this method is the name of the perceptual component. This string will appear in the map of controlling applications. The returned Main object is a handle for the connection to the daemon and the entry point for creating and subscribing to flows.

The video flow for the output of this component is created by the template method Main::create_video_flow<T>() in line 10. The method name indicates the type of the flow explicitly. The first string argument specifies the name of the flow, while the remaining arguments specify the parameters of the flow (color channels and resolution). The template parameter unsigned char defines the primitive data type representing the actual data. The data type is determined inside the client library by using run-time type information. In this case, the unsigned char data type means that each color channel in a pixel is represented by eight bits. The returned VideoOutputFlow<T> object stands for the source-side endpoint of the flow. There are other methods and classes to work with other types of flows, for example, Main::create_vector_flow() or AudioOutputFlow<T>.

The real work is done in the while loop in lines 12–17. The helper method Main::is_running() returns true as long as this component is not stopped by the user through a controlling application. Lines 13–14 retrieve an output buffer from the buffer history. This buffer will be filled with data in the fill_buffer() function. Providing this function is the developer's responsibility. Line 16 sends the buffer to all consumers subscribed to this flow. If any of the customers has selected only a subregion of the frames, then this Buffer::send() method will send only that selected subregion to reduce network load. If the buffer was not provided with a timestamp in the fill_buffer() function, the current time will be automatically assigned.

Once the component is stopped, it should exit gracefully. The flow is closed by calling VideoOutputFlow<T>::close() in line 18. This method will also send a message to all consumer components and the local daemon to notify them about the close event. Finally, the Main::finalize() method in line 19 disconnects the component from the local daemon and frees all resources used by the library.

Note that there is no indication of handling consumer components in any way. They are handled completely transparently in a background thread, as described in the previous section. Also note that both the type and data type of the flow are reflected throughout the code from the creation of the flow to the method signature of the fill_buffer() function (lines 3–4).

The structure of the consumer component shown in Fig. 24.3 is quite similar to the structure of the source component; hence, only the differences will be described.

The component subscribes to a flow for input by calling the template method Main::subscribe_video_flow<T>() in line 10. The method name indi-

```
1 #include <chilflow.h>
2 using namespace ChilFlow;
3 extern void process_buffer(
4     VideoInputFlow<unsigned char>::Buffer & buffer);
5 int main(int argc, char * argv[])
6 {
7     unsigned int width = 640, height = 480, history_size = 25;
8     Main * cf = Main::create(argc, argv, "sample consumer component");
9     VideoInputFlow<unsigned char> * flow
10        = cf->subscribe_video_flow<unsigned char>("sample video flow",
11            VIDEO_RGB, width, height, history_size);
12    while (cf->is_running()) {
13        VideoInputFlow<unsigned char>::Buffer * buffer
14            = flow->get_newest_buffer();
15        process_buffer(*buffer);
16        buffer->release();
17    }
18    flow->close();
19    cf->finalize();
20    return 0;
21 }
```

Fig. 24.3. An example consumer component with a video flow input.

cates the type of the flow explicitly. The first string argument specifies the name of the flow the component wants to receive. The `history_size` argument specifies the number of buffers that should be kept in the buffer history. The three remaining arguments specify what parameters the requested flow should have. Any of these parameters could be null, meaning that those parameters will equal the parameters specified by the source component. The template parameter `unsigned char` requests the primitive data type the data should be represented with. This single method call encapsulates all communication between the consumer component and the local daemon and the source component, respectively, that is required to establish the connection between the source and consumer components. If a flow with the given name does not exist, this method will block until one is created. If such a flow exists, but either its type or parameters or data type differ from what is requested, the component is not allowed to subscribe to the flow, and an exception is thrown. If everything matches, a `VideoInputFlow<T>` object is returned that stands for the consumer-side endpoint of the flow and provides methods for retrieving buffers from the buffer history easily.

Lines 13–14 retrieve the most recent buffer from the buffer history. If that buffer was already processed, this method will block until a new buffer arrives. The returned `Buffer` object not only contains the actual data sent by the source component, but also encapsulates brief information about the buffer itself, such as its timestamp or the dimensions of the selection in effect. Other methods are also available for retrieving buffers from the buffer history based on the buffer's timestamp or its position in the buffer history. The data in the retrieved buffer are processed by the `process_buffer()` function (lines 3–4). Providing this function is the developer's responsibility. After the processing is finished, the buffer must be released (line 16).

Note that there is no indication of any network address or the like where the component should connect to. Just like the source component, both the type and data type of the flow are reflected throughout the source code from the subscription to the flow to the method signature of the `process_buffer()` function. This way the accidental misinterpretation of received data is effectively prevented. Also note that merely two function calls were needed in each program to reach the point where the two components are connected.

When handling input on multiple flows simultaneously, the previously described `subscribe_*_flow()` and `get_*_buffer()` methods might well be inappropriate because of their blocking behavior. Therefore, ChilFlow's API also offers nonblocking methods and helper classes to facilitate handling multiple input flows. Both blocking `subscribe_*_flow()` and `get_*_buffer()` methods have nonblocking counterparts. These methods return immediately when the requested flow does not exist or the requested buffer has not arrived yet, respectively. One of the helper classes deals with synchronization of input on multiple flows. The consumer component can register which flows should be synchronized in a `Synchronizer` object. This object's `wait()` method will return when there are unprocessed buffers in the flows' buffer histories with the same timestamp. As the clocks of different computers are not perfectly synchronized, the exact match of timestamps is barely feasible. Hence, an optional threshold value can be specified that determines how large a timestamp difference is allowed to treat buffers that are synchronized.

A detailed description of ChilFlow's API can be found in [2].

24.4 Comparison with the NIST Smart Data Flow System

The NIST Smart Data Flow System (NSDFS) [1] is a data transfer system for smart rooms that was used in the CHIL project for bootstrapping the work on the low-level distributed data transfer layer and was deployed by several partners to facilitate the development of distributed perceptual components. We have chosen the NSDFS at early stages of the project, because it solved several important issues of distributing perceptual components over multiple computers. Although ChilFlow uses similar concepts, there are several important differences and improvements compared to the NSDFS.

ChilFlow's programming interface is very easy to use, is flexible, and offers strong type safety for flows. Merely around 20 lines of code are required to connect a simple source or consumer component in a network-transparent and type-safe way. When handling multiple input flows in a component simultaneously, developers can conveniently use the nonblocking methods and helper classes of the ChilFlow API; they don't have to resort to multiple threads for receiving flows and performing the actual processing. Also, consumer components are enabled to select a subset of the data sent by the source component.

Furthermore, ChilFlow offers great flexibility in managing the distributed system; e.g., it supports multiple simultaneous users, it allows hosts to be added to or removed from the running system, and it also allows components to be started from scripts or directly from the command line.

References

1. NIST Smart Data Flow System. `http://www.nist.gov/smartspace/nsfs.html`.
2. G. Szeder. The ChilFlow API. `http://www.ipd.uni-karlsruhe.de/CHIL/ChilFlow/Documentation/libchilflow/html/`.
3. G. Szeder and W. F. Tichy. A communication middleware for smart room environments. In *AmI'07: European Conference on Ambient Intelligence*, pages 195–210, 2007.

25

Perceptual Component Data Models and APIs

Nikolaos Dimakis[1], John Soldatos[1], Lazaros Polymenakos[1], Jan Cuřín[2], Jan Kleindienst[2]

[1] Athens Information Technology, Peania, Attiki, Greece
[2] IBM Research, Prague, Czech Republic

Perceptual components can be defined as sophisticated software processing units that extract context from audio and visual sensor streams. This chapter starts by outlining the importance of interface specifications for perceptual components in a multivendor environment, i.e., an environment with multiple perceptual component providers. Interface specifications are discussed in terms of input, output, and configuration parameters of perceptual components. Accordingly, this chapter introduces a set of structuring principles for integrating perceptual components that leverage such specifications into context-aware applications. Concrete interface specification examples for body trackers are presented. Based on these specifications, service developers in the CHIL project were able to select among a wide range of multivendor components for building their services. Furthermore, different partners in CHIL were able to exchange perceptual components of the same type (e.g., body trackers and person identification components). This first-of-a-kind effort toward building applications using multivendor perceptual components is thoroughly discussed in the scope of the concluding remarks of this chapter.

25.1 Plug-and-Play Perceptual Components

The CHIL project brought together a team of perceptual component developers who were all using customized tools, different operating systems, a wide variety of signal processing (and other) libraries, and in general a plethora of equipment that was not agreed upon from the early stages of the project. However, the vision of the CHIL project, as expressed from the initial stages, was to be able to combine the developed technologies and deliver state-of-the-art services that would exploit the whole underlying sensor and actuator infrastructure and, ideally, would be independent of the technology provider. This need introduced the concept of a "plug-and-play" perceptual component.

The characteristics of the plug-and-play perceptual component do not define the parameters of the core signal processing algorithm, but pertain to the input and output data modeling aspects of it; they consider the perceptual component as a "black box".

25.2 The CHIL Compliance

The CHIL compliant perceptual components provide the developer with specific APIs that enable the rapid plugability of the perceptual component with the rest of the framework. This API has been designed to cover the basic needs of the perceptual component (as a context provider to the system, without focus on the algorithmic part), which span from perceptual component control, sensor data processing, to contextual formulation, etc. This API is applied by a controlling component that controls the perceptual component and is described in more detail below:

initialize(): Initializes the perceptual component. The initialization includes the subscription to the ChilFlow system but not starting the data reception from the sensor. The component is not yet ready for operation, and so there is no need to introduce additional burden to the system.

register(): Registers the perceptual component to the IPD knowledge base. The registration of the component to the IPD knowledge base is perhaps the key step of the overall process as it enables the perceptual component to advertise itself to the rest of the CHIL framework. This registration goes beyond simple IP and port specifications but also includes information about the perceptual component itself, the vendor, and the communication "dialect" it uses, which is crucial information for the decoding of the produced context by the subscribing components.

query(): Initiates information requests (queries) from the knowledge base for environment-specific parameters. There are perceptual components that require information about, for instance, the dimension of the room, the location of the sensors, the flow identities, etc. This information is provided by the query() function which instantiates a querying process with the IPD knowledge base about any information that the component requires (provided that the knowledge base already contains it).

start(): Starts the perceptual component. The controlling component starts the perceptual component as soon as it needs contextual information. All the previously mentioned steps must be completed before the procedure is started. The nature of the communication middleware used was designed to be as light-weight as possible and thus does not introduce significant delays. The starting process automatically starts the data stream communication, which leads to the algorithm being applied to the incoming data.

stop(): Stops the perceptual component. As soon as the controlling component no longer needs the contextual information, it stops the perceptual component. This does not force the perceptual component to deregister from the ChilFlow system or the knowledge base, as further (re)starting of the component is possible. These steps are completed when the perceptual component terminates.

set/getParameter(): Provides mechanisms to set/get parameters by the consumer component. Each controlling component can request internal parameters (which can be available by the perceptual component developer), using the getParameter method. Such information could be the data of developing, the developer's email, etc. Similarly, each perceptual component can set a parameter using the setParameter method.

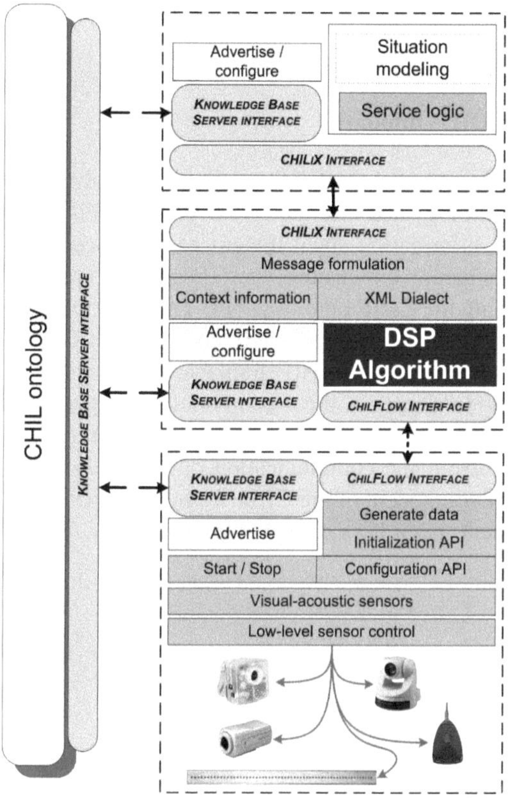

Fig. 25.1. The structure of the CHIL compliant perceptual component, how it is positioned in the CHIL Architecture, and how it encapsulates independent pieces: (a) original algorithm-specific code (in dark blue); (b) ontology communication front end (dashed arrow); (c) situation modeling connectivity (solid arrow); (d) sensor data retrieval (dotted arrow); and (e) context information formulation (final step prior to forwarding the context to the top tier).

25.3 Data Models and Interfaces

The CHIL compliant perceptual component is a significantly enhanced form of a perceptual component, and is illustrated in Fig. 25.1. It encapsulates the algorithm itself and orchestrates all operations by triggering the appropriate interfaces, which interact with other key points of the CHIL architecture. The CHIL compliant perceptual component combines three different middleware interfaces:

1. the CHIL ontology interface for communicating with the CHIL ontology using XML-over-TCP messages;
2. the *CHiLiX* library, which undertakes the communication of the perceptual component with the situation modeling components; and

3. the ChilFlow interface, which is responsible for providing constant flow of the sensor data to the perceptual component.

In the following sections, we elaborate on these interfaces and describe how they facilitate the plugability of the CHIL compliant perceptual component to the CHIL framework.

25.3.1 Perceptual Component to CHIL Ontology

The CHIL compliant perceptual component uses the CHIL ontology excessively for two reasons: to register itself to the ontology and essentially "advertise" itself to the rest of the framework, and to obtain information about the environment regarding the availability of sensors, the environment's dimensions, etc. These steps are done using an XML-over-TCP schema, using the CHIL ontology API. The advertisement of the perceptual component is an important action of the CHIL compliant perceptual component as it enables other components of the CHIL framework to connect to it and obtain its contextual information. Moreover, these components can parse the output of these components as each CHIL compliant perceptual component registers, in addition to all information about itself, the communication dialect it uses (see Table 25.1). It is thus possible for any component of the CHIL framework to use any CHIL compliant perceptual component, regardless of technology provider. This is

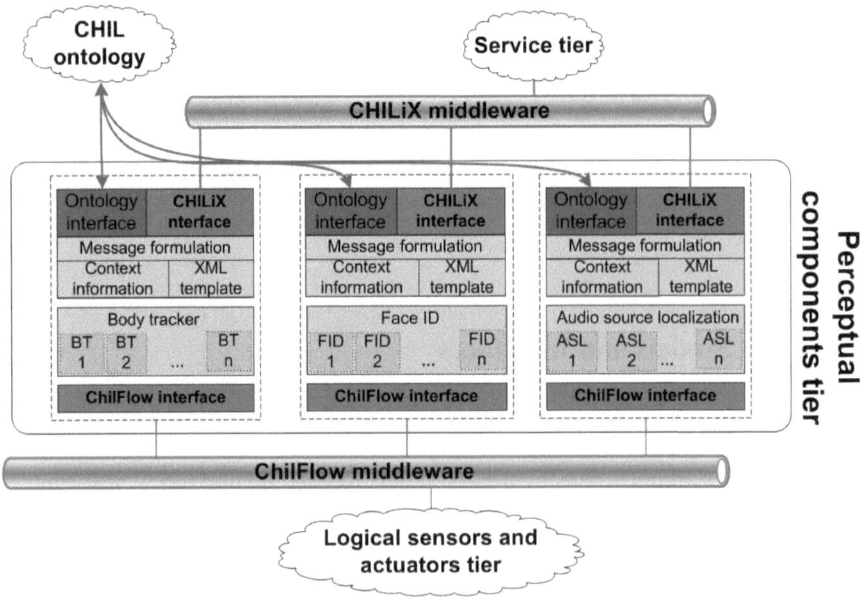

Fig. 25.2. Combining perceptual components using the CHIL compliant perceptual components guidelines [examples for the case of body tracker (BT), face ID (FID), and audio source localization (ASL) perceptual components].

also made easier by the use of the *CHiLiX* library, which is described in following paragraphs.

25.3.2 Perceptual Component to ChilFlow

The CHIL compliant perceptual component subscribes to the ChilFlow as soon as it is started by the consumer component. This subscription enables the CHIL compliant perceptual component to receive sensor data from the corresponding sensor to whose flow it has subscribed and apply its signal processing algorithm to extracting context. As soon as this context is not needed by the application, the CHIL compliant perceptual component can unsubscribe to the flow and pause its processing. Similarly, it can resume its operation when needed.

25.3.3 Perceptual Component to Situation Modeling

Forwarding the contextual information to the situation modeling component is the last part of the journey, and the IBM *CHiLiX*library plays a vital role in achieving this effort. Each perceptual component (following the CHIL compliant perceptual component guidelines) implements this interface and sets itself as a "producer" (i.e., a context provider) and waits for incoming connections from situation modeling, or other, components, which bind the connection by stating themselves as "consumers". As soon as the connection is set, the perceptual component initiates the transmission of contextual information to the consumers, following a predefined dialect. An example of such a dialect is illustrated in Table 25.1.

In addition to receiving the contextual information from the producer, the context consumer has the capability to control the component by using the API illustrated in the beginning of this chapter. For example, it can stop the component and restart it, or stop it and start another one, etc. This option was used in our evaluation of this approach during the exchange of three different body trackers, as described next.

25.4 Body Tracker Exchange Experience

Our approach for a CHIL compliant perceptual component was evaluated during an attempt to exchange three different perceptual components, developed by different development teams in different labs. These perceptual components were extended to meet the requirements of the CHIL compliant perceptual component and were tested during the CHIL EC review meeting in Karlsruhe in May of 2006. During this evaluation, the body trackers exchanged were

1. a 2D body tracker developed by Athens Information Technology (AIT), Greece, requiring one panoramic camera,
2. a 2D body tracker developed by the Interactive Systems Laboratory of the Universität Karlsruhe (UKA-ISL), which operated on a panoramic camera as well, and

The DTD XML and an example for the FaceID-type perceptual component

```
<!ELEMENT ChilixMessage (FaceID)>
<!ELEMENT FaceID (Target*)>
<!ATTLIST FaceID
   tsImage CDATA #REQUIRED>

<!ELEMENT Target EMPTY>
<!ATTLIST Target
   id CDATA #REQUIRED
   username CDATA #REQUIRED>

<?xml version="1.0" encoding="utf-8"?>
<!DOCTYPE ChilixMessage SYSTEM>
<ChilixMessage>
 <FaceID tsImage="TIMESTAMP">
   <Target id="id1" username="uname1"/>
 </FaceID>
</ChilixMessage>
```

Table 25.1. Perceptual component dialect

3. a 3D body tracker developed by the Institut National de Recherche en Informatique et en Automatique (INRIA), which used four corner cameras.

Despite the fact that this was a very ambitious problem, the perceptual component development teams and the architecture teams managed to collaborate closely and operate in parallel as the CHIL compliance enables this separation of tasks. Factors such as operating system, development platform, development language, algorithm details, etc., are not visible by the other, higher-level, components. Thus, the communication is independent and unaffected by them. As soon as the perceptual component was prepared (as a signal processing component, and not necessarily as a CHIL compliant perceptual component), the architecture teams had to enforce them by plugging the component into the ChilFlow (using the ChilFlow interface), enable the component-ontology communication by using the ontology interface, and, eventually, integrate the *CHiLiX*library to forward the contextual information which would be extracted by the perceptual component to the consumer component (which could be a CHIL agent, another component, or, in our case SitCom).

The evaluation setup was a typical setup commonly used in the project, having five cameras connected in different computers and operating in parallel, and using ChilFlow to stream their output, three body trackers (the AIT, UKA-ISL, and INRIA) in different computers, the CHIL knowledge base and sitcom running on two different computers. During the setup, all body trackers were instantiated and followed the CHIL compliant perceptual component steps to register to the ontology, get information about the current room, and adjust their internal parameters, which were room-dependent, subscribe to a camera flow, and wait for Sitom to control them.

The result was that all three body trackers were successfully instantiated and managed to receive input from the cameras. Finally, after requesting from the knowledge base the list of currently registered perceptual components in the CHIL system, SitCom started all three components. All of the body trackers managed to operate fully autonomously. During this evaluation, we had the chance to have all body trackers operating in parallel and visualize output, start, and stop on demand for any of them and have either one or two continue their operation, etc. Despite the fact that the output of the perceptual component did not match 100%, which was expected as all these body trackers were developed at different sites having different algorithmic approaches, this exchange proved that the CHIL guidelines to have a common interface to exchange perceptual components was correct. The CHIL compliant perceptual component guidelines have also been incorporated for other perceptual components, including components for face identification, audio source localization, and others.

Situation Modeling Layer

Jan Kleindienst[1], Jan Cuřín[1], Oliver Brdiczka[2], Nikolaos Dimakis[3]

[1] IBM Czech Republic, Praha, Czech Republic
[2] INRIA Rhône-Alpes, Saint Ismier Cedex, France
[3] Athens Information Technology, Peania, Attiki, Greece

CHIL services require observation of human activity. The observation of humans and their activities is provided by perceptual components. For most human activities, a potentially infinite number of entities could be detected, and an infinite number of possible relations exist for any set of entities. The appropriate entities and relations must be determined for a task or service to be provided. This is the role of the situation model. Situation models allow focusing attention and computing resources to determine the information required for operation of CHIL services. In this chapter, we introduce concepts and abstractions of situation modeling schema used in the CHIL architecture.

The concept of ubiquitous computing [1] opened a new era of computer science. Embedding computers into mobile or wearable appliances and blending them into the physical environment broke the paradigm of static deskbound computers. This led to a new quality of relationship between the user and the machine. Situation-awareness—the ability to sense relevant states of environment—plays an important role in facilitating this new relationship. Situation-aware applications are prone to better address user needs given the time, space, and situation constraints.

In the CHIL layered architecture, the situation modeling layer facilitates, mediates, and processes the flow of contextual information between multimodal perceptual components and services. The main goal of this layer is to filter high-bandwidth data flow from perceptual components and infer more abstract events that can help CHIL services fulfill their tasks.

The *situation modeling* layer is the place where the situation context received from audio and video sensors is processed and modeled. The context information acquired by the components of this layer enables services to respond correctly to varying user activities and environmental changes. For example, the situating modeling answers questions such as, "Is there a meeting going on in the smart room?" "Who is the person speaking at the whiteboard?" "Has this person been in the room before?"

In this sense, the situation modeling layer is a collection of abstractions representing the environment context in which the user interacts with the application. Ideally, it acts as a database that maintains up-to-date state of objects (people, artifacts, situations) and their relationships. The situation model itself acts as a directed inference engine that watches for the occurrence of certain *situations* in the current environment. The searched-for situations are dictated by the needs of active services, such as the detection of the interruptibility level of a particular person in a room for whom a service has to monitor incoming calls.

The general introduction to the situation modeling task with definitions of basic terms is given in Chapter 12 of this book.

26.1 Principles of Situation Modeling

In this section, we identify key usage principles that drove the architecture design of the situation modeling layer. A CHIL system exists in order to provide services. Providing services requires the system to perform actions, including modifying its internal configuration. The selection and execution of actions has long been studied in the field of artificial intelligence under the theme of planning. Almost all planning techniques since the 1950s have been formally defined using the concept of a state space. A state is defined as a conjunction of truth functions or predicates. Predicates express logical tests based on a priori axioms or perceptual observations.

In the CHIL project, we propose to adapt a "state-space" approach to formalize the way in which CHIL services use the perceptual components to determine actions. The current state, or situation, of the environment is expressed as a conjunction of predicates based on the observations provided by perceptual components. Predicates may express information observed in the environment, including the position, orientation, and activity of actors and props, as well as the state of electronic equipment and communications. This information is provided by perceptual components. The truth values of such predicates are determined by applying logical and probabilistic tests to properties provided by the perceptual components. The result is a logical description of the current situation (or state) of the CHIL users and their surroundings. Examples of actions for CHIL services include adapting the ambient illumination in a room, or displaying a user's "availability for interruption", configuring an information display at a specific location and orientation, or providing information or communications services to a group of people working on a common task.

The first step in building a situation model is to specify the desired system behavior. For a CHIL environment, this corresponds to specifying the set of actions that can be taken by the CHIL service, and formally describing the conditions under which such actions can or should be taken. For each action, the CHIL service designer lists a set of possible situations, where each situation is a configuration of entities and relations to be observed in the environment. Situations form a network, where the arcs correspond to changes in the roles or relations between the entities that define the situation. Arcs define events that must be detected to observe the environment and to enable the system to take appropriate actions. In real examples, we have noticed that there is a natural tendency for designers to include entities and relations in the situation model that are not really relevant to the system task. It is important to define the situations in terms of a minimal set of relations to prevent an explosion in the complexity of the system. This is best obtained by first specifying the system output state, then for each state specifying the situations, and for each situation specifying the entities and relations. Finally, for each entity and relation, we determine the configuration of perceptual processes that may be used.

26.1.1 The Process of Designing a Situation Model

One of the challenges of specifying a context model is avoiding the natural tendency toward complexity. Over a series of experiments we have evolved a method for defining situation models for context-aware observation using process federations. Our method is based on several principles:

Principle 1: Keep it simple.

The idea behind this principle is to start with the simplest possible network of situations, and then gradually add new situations. This leads to avoiding the definition of perceptual processes for unnecessary entities. The initial specification of a simple network of situations leads to a simple federation of processes. This federation can then be extended by the addition of situations and processes as needed. This process of incremental design is currently performed by hand. We are currently looking for means to automatically acquire new situations by observation.

Principle 2: Action drives design.

The idea behind this principle is to drive the design process from a specification of the actions that the system is to take. From the actions, we pass through a series of phases in the design process.

Phase 1: Map actions to situations.

The actions to be taken by the system provide the means to define a minimal set of situations to be recognized. The mapping from actions to situations need not be one-to-one. It is perfectly reasonable that several situations will lead to the same action. However, there can only be one action list for any situation.

Phase 2: Identify the roles and relations required to define each situation.

A situation is a unique set of roles and relations between the entities playing roles. Roles act as a kind of variable so that multiple versions of a situation played by different persons are equivalent. Determine a minimal set of roles and the required relations between entities for each situation.

Phase 3: Define acceptance tests for roles.

Define the predicates that must be true for an entity or agent to be assigned to a role. Currently these are logical predicates. We plan to move to probabilistic predicates, with the most likely entity being assigned to each role.

Phase 4: Define the processes for observation.

Define a set of perceptual processes to observe the entities required for the roles and to measure the properties required for the relations. Define processes to assign entities to roles and to measure the required properties.

Phase 5: Define the events.

Changes in situations are defined by events. Events are the results of changes in the assignment of agents to roles or changes in relations between the agents that play roles.

Phase 6: Implement and then refine.

Given a first definition, implement the system. Extend the system by seeking the minimal perceptual information required to appropriately perform new actions. To implement a situation modeling system, different kinds of developers are required. We introduce the developer roles and their required skills in the next section.

26.2 Perspective of Situation Modeling Developers

The construction of contextual services demands an interdisciplinary effort because the context modeling layer typically sits between two infrastructure parts: the environment sensing part (*perceptual components* layer) and the application logic (*services* layer). Thus, even a simple service built for an intelligent room requires at least three different roles, each with a different set of skills:

- **Perceptual technology providers** supply sensing components such as person trackers, sound and speech recognizers, activity detectors, etc. to see, hear, and feel the environment. The skills needed here include signal processing, pattern recognition, statistical modeling, etc.
- **Context model builders** make models that synthesize the flood of information acquired by the sensing layer into semantically higher-level information suitable to user services. Here, the skills are skewed toward probabilistic modeling, inferencing, logic, etc.
- **Service Developers** construct the context-aware services by using the abstracted information from the context model (the situation modeling layer). Needed skills are application logic and user interface design; multimodal user interfaces typically require yet an additional set of roles and skills.

How are these roles exercised during the process of service construction? Imagine for a moment a service developer creating a new context-aware service with this goal: *Once people have gathered in a meeting room, determine when they are taking a break for coffee.* This would, for example, let staff enter the room only when they do not disturb the meeting participants.

The service developer does not, typically, have access to a fully equipped room with sensors, nor does she have the technology for detecting people and meetings. But for an initial prototype of such a service, prerecorded input data may suffice.

A service developer then talks to the context model builder to define what new contextual abstractions will be needed, and which can be reused from an existing catalog. The context model builder in turn talks to technology providers to verify that capturing the requested environment state is within state-of-the-art of the sensing technology. If not, there is still a possibility to synthesize the requested information from other sensors under a given quality of service; for example, the number of people in a room may be inferred through sound and speech analysis if no video analysis technology is available.

26.2.1 Perceptual Technology Providers

Researchers working on sensing technologies sometimes suffers from a lack of use cases to understand service requirements toward technologies. The ability to simulate and visualize the scenarios would help in specifying and understanding what functionality and capability are sought from the technologies. For example, the coffee break scenario indicates to the technology provider that the focus in not on exact tracking of the number of people in the room, but rather on detecting the change in the interaction pattern in two regions of interest in the room (i.e., whiteboard area and audience).

Moreover, since the components of the situation modeling layer use full-fledged sensing technology APIs, it is possible to plug in the respective component (e.g., a body tracker) and test it together with the service logic. For example, the simulated people activity detector can be replaced with the real component without any change of APIs. The other sensing technologies can still only be simulated (we call this semi-automatic mode). In this case, the technology provider gets the end-to-end picture of how the sensing component behaves in tandem with the service. When the sensing component is plugged in, the context model developer can refine the set of context models and iteratively feed back to both technology and service developers.

26.2.2 Context Model Builders

The job of the context model developer is to ensure that information received from the perceptual components is sufficiently modeled by the contextual abstractions. This developer needs an environment that serves both as testbed that allows him to model and test various context acquisition algorithms, and as an observer of their "real" performance in the simulated (or semisimulated) environment. In the case of the coffee break scenario, the meeting would typically be modeled as a sequence of states, where *coffee break* would be one of them. The context model makes the decision about what the current state of the meeting is, based on the events from the sensing technology APIs.

26.2.3 Service Developers

It is important for service developers to have the context information available as if it were collected by processing real-time data sensed in a real environment – in other words, to simulate the function of various recognizers and detectors running at the multimodal sensing layer. This brings significant advantages to service developers. They can focus on crafting the application logic in a lightweight configuration represented by a "virtual room", where people's movements, interactions, and room situations are simulated.

At the same time, the application should receive information from the context model as if the service were running in a real room. In addition, the developer can play scenarios repeatedly (e.g., in normal, slow, or fast motion) and easily create new scenarios, e.g., to broaden the collection of functional tests. Via simulation, the developer may debug the service without needing to deal with the plethora of sensing hardware and software that typically needs nontrivial setup and maintenance, and without needing a bunch of student volunteers willing to play the crowd in the smart room. The time savings (and saving of lunch coupons) are obvious. Clearly, near the end of the development cycle, the service must be deployed in the real setting. But since the same set of contextual abstractions is used, substituting simulation for real-time data processing should be rather transparent.

The situation modeling framework introduced in further sections is the development environment where all these roles meet. This framework was designed to alleviate the above process and make the communication between those roles more efficient and straightforward.

26.3 Architecture and Key Abstractions

What exactly is the role of the SM layer during CHIL run time? And what are key implementation concepts of this layer? We will discuss such questions in this section.

In the CHIL system, sensors stream the acquired audio and visual data to perceptual components, which process the data to extract meaning from the signals. The result of the processing is delivered to the upper layers of the CHIL stack as a stream of events. For example, a body tracker (i.e., a perceptual component) receives a continuous stream of a video signal on its input, and on its output continuously updates the information about the number of detected subjects, including their location. In another example, a room noise classifier processes an audio stream to generate a stream of events bearing the detected noise label. Both such event streams are then provided to the situation modeling layer and/or services layer for further processing. In that sense, the data flow through the layers as streams exchanged between logical entities. The concept of streams and entities is presented in the next section.

26.3.1 Entities and Streams

The basic unit visible from the perceptual component layer (introduced in Chapter 25) is *entity* (see definitions in Chapter 12). The entity is defined as an entity type (such as person, whiteboard, or room) and a set of property streams as shown in Fig. 26.1. The property stream represents a collection of events describing (and updating) a particular attribute of an entity.

Examples of such properties for an entity of the type person are *Location* (indicating the physical location of the person), *Heading* (tracking direction the person is looking at), or *Person ID* stream (sending events about the identification process of the person). Schema of the entity-based communication between perceptual components and situation modeling layers is shown in Fig. 26.1. In the situation modeling

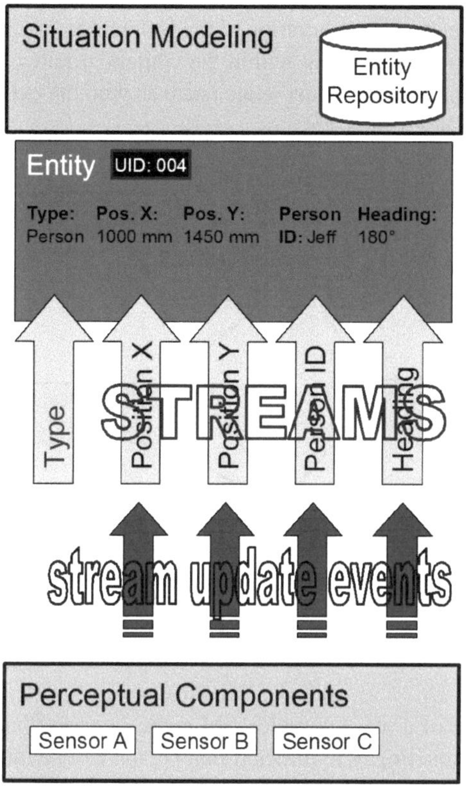

Fig. 26.1. The communication between perceptual components and situation modeling layers uses concept of entities and streams. Entity instance created on the perceptual components layer is registered in Entity Repository and then all events belonging to this particular entity are reflected as its stream updates.

layer, entities are stored in a container called *entity repository*. The entity repository notifies registered listeners about newly created entities, and a subscribing mechanism is used for notification about changes for a particular entity or stream.

26.3.2 Situation Machines

The concept of situations in the situation modeling layer is implemented by *situation machines*. A situation machine (SM) infers information by combining outputs of various entity streams, optionally also including the states of other situation machines. Internally, SMs may use different algorithms and techniques for information fusion, ranging from rule-based approaches to statistical training.

An example of a situation machine is the *whiteboard talker* SM, whose task is to infer which of the meeting participants currently plays the role of a whiteboard talker. The state of such a situation machine is implemented as a binary function – "a talker is present" or "a talker is not present". The algorithm for inferring this information may rely, for example, on the observation of the following: the distance of the person from the whiteboard, the time spent within the whiteboard area, whether a person is talking or not, some pre-entered knowledge (such as who the expected speaker of the meeting is), a person's gestures, etc.

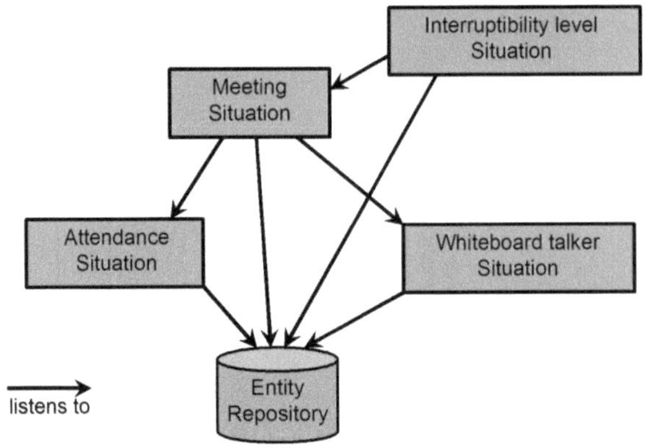

Fig. 26.2. Hierarchical tree of situation machines. The *listens to* arrows lead from the requester to the provider of the information.

Another example of a more complex SM is the *interruptibility level* SM, which for each person in a meeting determines if that person can be interrupted. For example, a person in the meeting audience might be interrupted by an urgent call, whereas the lecture presenter has his interruptibility level automatically set to zero during the entire presentation. In the current implementation, such a situation machine infers this information from the current state of other SMs, as shown in Fig. 26.2.

As we can see from this figure, the "interruptibility" of a particular person can be decided by observing the current state of the meeting and the presenter role (both provided by the Meeting SM), while the current state of the meeting is determined from the information about the whiteboard talker role, the number of attendees, and other relations among entities. Figure 26.3 shows the state diagram of the Meeting SM.

26.4 Situation Modeling Framework

We designed the CHIL situation modeling layer as a framework with well-defined abstractions, interfaces, and data contracts that enables one to plug in third-party components and ensures portability of the components to different CHIL environments. The APIs are based on the concepts of entities, streams, and situation machines introduced in the sections above. Typically, the framework supports both synchronous polling and asynchronous notifications, the latter commonly implemented via subscriber/listener patterns.

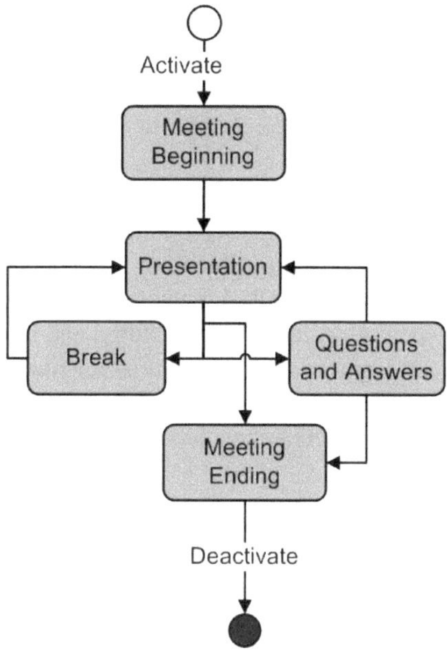

Fig. 26.3. Meeting situation machine states.

The situation modeling framework must fulfill several key tasks to address the needs identified by different developer roles as introduced in previous sections. One of them is to handle the life cycle of the perceptual components, i.e., to keep track of available and running perceptual components as well as to provide means for controlling their state and possibly adjust the performance parameters. In the ideal sense, such a framework would dynamically direct the CHIL perceptual resources to understanding that part of physical surroundings where it is most desired by the CHIL services.

The ability of the framework to "understand" the information coming from a plethora of perceptual components that themselves process a plethora of audio, visual, etc. signals places a strong emphasis on data formats, data standards, and perceptual component compliance with CHIL data interfaces. The framework is thus a logical place for cultivating respective data contracts between services and sensing layers, as described by principle 2 in Section 26.1.

Having well-defined APIs for the different layers not only benefits the easy exchange of components for cross-evaluation or cross-validation, but also enables the framework to actually plug into a running system seamlessly. Using the perceptual component API, the framework may provide the situation model with some additional simulated entities, which will then be received at the situation model as regular entities.

The standardized interface is also of importance if automatic verification against a labeled set of data is to be performed, so that quality metrics can be computed automatically. The verification process will simulate a given scenario, check the output of the model being evaluated, and compute the quality metric against the labeled set of output data. This can be done for several different models of situations and services, enabling their comparison.

It is still unclear to us to what extent the framework should play the role of the universal database for storing the situation modeling knowledge related to short-term (session) as well as long-term context acquisition tasks (world knowledge). We have addressed these open issues with the CHIL knowledge base server, which is the place where the majority of such information can be stored and universally accessed.

In CHIL, we have put together a particular implementation of the situation modeling framework that is called SITCOM, which is Java run time and IDE for writing situation models and inference hierarchies based on situation machines as described in Chapter 29.

References

1. M. Weiser. The computer for the twenty-first century. *Scientific American*, 256(3):94–104, 1991.

27

Ontological Modeling and Reasoning

Alexander Paar and Jürgen Reuter

Institute for Program Structures and Data Organization (IPD), Universität Karlsruhe (TH), Karlsruhe, Germany

A major challenge in putting the CHIL vision into practice was to integrate contributions from project partners all over Europe and the United States. While standard Web technologies such as HTTP support core data exchange, perceptual component integration places higher demands on distributed intercommunication. Interfaces have to be defined for the type-safe exchange of structured data between components that are implemented in diverse computer languages.

Rather than defining component interfaces in a particular programming language (e.g., C++, C#, Java), we prefer a programming language-independent *knowledge representation* (KR) for system and component modeling. KR can exploit formal mathematical notion for automatic *reasoning*, which is useful for checking the consistency of the model, querying the model, and deriving implicit knowledge as we will demonstrate later.

Knowledge representation was probably first implemented through *semantic networks* in the late 1960s [20] and later with *frame*-based approaches [12]. Semantic networks introduced semantics for describing *entities* and *relations* between these entities. The use of first-order logic *predicates* established a formal mathematical foundation for logically describing semantic networks. In the course of time, *description logics* (DL) were refined to support set-related features such as hierarchies of *classes* of entities and relations. At the same time, decidability had to be preserved for essential querying tasks such as satisfiability of the model or membership of an entity in a class. Compared to deductive relational database systems, description logics typically match the needs of domain representation much better and hence result in a more natural and compact knowledge representation [21].

The Web Ontology Language OWL [11] combines well-proven Web technologies with state-of-the-art description logics [1]. OWL is available in three versions: OWL Lite as the least expressive subset; OWL DL with decidable description logics; and the undecidable OWL Full with support for full first-order logic. In CHIL, we deploy OWL DL for modeling the CHIL domain of discourse.

Designed for interoperability, an OWL ontology is represented as a globally accessible Web resource. Since OWL is independent of programming languages, domain experts can focus on the modeling of the entities in their domain. Similarly to object-oriented programming languages, OWL's description logics deploy a hierarchy of *classes* for modeling. *Instances* represent incarnations of classes. Unlike object-oriented languages, however, *properties* in OWL are not defined as part of classes but form a hierarchy of their own. The two orthogonal hierarchies allow for distributed extension of the ontology better than object-oriented programming languages do. In fact, OWL uses *open world semantics*.

An ontology can be split into separate Web resources through OWL's *import* mechanism. This way, complex ontologies can be composed on demand from smaller units. Conflicting ontology fragments may result in logical *inconsistency*. OWL's formally specified semantics enable tools to detect and report inconsistencies. For this purpose as well as for managing the ontological data, access to an ontology is typically backed by a central knowledge base management system that performs reasoning and management of changes to the ontology. The management system can also exploit OWL's description logic features to make implicit knowledge explicit. Reasoning typically includes testing for *satisfiability* of the defined concept descriptions.

Deploying OWL in the CHIL project was not only a technological challenge but also a social one. Why should you be concerned with the formal overhead of ontological descriptions in OWL when the technology works fine, asked many of the technology providers. From their perspective of working on the C code level (or other languages on a similar abstraction level), the overhead of adapting their high-speed implementations to some formally specified interfaces was counterproductive to the overall system performance. The answer to their question is similar to what one would respond to an assembly language programmer in order to convince him to use high-level languages: While the ad hoc approach works perfectly for small sizes, it does not scale in the large. That is, as long as the technology providers stayed within the bounds of their domain of discourse and only had to interface with their floor neighbors' implementations, they got impressive results. However, one of the vital goals of the CHIL project was to bring together technologies from the most different partners. In fact, the process of formally modeling interfaces in fact forced the technology providers to make the features and limitations of their technology transparent for partners. They had to decide what is part of the black box and what should be visible and controllable from outside.

While the technology providers understood that formally specified interfaces are essential for flexible composition of the most diverse technologies, some still wondered why not using more traditional tools (e.g., deductive DBMS) was sufficient. The CHIL ontology (Section 27.1) proved that the description logics of OWL are suitable for compact modeling of the entities and relations in the CHIL domain of discourse. In the following section, we will present the structure of the CHIL ontology and introduce an example scenario to demonstrate some reasoning features of the Web Ontology Language OWL.

While the purpose of OWL is to formally specify ontologies as XML documents, ontologies usually need to be maintained. For example, instances of classes may be created and disappear over time and class memberships of particular instances need to be checked. Until now, there are no widely accepted standards yet that define

APIs to manage ontological data have existed. Processing ontological information still suffers from the heterogeneity imposed by the plethora of available ontology management systems. The CHIL knowledge base server (Section 27.2) exposes the functionality of arbitrary off-the-shelf ontology management systems via the formally specified and well-defined CHIL OWL API. As such, it provides unified access to a central ontological knowledge base for heterogeneous platforms, programming languages, and communication protocols.

Finally, while having the well-defined CHIL OWL API is highly desirable, smooth integration of knowledge base management into programming languages as first-level language constructs is even better. The Zhi# programming language is an extension of ECMA standard C#, which boasts compiler support for XML Schema Definition data types and Web Ontology Language OWL DL concept descriptions (Section 27.3).

27.1 The CHIL Ontology

The role of the CHIL ontology is to provide a unique vocabulary and description logics-based model of CHIL entities for rudimentary reasoning. The ontology comprises several modules that are physically represented by separate Web resources with distinct URLs. The idea of modularization is that software developers need to reference only those parts of the ontology that are relevant to them. Additionally, modularization increases performance, for example, when deploying the communication agents subset of the ontology in order to generate software agents code.

The CHIL ontology consists of the following currently actively deployed modules.

- **`chil.rdf`** is the main file of the full CHIL ontology. It imports all the domain-specific modules and thus can be considered as a kind of main setup file. It does not contribute otherwise to the CHIL ontology.
- **`chil-core.rdf`** contains the axiomatic knowledge shared by many architectural layers of CHIL that are common to all smart-room installations. It can be considered as the core of the CHIL ontology.
- **`chil-pc.rdf`** contains the descriptions of all perceptual components in CHIL. It consists of a generic perceptual component model, a set of standard categories of perceptual components, and finally descriptions of all vendor-specific components.
- **`chil-ca.rdf`** contains the communication agents part of the ontology, originally contributed by Fraunhofer IITB. It is put into a namespace of its own and has been editorially adapted (with respect to naming conventions, etc.) to the remainder of the CHIL ontology. Concepts in the `chil-ca` module that semantically overlap with concepts in the `chil-core` module either were moved to `chil-core.rdf`, if they are of general interest, or were left in `chil-ca.rdf` and may appropriately extend concepts of the `chil-core` module.

- **`chil-isl.rdf`** finally gives an (imaginary) example of the factual knowledge of an actual CHIL installation.

Another module, which is not actively used (experimental or obsolete code), is **`chil-metadata.rdf`**, which provides algorithmic support for metadata management such as knowledge base facts, timestamps, or access rights. This module is in an experimental state. WP2 has decided to defer full implementation of access control until a future CHIL implementation.

27.1.1 Example Scenario

An ontological knowledge base comprises a terminology (TBox) and contingent assertional knowledge about a particular state of the described world (ABox). The following knowledge base contains information about an exemplary CHIL scenario, which may illustrate applications of OWL modeling and reasoning features in the CHIL domain of discourse. The TBox and ABox are given in description logics notation (i.e., the symbol $\top$ denotes the top-level concept, the relational operator $\sqsubseteq$ denotes a subconcept/subrole relation, and $\equiv$ denotes equivalence).

The given ABox declares three persons, Alice, Bob, and Charlie. A project meeting is taking place at a certain time (given as an XML schema definition `dateTime` value) in smart room no. 248. The meeting is moderated by Alice. Bob and Charlie also participate.

TBox

$Moderator \sqsubseteq Participant \sqsubseteq Person \sqsubseteq \top$, $Room \sqsubseteq \top$, $Meeting \sqsubseteq Event \sqsubseteq \top$,
$ActivityLevel \sqsubseteq \top$, $\geq 1 hasActivityLevel \sqsubseteq Meeting$,
$\top \sqsubseteq \forall hasActivityLevel.ActivityLevel, \geq 1 takesPlaceIn \sqsubseteq Event$,
$\top \sqsubseteq \forall takesPlaceIn.Room$, $hosts \equiv takesPlaceIn^{-}, ActivityLevel(HIGH)$,
$ActiveMeeting \equiv Meeting \sqcap \exists hasActivityLevel.HIGH$,
$ActivityLevel(LOW)$, $\geq 1 scheduledAt \sqsubseteq Event, \top \sqsubseteq \forall scheduledAt.\texttt{xsd\#dateTime}$,
$\geq 1 hasModerator \sqsubseteq Meeting, \top \sqsubseteq \forall hasModerator.Moderator$,
$moderates \equiv hasModerator^{-}, \geq 1 hasParticipant \sqsubseteq Event$,
$\top \sqsubseteq \forall hasParticipant.Participant, participatesIn \equiv hasParticipant^{-}$,
$hasModerator \sqsubseteq hasParticipant$, $\mathbf{func}(hasModerator)$

ABox

Person(ALICE), *Person*(BOB), *Person*(CHARLIE), *Room*(ROOM_248),
Event(PROJECT_MEETING), *takesPlaceIn*(PROJECT_MEETING, ROOM_248),
scheduledAt(PROJECT_MEETING, `2008-06-27T13:00:00Z`),
moderates(ALICE, PROJECT_MEETING), *participatesIn*(BOB, PROJECT_MEETING),
participatesIn(CHARLIE, PROJECT_MEETING).

The following list contains tell and ask operations on the given knowledge base. The input (⇥) and output knowledge (↦) are given in natural language (NL) and

RDF triplesyntax (RDF). Note how much the RDF triple notation resembles the natural-language sentences.

1. ↦ NL: Room no. 248 hosts the project meeting.
RDF: ROOM_248 *hosts* PROJECT_MEETING
2. ↦ NL: The project meeting is scheduled at 27 June 2008, 1:00 PM.
RDF: PROJECT_MEETING *scheduledAt* `2008-06-27T13:00:00Z`
3. ↦ NL: The project meeting has the participants Alice, Bob, and Charlie.
RDF: PROJECT_MEETING *hasParticipant* {ALICE, BOB, CHARLIE}
4. ↦ NL: Alice is a moderator.
RDF: ALICE *rdf:Type Moderator*
5. ↦ NL: The project meeting event is a meeting.
RDF: PROJECT_MEETING *rdf:Type Meeting*
6. ⇥ NL: Elsie moderates the project meeting.
RDF: ELSIE *moderates* PROJECT_MEETING
7. ↦ NL: Alice and Elsie are the same persons.
RDF: ALICE *owl:sameAs* ELSIE
8. ⇥ NL: The project meeting has a high activity level.
RDF: PROJECT_MEETING *hasActivityLevel* HIGH
9. ↦ NL: The project meeting is an active meeting.
RDF: PROJECT_MEETING *rdf:Type ActiveMeeting*

The storyline develops as follows. A meeting summarization perceptual component queries all events that are hosted in room no. 248. Because the ontological role *hosts* is defined as the inverse of *takesPlaceIn*, the reasoner automatically determines that room no. 248 hosts the project meeting (1: *inverse roles*). The meeting summarizer queries the scheduled start time of the project meeting (2) and starts recording. It needs to assemble a list of participants and queries the ontological knowledge base. Bob and Charlie are participants of the project meeting since the ontological role *hasParticipant* is declared to be the inverse of *participatesIn*. Alice is a meeting participant, too, because the ontological role *hasModerator* is a subproperty of *hasParticipant* (3: *inverse roles*, *subroles*).

In particular, the RDF type of individuals Bob and Charlie is inferred to be *Participant* because of the domain restriction of the *participatesIn* role. Analogously, Alice is inferred to be a *Moderator* because of the domain restriction of *moderates* (4: *property domain restrictions*). Due to the range restriction of the *moderates* role, the project meeting *Event* is classified as a *Meeting* (5: *property range restrictions*). Note how the domain and range restrictions of ontological roles are different from class definitions in object-oriented programming languages.

Another example for intuitive ontological reasoning is functional roles such as *hasModerator* along with several property values. Suppose in addition that Elsie moderates the project meeting under consideration (6). Because there can be only one value for the functional role *hasModerator*, the reasoner concludes that the ontological individuals ALICE and ELSIE must refer to the same entities in the described world (7: *inverse roles*, *functional roles*, *equivalent individuals*). Note that because in OWL there is no *unique name assumption* with additional information, a reasoner will not deduce that two individuals are distinct.

During the project meeting, the activity level, which is detected by the smart-room technologies, changes from low to high (8). Because of the nominal definition in the ontology, the project meeting is inferred to be an *ActiveMeeting* (9: *nominals*).

For a complete list of $\mathcal{SHOIN}(\mathbf{D})$ concept constructors, see [7]. More examples of ontological reasoning can be found in /it The Description Logic Handbook [1].

27.2 The CHIL OWL API

As the underlying Semantic Web standards such as RDF(S) [10, 4], DAML+OIL [8], and their common description logics (DL) [1] -based successor OWL DL [11] have matured, tools for ontology engineering have emerged in both commercial as well as academic fields.

Common to all widely used ontology management systems [9, 14, 19, 23, 13, 15] and knowledge base interface specifications [2] is the lack of formal specifications that define the semantics of the ontology management APIs. This is particularly problematic since typical interface methods (e.g., *listSubclasses*, *addIndividual*) are closely related to the semantics of the formal foundations of ontology languages (i.e., description logics). Also, existing off-the-shelf ontology management systems provide only limited connectivity with respect to native support for programming languages and remoting protocols. Hence, it is particularly difficult to use ontology management systems remotely or along with a variety of different programming languages (i.e., in heterogeneous distributed computing environments). Consequently, it may be unfeasible to replace an ontology management system by alternative products without rewriting significant parts of client code.

27.2.1 The CHIL Knowledge Base Server

The authors devised a pluggable architectural model of an ontological knowledge base server, which can aggregate arbitrary combinations of ontology management systems and remoting protocol servers. In particular, the CHIL knowledge base server can adapt off-the-shelf ontology management systems and expose their functionality by means of remoting technologies.

In the current implementation, the Jena Semantic Web Framework [9] is adapted and configured to use the Pellet OWL DL reasoner [19] with in-memory and database-backed ontology models. The formally specified CHIL OWL API [17] is exposed via an XML-over-TCP interface.

The CHIL knowledge base server can be run as a standalone application and as an Eclipse plug-in. The Eclipse plug-in provides a GUI that can be used to control the CHIL knowledge base server and to browse managed ontologies.

The definition of the programming language-independent CHIL OWL API is twofold. The first part provides a schema for defining interfaces and methods on the meta-level. The second part defines the format of request and response messages on the object level (i.e., the format of the messages that are sent over the wire). Both the XML-over-TCP server component of the CHIL knowledge base server as well as the client libraries were generated automatically from the XML-based API definition. The code generation approach helped a great deal during the development phase of the CHIL OWL API. Client libraries for Java, C#, C++, and Python could be automatically updated to instantaneously reflect changes in the API definition. Note that the XML-based definition of the CHIL OWL API does not make any assumptions about the programming language data types that may be used to denote the elements of an ontology (e.g., concept names). Instead, interpretation information is given that identifies formal parameters and returns values as particular ontology elements. Code generators may use arbitrary data types to represent these elements in different programming languages.

27.2.2 A Formally Specified OWL API

Complementary to the XML-based API definition, a Floyd-Hoare logic-based formal specification of the CHIL OWL API was devised in order to make it possible to consistently adapt off-the-shelf ontology management systems. The methods of the CHIL OWL API are specified as Hoare triples [6] of the following form.

$$\begin{array}{ll} \{P\}\text{:} & \{\eta(\mathcal{O}) \wedge (\bigwedge \mathcal{O} \vdash ((\mathcal{SHOIN}(\mathbf{D})\ \textit{semantics}) \oplus \texttt{Exception}))\} \\ Q\text{:} & \qquad\qquad \mathcal{O}, \mathrm{m}(p_1, .., p_n) \\ \{R\}\text{:} & \{(\bigwedge \mathcal{O} \vdash (\mathcal{SHOIN}(\mathbf{D})\ \textit{semantics})) \wedge (\textit{return value semantics})\} \end{array}$$

Preconditions guarantee the ontology to be consistent and particular axioms and facts to hold; the program terminates with the given exceptions otherwise. We define the partial function $\eta(\mathcal{O})$, which returns true for consistent ontologies; it is undefined for inconsistent ontologies. Next, we define a total function $\overline{\eta(\mathcal{O})}$, which additionally evaluates to the symbol `InconsistentOntologyException` for all arguments for which $\eta(\mathcal{O})$ is undefined. For the sake of simplicity, we still write $\eta(\mathcal{O})$ instead of $\overline{\eta(\mathcal{O})}$. Thus, the function $\eta(\mathcal{O})$ is to be interpreted as a consistency check of the ontology $\mathcal{O}$.

In the Hoare triples of the CHIL OWL API specification, $\mathcal{SHOIN}(\mathbf{D})$ semantics are used since the $\mathcal{SHOIN}(\mathbf{D})$ Description Logic constitutes the formal foundation of the Web Ontology Language OWL DL. $\mathcal{SHOIN}(\mathbf{D})$ semantics are denoted in standard description logics terminology. Note that the given preconditions are positional (i.e., evaluated in the given order).

The comma operator applies a method to the ontology. If the program terminates, particular axioms and facts and return-value semantics can be asserted.

The chosen API formalization allows for reasoning on a meta-level about the effect of subsequent method calls on an ontology. For example, assuming a previous method call $\mathcal{O}, \textit{addSubclass}(S, T)$, the corresponding Hoare triple HT-OwlApi-AddSubClass asserts only that $S^{\mathcal{I}} \subseteq T^{\mathcal{I}}$ (i.e., the extension of S is a subset of the extension of T), while a method call $\mathcal{O}, \textit{addIndividual}(s, S)$ requires the concept S to be declared in the ontology as a subconcept of the top-level concept $\top$

(i.e., $S^{\mathcal{I}} \subseteq \Delta^{\mathcal{I}}$). Using the Floyd-Hoare logic rule HR-PRESTRENGTH, one can infer that the precondition $S^{\mathcal{I}} \subseteq \Delta^{\mathcal{I}}$ is met by the postcondition $S^{\mathcal{I}} \subseteq T^{\mathcal{I}}$ of the previous method call $\mathcal{O}, addSubclass(S,T)$.

The CHIL OWL API comprises 37 tell operations to modify the TBox, 16 tell operations to modify the ABox, 26 ask operations to query the TBox, and 12 ask operations to query the ABox of OWL DL knowledge bases. In addition, there are 15 methods to query and modify the annotations of ontology elements and 17 auxiliary methods, for example, to load an ontology from a file or to serialize an ontology model to its RDF/XML syntax. These methods are not included in the formal specification of the CHIL OWL API since their semantics cannot be founded on Description Logics.

Regression test cases were automatically generated from the formal specification. The CHIL OWL API adapter code and the Jena Semantic Web Framework were instrumented using the code coverage tool Emma [22]. In the adapter component, the measured method-, block-, and line coverage based on the purely automatically generated test code was 100%, 97%, and 96%, respectively. For the 10 most often used Jena Semantic Web Framework Java packages, a class-, method-, block-, and line coverage of 68%, 51%, 48%, and 49%, respectively, could be achieved. For the five most often used packages, the class-, method-, block-, and line coverage was 75%, 58%, 57%, and 55%, respectively.

27.2.3 Example Scenario Implementation

Using the CHIL OWL API, knowledge base queries and modifications as they occur in the example scenario described in Section 27.1.1 can be implemented. The project meeting of the example scenario can be scheduled and the set time queried in C# as follows (the object o refers to an instance of the `ICHILOWLAPI` interface).

```
1 string ns = "http://chil.server.de/ontology#";
2 string xs = "http://www.w3.org/2001/XMLSchema#";
3 o.AddDatatypePropertyValue(ns + "PROJECT_MEETING",
4   ns + "scheduledAt", "2008-06-27T13:00:00Z", xs + "dateTime");
5 string[] arrTimes = o.ListDatatypePropertyValuesOfIndividual(
6   ns + "PROJECT_MEETING", ns + "scheduledAt");
```

The CHIL knowledge base server ships with a full-fledged XSD validator. Still, using an API to modify external data is inherently unsafe and error-prone. In the code snippet above, there is no compile-time support to (1) check that the referenced ontological individual and data-type property actually exist in the knowledge base and (2) type-check the string literal that denotes the `xsd#dateTime` value. In line 5, the burden to parse the string representation of the returned `xsd#dateTime` value is put on the programmer. A well-typed C# program may be invalid with respect to a particular ontology and XML schema.

In programming languages such as C#, the `is` operator can be used to check the run-time type of an object. For ontological individuals, such a type check has to be devised manually as follows. Again, undetected syntax errors may lead to program failure at run time.

```
string ns = "http://chil.server.de/ontology#";
string[] arrTypes =
 o.ListRDFTypesOfIndividual(ns + "PROJECT_MEETING");
if (arrTypes.Contains(ns + "ActiveMeeting")) { [...] }
```

In statically typed programming languages, the structure of input objects that can substitute a formal method parameter can be restricted by its declared type. This feature is not applicable for external types since there is no isomorphic mapping from ontological concept descriptions and XSD type definitions to C# classes. As shown below, a method that returns the start time of an example scenario *Event* object can only be declared with a string parameter that references the individual in the knowledge base. There is, however, no guarantee that at run time the referenced individual actually is in the extension of the *Event* concept. A manual "type check" as shown above would be necessary.

```
public int getStartTime(string Event) { [...] }
```

The authors developed the Zhi# programming language as introduced in the following section per request of the CHIL project partners in order to make OWL concept descriptions and XSD type definitions first-class citizens of an object-oriented programming language.

27.3 The Zhi# Programming Language

The applicability of the Web Ontology Language OWL to describe complex data and to automate reasoning on ontological knowledge bases has been widely acknowledged in recent years. However, processing ontological information programmatically is still laborious and error-prone. Until now, in general-purpose programming languages such as Java or C#, there has been no type checking for XML Schema Definition type definitions, which may be the range of OWL data-type properties, and ontological knowledge bases.

Ontology management systems merely provide APIs, which have to be used in an explicit manner. The burden to wisely manipulate ontological data is put on the programmer. Naïve approaches to simply use wrapper classes for XSD and OWL type definitions in order to benefit from built-in type systems of object-oriented programming languages have proved to be insufficient or even incorrect.

In XML Schema Definition, atomic types can be derived through the application of lexical and value-space constraints (e.g., an integer data type `meetingRoomCapacity` can be derived from the built-in data type `xsd#nonNegativeInteger` by constraining its value space to numbers less than 30 using the constraint `xsd:maxExclusive`).

In the Web Ontology Language OWL, ontological concepts reveal a behavior in terms of type inference and concept subsumption that is different – if not antithetic – to object-oriented programming languages. In C#, properties are declared as class members. The domain of a property corresponds to the type of the containing host object. Only instances of the domain type can have the declared property. The range of a property (i.e., class attribute) is given by an explicit type declaration. This type declaration is authoritative, too. All objects that are declared to be values of a property must be instances of the declared type.

For ontological concepts and properties, which are subject to automatic reasoning, domain and range constraints are interpreted differently. In particular, if an ontological property relates an individual to another individual, and the property has a class as its range, then the other individual is inferred to belong to this range class. For example, the property *moderates* may be stated to have the range *Meeting*. From this a reasoner can deduce that if ALICE is related to a PROJECT_MEETING by the *moderates* property (i.e., the project meeting is moderated by Alice), then the *Event* PROJECT_MEETING is a *Meeting*. In the same way, if *moderates* is stated to have the domain *Moderator*, the reasoner can deduce that ALICE is a *Moderator*.

In contrast to C#, in OWL ontologies, property assignments do not fail immediately. Instead, the reasoner considers the declared piece of knowledge for subsequent inference and querying steps. These different property domain and range semantics alone make it particularly difficult – if not impossible – to simply substitute an OWL ontology by an isomorphic set of C# wrapper classes.

In the course of the CHIL research project, the Zhi#[1] programming language was developed. It implements programming language inherent support for both XML Schema Definition and the Web Ontology Language OWL DL. The Zhi# programming language is a pluggable extension of conventional C# 1.0 [5]. XML and OWL compiler components can be activated to operate separately with standard C# code or to cooperate. External types can be included using the keyword `import`, which permits the use of external types in a Zhi# namespace such that one does not have to qualify the use of a type in that namespace. Type system specifiers such as `XML` and `OWL` indicate which external type system (i.e., compiler plug-in) is responsible for processing the imported namespaces.

```
import XML xsd = http://www.w3.org/2001/XMLSchema;
import XML chx = http://chil.server.de/schema;
import OWL cho = http://chil.server.de/ontology;
```

The Zhi# compiler framework supports the usage of arithmetic, relational, and logical operators with external types. The dot operator '`.`' can be used to access members of external types. Types of different type systems may be used cooperatively in one single statement. For example, a .NET `System.Int32` variable can be assigned the XML value `meetingRoomCapacity`, which may be defined as a *hasCapacity* property value of an instance of the ontological concept *Room*.

```
#cho#Room r = [...];
int i = r.#cho#hasCapacity;
```

In Zhi#, external type definitions can almost unrestrictedly be used with almost all C# programming language features. For example, methods can be overridden using external types, user-defined operators can have external input and output parameters, and arithmetic and logical expressions can be built up using external objects.

[1] Zhi (Chinese): knowledge, information, wisdom.

Zhi# programs are compiled into conventional C# code (i.e., standard .NET assemblies), which uses the extensible Zhi# run-time library in order to implement external type system functionality (e.g., create an ontological individual in a knowledge base using the CHIL OWL API).

27.3.1 XSD Aware Compilation

The Zhi# programming language boasts *XSD aware compilation* [18]. The Zhi# compiler takes as input both Zhi# source files as well as XML Schema Definitions [3].

The authors formally devised and fully implemented a constraint-based type system in order to facilitate the correct, complete, and concise handling of XML Schema Definition type definitions. The soundness of the developed type system was formally proved in the context of an extension of the typed lambda calculus with subtyping ($\lambda_{<:}$).

In the developed λ_C-calculus, constrained data types are computational structures. The only operations on atomic types are constraint applications. Constraining facets can be modifying (i.e., string literals that are assigned to instances of constrained types are implicitly modified upon assignment) or enforcing (i.e., certain value-space constraints must hold for the interpretations of string literals that are assigned to instances of constrained types). Types can be derived from a number of built-in primitive types through the application of lexical and value-space constraints. This form of type construction can be used to mimic the semantics of XML Schema Definition. Also, it is possible to infer transient constraints that hold for the instances of constrained types within only a limited scope of a program. Types of binary expressions that involve constrained types are inferred using constraint arithmetic.

The λ_C-type system was fully implemented in C#. In particular, an XML Schema Definition implementation of the λ_C-type system interface was devised that can be used to load type definitions from XSD files and classify these types in a hierarchy.

The XSD compiler plug-in of the Zhi# programming language provides static and dynamic type checking of XML Schema Definition type references. At compile time, the types of XSD objects are checked according to the constraint-based typing rules of the λ_C-type system. In particular, the Zhi# compiler is able to detect exactly which constraint is violated by an assignment instead of merely reporting generic type incompatibilities. XSD types can be used along with C# programming language features such as method overriding, user-defined operators, and run-time type checks.

The Zhi# compiler plug-in for XML Schema Definition infers types of variables based on control and data flow analysis. The type inference rule for `if` statements in the λ_C-calculus was complemented with type inference rules for `for` and `while` statements. Types of literals are inferred based on a literal expression's particular value. Types of binary arithmetic expressions are inferred based on the constraint arithmetic of the λ_C-type system. A number of relational operators were added to the Zhi# language grammar in order to cope particularly well with XSD constraining facets.

The generated C# code contains dynamic type checks in order to allow for schema modifications after the compilation of a Zhi# program and safe usage of Zhi# assemblies from conventional .NET programs.

27.3.2 OWL Aware Compilation

Based on the formally specified CHIL OWL API, an OWL plug-in [16] for the Zhi# compiler framework was developed that enables the usage of OWL concepts, roles, and individuals in Zhi# programs. In particular, the OWL plug-in can be used co-operatively with the XSD plug-in described in the previous subsection in order to facilitate the handling of OWL data type properties.

The most fundamental compile-time feature that the OWL plug-in provides is checking the existence of referenced ontology elements in the imported terminology. The C# statements below declare the ontological individuals a and b. Individual b is added as a property value of property R of individual a. For the sake of brevity, in this chapter, the URI fragment identifier "#" is used to indicate ontology elements in Zhi# programs instead of using fully qualified names. The object `o` shall be an instance of the `ICHILOWLAPI` interface.

```
o.addIndividual("#a", "#A");
o.addIndividual("#b", "#B");
o.addObjectPropertyValue("#a", "#R", "#b");
```

The given code is a well-typed C# program. It may, however, fail at run time if, in the TBox of the referenced ontology, classes A and B and property R do not exist. In Zhi#, the same declarations can be rewritten as shown below. As a result, the Zhi# compiler statically checks if concept descriptions A and B and property R exist in the ontology.

```
#A a = new #A("#a");
#B b = new #B("#b");
a.#R = b;
```

C# is a statically typed programming language. Type checking is performed during compile time as opposed to run time. As a consequence, many errors can be caught early at compile time (i.e., fail-fast), which allows for efficient execution at run time. Unfortunately, the non contextual property-centric data modeling features of the Web Ontology Language OWL render compile-time type checking only a partial test on Zhi# program text. In a well-typed C# program, every expression is guaranteed to be of a certain type at run time. In Zhi#, the same is not generally true for instances of OWL concept definitions such as the definition of a small meeting ($SmallMeeting \equiv Meeting \sqcap \leq 3 hasParticipant$).

```
import OWL cho = http://chil.server.de/ontology;
class C { public static void Main() {
  #cho#SmallMeeting m = [...];
  #cho#Person p1 = [...], p2 = [...], p3 = [...], p4 = [...];
  m.#cho#hasParticipant =
```

```
6       new #cho#Participant[]{p1, p2, p3, p4};
7     #cho#Participant[] arrParticipants = m.#cho#hasParticipant;
8   }}
```

Given an instance m of a *SmallMeeting*, further persons may be added as participants of the meeting as shown in line 5 in the previous code snippet. Eventually, the ontological individual referred to by instance m becomes a standard *Meeting* while for the assignment expression in line 7, the instance m is still declared to refer to a *SmallMeeting*. Accordingly, a programmer will assume that – given the definition of the *SmallMeeting* concept in the ontology – there may be at most three participants in meeting m.

This is why Zhi# implements a form of dynamic typing. Every access of an ontological individual is dynamically type-checked such that the individual must be of the declared type in the Zhi# program. These type checks are enforced by the Zhi# run time, which raises an exception if a given individual is not of the required type. In this way, Zhi#'s dynamic type checking of OWL individuals is similar to structural typing and different to "duck typing" since not only the part of a type's structure that is accessed is checked, but, rather, the complete type of the object is checked. As a consequence, typing errors at run time are detected not only when a particular object property is accessed but also when the object itself is referenced. Accordingly, the Zhi# program above terminates with an exception in line 7 even before the property *hasParticipant* is accessed since instance m does not refer to an individual of the declared type *SmallMeeting* anymore.

For OWL object and data-type properties, the property assignment semantics are *additive*. A Zhi# assignment statement such as `a.#R = b` *adds* the individual *b* as a value of property *R* of individual *a*; it does not replace existing triples in the ontology model. In order to remove property-value assertions from the ontology, the Zhi# OWL plug-in provides the auxiliary methods `Remove` and `Clear` for OWL properties to remove one particular value and all values from an OWL property of the specified individual, respectively. The following statement removes all values of property *R* of individual *a*.

```
a.#R.Clear();
```

The auxiliary properties `Types` and `EquivalentIndividuals` that are defined for ontological individuals in Zhi# yield the RDF types and equivalent individuals for the given individual. The `Individuals` property defined for static ontological type references yields an array of individuals that are in the extension of the given OWL concept. Note that the `EquivalentIndividuals` and `Individuals` property are generic with respect to the RDF type of the host object.

27.3.3 Example Scenario Implementation

In Zhi#, the example scenario project meeting can be scheduled as shown ahead. The Zhi# compiler statically checks that the used concepts and properties are defined in the ontology. In line 6, the string literal is statically checked to denote a valid `xsd#dateTime` object. Values that are assigned to OWL data-type properties as shown in line 7 are statically checked, too, to be compatible with the data-type property's range restriction. Following the assignment of the ontological individual ALICE to the project meeting's *hasModerator* object property in line 8, the reasoner infers ALICE to be in the extension of the property's range restriction (i.e., Alice must be a moderator). The tight integration of OWL with C# programming language features makes it possible in Zhi# to use the `is` operator to dynamically check the RDF type of ontological individuals as shown in line 10. Also, OWL concepts and XSD type definitions can be used to define fields, properties, operators, and methods as shown in line 12.

```
1  import XML xsd = http://www.w3.org/2001/XMLSchema;
2  import OWL cho = http://chil.server.de/ontology;
3  class C {
4   public static void Main() {
5    #cho#Event e = new #cho#Event("#PROJECT_MEETING");
6    #xsd#dateTime dt = "2008-06-27T13:00:00Z";
7    e.#scheduledAt = dt;
8    e.#hasModerator = new #cho#Person("#ALICE");
9    [...]
10   if (e is #cho#ActiveMeeting) { [...] }
11  }
12  public #xsd#dateTime getStartTime(#cho#Event e) { [...] }
13 }
```

27.4 Conclusion

The authors developed an ontology in order to provide a formal high-level description of the CHIL domain of discourse that can be efficiently used to build intelligent applications. By using the description logics-based Web Ontology Language OWL DL for modeling concepts, properties, and individuals of the CHIL domain of discourse, it is possible to use a reasoner to automatically check the consistency of the knowledge base and make implicit knowledge explicit, which can be considered a form of artificial intelligence.

The authors devised and implemented a pluggable architectural model of an ontological knowledge base server in order to make the CHIL ontology available in the distributed heterogeneous CHIL computing environment. The CHIL knowledge base server can adapt off-the-shelf ontology management systems and expose their functionality by means of remoting technologies.

The CHIL knowledge base server implements the formally specified CHIL OWL API, which was defined based on a combination of Floyd-Hoare logic and formal description logics terminology. The formal specification of the CHIL OWL API was devised in order to make it possible to consistently adapt off-the-shelf ontology management systems and to provide knowledge base clients with well-defined programming language-independent semantics of the OWL DL API. Regression test cases and client libraries for Java, C#, C++, and Python can be automatically generated from the CHIL OWL API specification.

The authors developed a compiler framework that makes programming language features of the host language extensible with respect to external typing mechanisms (e.g., subsumption, type derivation, type inference) in order to regain compiler support for XSD and OWL type definitions. In the current implementation, the Zhi# compiler framework extends the C# programming language with two external compiler plug-ins, which provide type-checking and program transformation functionality for XML Schema Definition [3] and the Web Ontology Language OWL [11].

Zhi# programs are compiled into conventional C# code. In contrast to approaches based on wrapper classes or additional code generation, the program overhead of compiled Zhi# programs is constant and does not grow with, for example, the number of referenced XSD or OWL types.

The presented work demonstrates that it is possible to devise and use ontological domain models in order to benefit from automatic reasoning while at the same time preserving compiler support such as type checking of ontological concept descriptions.

References

1. F. Baader, D. Calvanese, D. McGuiness, D. Nardi, and P. Patel-Schneider, editors. *The Description Logic Handbook*. Cambridge University Press, Cambridge, UK, 2003.
2. S. Bechhofer. The DIG description logic interface: DIG/1.0. Technical report, University of Manchester, 2002.
3. P. V. Biron and A. Malhotra. XML Schema Part 2: Datatypes Second Edition. Technical report, World Wide Web Consortium (W3C), Oct. 2004. `http://www.w3.org/TR/xmlschema-2/`.
4. D. Brickley and R. Guha. RDF Vocabulary Description Language 1.0: RDF Schema. Technical report, World Wide Web Consortium (W3C), Feb. 2004. `http://www.w3.org/TR/rdf-schema/`.
5. A. Hejlsberg, S. Wiltamuth, and P. Golde. C# language specification. Technical report, ECMA International, Jun. 2006. `http://www.ecma-international.org/publications/standards/Ecma-334.htm`.
6. C. A. R. Hoare. An axiomatic basis for computer programming. *Communications of the ACM (CACM)*, 12(10):576–580, 1969.
7. I. Horrocks and P. F. Patel-Schneider. Reducing OWL entailment to description logic satisfiability. *Journal of Web Semantics*, 1(4):345–357, 2004.
8. I. Horrocks, F. van Harmelen, and P. Patel-Schneider. DAML+OIL. Technical report, DARPA's Information Exploitation Office and European Union's Information Society Technologies, Mar. 2001. `http://www.daml.org/2001/03/daml+oil-index.html`.
9. HP Labs. Jena Semantic Web Framework, 2004. `http://www.hpl.hp.com/semweb/jena.htm`.
10. F. Manola and E. Miller. RDF Primer. Technical report, World Wide Web Consortium (W3C), Feb. 2004. `http://www.w3.org/TR/rdf-primer/`.

11. D. L. McGuinness and F. van Harmelen. OWL Web Ontology Language Overview. Technical report, World Wide Web Consortium (W3C), Feb. 2004. http://www.w3.org/TR/owl-features/.
12. M. Minsky. *Mind Design*, chapter A framework for representing knowledge. MIT Press, Cambridge, MA, 1981. A longer version appeared in *The Psychology of Computer Vision* (1975).
13. R. Möller and V. Haarslev. RACER: Renamed ABox and Concept Expression Reasoner, 2004. Technische Universität Hamburg-Haburg.
14. B. Motik. KAON2, 2006. http://kaon2.semanticweb.org.
15. Open RDF. Sesame RDF Database, 2006. http://www.openrdf.org.
16. A. Paar. Zhi# – programming language inherent support for ontologies. In J.-M. Favre, D. Gasevic, R. Lämmel, and A. Winter, editors, *ateM '07: Proceedings of the 4th International Workshop on Software Language Engineering*, pages 165–181, Mainz, Germany, Oct. 2007.
17. A. Paar, J. Reuter, J. Soldatos, K. Stamatis, and L. Polymenakos. A formally specified ontology management API as a registry for ubiquitous computing systems. *The International Journal of Artificial Intelligence, Neural Networks, and Complex Problem-Solving Technologies (Applied Intelligence)*, 2007.
18. A. Paar and W. F. Tichy. Zhi#: Programming language inherent support for XML Schema Definition. In W.-T. Tsai and M. Hamza, editors, *SEA '05: Proceedings of the 9th IASTED International Conference on Software Engineering and Applications*, pages 407–414, Anaheim, CA, Nov. 2005.
19. Pellet, 2006. http://pellet.owldl.com.
20. M. R. Quillian. Word concepts: A theory and simulation of some basic capabilities. *Behavioral Science*, 12:410–430, 1967.
21. J. Reuter. Ontological processing of sound resources. In *Proceedings of the 4th International Linux Audio Conference (LAC2006)*, pages 97–104. Zentrum für Kunst und Medientechnologie (ZKM), Apr 2006.
22. V. Roubtsov. Emma: A free Java code coverage tool, 2007. http://emma.sourceforge.net.
23. Stanford University School of Medicine. Protégé knowledge acquisition system, 2006. http://protege.stanford.edu.

28

Building Scalable Services: The CHIL Agent Framework

Axel Bürkle[1], Nikolaos Dimakis[2], Ruth Karl[1], Wilmuth Müller[1], Uwe Pfirrmann[1], Manfred Schenk[1], Gerhard Sutschet[1]

[1] Fraunhofer Institut für Informations- und Datenverarbeitung, IITB, Karlsruhe, Germany
[2] Athens Information Technology, Peania, Attiki, Greece

The services realized within the CHIL project are implemented by a set of collaborative software agents communicating with each other on a semantic level. In order to ensure this collaboration as well as a scalable service composition, coordination, and configuration, an agent framework and infrastructure was designed. A special feature of the CHIL agent infrastructure is the "pluggable behaviors" mechanism. This concept allows implementing service-specific code in agent behaviors, which will be plugged into the agents. It keeps the agent free from service functionality and enables a service-oriented scalable configuration. Service-specific communication ontologies can be plugged into the system without recompiling the source code. Furthermore, the autonomy feature of the CHIL architecture facilitates self-healing and restarting of agents, both in a stateless mode and in a stateful mode, while a directory service leverages a knowledge base that services requests and handles registration of any component in the architectural framework.

28.1 The CHIL Agent Infrastructure

Ubiquitous services are usually based on complex heterogeneous distributed systems comprising sensors, actuators, perceptual components, and information fusion middleware. In projects like CHIL, where a number of service developers concentrate on radically different services, it is of high value that a framework ensures reusability in the scope of a range of services. To this end, we have devised a multiagent framework that meets the following target objectives:

- facilitates integration of diverse context-aware services developed by different service providers;
- facilitates services in leveraging basic services (e.g., sensor and actuator control) available within the smart rooms;
- allows augmentation and evolution of the underlying infrastructure independent of the services installed in the room;
- controls user access to services;

- supports service personalization by maintaining appropriate profiles;
- enables discovery, involvement, and collaboration of services.

28.1.1 Software Agents

> *An agent is a computer system, situated in some environment, that is capable of flexible autonomous action in order to meet its design objectives.*

This definition by Jennings et al. [6] emphasizes three key concepts – situatedness, flexibility, and autonomy – which perfectly meet the main requirement for the CHIL architecture: to support the integration and cooperation of autonomous, context-aware services in a heterogeneous environment. Summarized, the major goal for the infrastructure incorporates the discovery, involvement, and collaboration of services as well as competition between services in order to perform a certain task the best way possible. Standardized communication mechanisms and protocols have to be considered to raise information exchange onto a semantic level and to ensure location transparency.

The following sections describe the CHIL agent infrastructure and how we achieved these objectives. Moreover, they demonstrate how we realized a multiagent system that is capable to "solve problems that are beyond the individual capabilities or knowledge of each problem solver" [6].

28.1.2 Agent Description

The CHIL software agent infrastructure covers the upper two layers of the CHIL architecture. Agents and components close to the user are situated in the *user front-end* layer, e.g., the user's personal agent and the device agents, whereas the *services and control* layer contains the basic agents and communication as well as the service agents. Figure 28.1 shows an excerpt of the agent infrastructure.

Basic Agents

CHIL agent: CHIL agent is the basic abstract class for all agents used in the CHIL environment. It provides methods for basic agent administrative functionality (setup, takedown), directory facilitator service (*DF Service*) functions (register/deregister agents, modify agent descriptions, search service-providing agents based on a semantic service description ontology), and additional supporting utility functions for creating and sending messages, extracting message contents, and logging. Special importance is attached to keep the agent communication conforming to FIPA (*Foundation for Intelligent Physical Agents*) standards [3], i.e., to comply with the FIPA Interaction Protocols and the FIPA Communicative Acts. The message transfer is based on a well-defined communication ontology including agent actions, content concepts, and predicates.

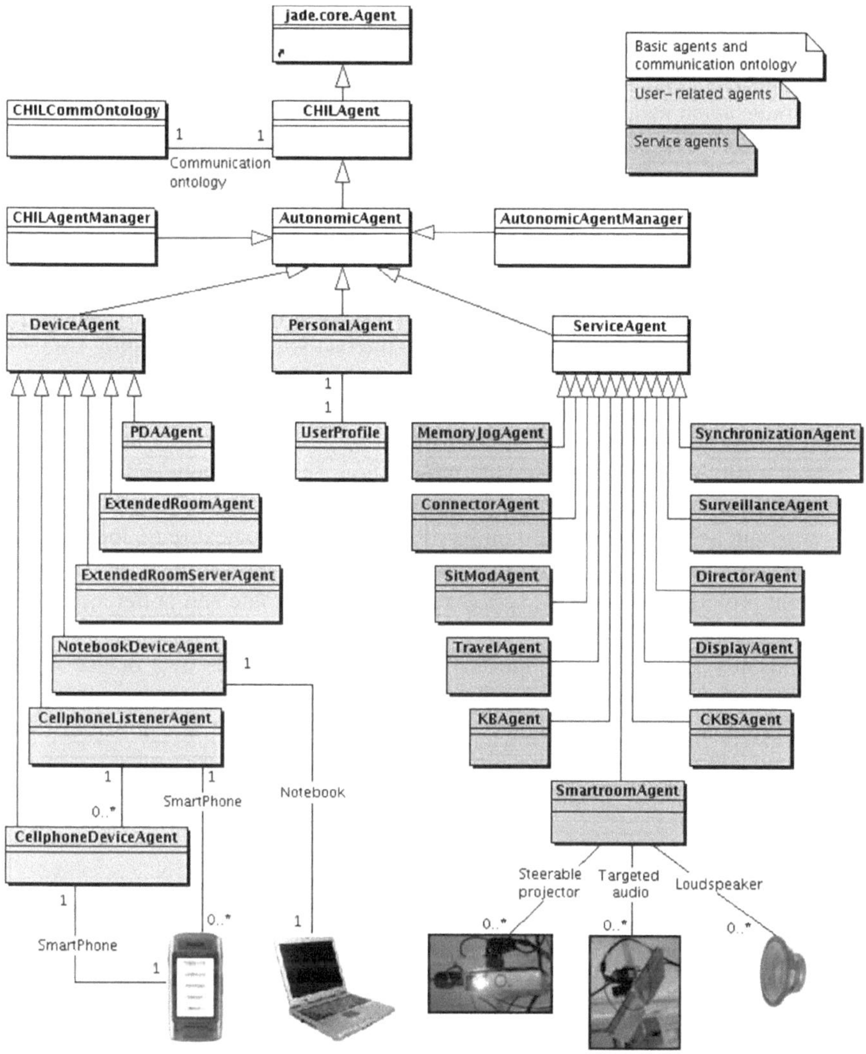

Fig. 28.1. The CHIL agent infrastructure.

CHIL agent manager: The CHIL agent manager is a central instance encapsulating and adding functionality to the JADE [5] Directory Facilitator (DF). Other CHIL agents register their services with the agent manager, including required resources for carrying out that service. The CHIL agent manager can act as a matchmaker or as a broker. As a matchmaker, it provides a requesting agent with a handle to appropriate service agents that are capable of satisfying the request. As a broker, it divides a request into subrequests, generates a task schedule, forwards the subrequests to appropriate service agents, composes the results provided by the service agents into an overall result, and forwards this result to the initial requestor.

Autonomic agent and autonomic agent manager: The autonomic agent and the autonomic agent manager provide self-healing control. The autonomic agent is the base class for all agents requiring this feature. A detailed description of the agent autonomy is given in Section 28.5.

User-Related Agents

Personal agent: Every person in the CHIL environment has his own personal agent, which acts as a personal secretary. Users interact with the system only via their self-adapting personal agents assigned during the login procedure. The personal agent manages (via dedicated device agents, bound to specific devices) interactions with its master: It knows what front-end devices its master has access to, how it can best receive or provide information, and what input and notification types he prefers. Moreover, the personal agent communicates with the situation watching agent to be updated on its master's current context (e.g., location, activity) in order to act or react in an appropriate way. Furthermore, the personal agent provides and controls access to its master's profile and preferences, thus ensuring user data privacy.

Device agents: Each device in the CHIL environment has its own device agent, which manages communication between the device and a user's personal agent. For example, the notebook device agent is a concrete implementation of a device agent handling the communication between the graphical user interface on the user's notebook and his personal agent.

Service Agents

Service agents implement the services provided by CHIL. Based on the common ontology, they map the syntactical level of the services to the semantical level of the agent community. They may engage other service agents to supply their own service.

Connector agent: The connector agent comprises the core connector functionality: manage intelligent communication links. It can handle both bilateral and multilateral connection and notification requests. This includes having the connector agent store all pending connections and finding a suitable point of time for these connections to take place.

Travel agent: The Travel Agent can arrange and rearrange itineraries according to the user's profile and current situation. It uses a simulated Semantic Web interface to gain information about flights and train connections. The simulated interface gathers the required data from several online resources on the Internet.

Memory Jog agent: The Memory Jog agent maintains and controls a data model of the current scenario in the smart room. This model holds information regarding the current event, the agenda, the participants, their locations and IDs, etc. The

Memory Jog accepts registrations by personal agents of participants of the current event and broadcasts this data model to all these registered personal agents, which forward the changes to the appropriate device agents.

Smart room agent: The smart room agent controls the service-to-user notification inside the smart room using appropriate devices (loudspeaker, steerable video projector, targeted audio).

Situation watching agent: The situation watching agent wraps the situation modeling layer. It manages both access to the situation model and subscriptions to special events from other agents using an access API and a subscription API to the situation model.

28.2 Intelligent Messaging

In order to ensure that the semantic content of tokens is preserved across agents, the information exchange between agents is based upon a well-defined communication ontology, as proposed in the FIPA Abstract Architecture Specification [3]. The importance of such a common semantic concept is heightened by the facts that the services are implemented by various service developers in different places and these service developers must understand each other correctly.

The CHIL Communication Ontology is completely defined using the Web Ontology Language OWL [8] and fully integrated in the overall CHIL domain ontology. It is based upon the "Simple JADE Abstract Ontology", an elementary ontology provided by JADE, which must be used as root for every ontology-based message exchange within the JADE agent management system. Built upon the basic JADE classes AID (agent identifier), Concept, AgentAction, and Predicate, the CHIL communication ontology extends the core CHIL ontology by tokens, which are necessary for the agent communication, particularly agent actions for requesting services from other agents and concepts such as answers to these requests if they are not already defined in the CHIL ontology.

The ontology concepts are used in the Java environment by means of the JADE *abs* package, found in *jade.core.abs*. This package provides several classes for "abstract" concepts, agent actions, predicates, etc., whereas "abstract" in this sense means that the concepts are not transformed to concrete Java classes. The handling of these semantic entries is implemented by the central *CHILAgent*, a basic abstract class for all agents in the CHIL system (cf. Fig. 28.1), providing them for the complete CHIL agent community. Besides the methods already described in Section 28.1, the CHILAgent particularly provides ontology-based messaging methods for encoding, sending, receiving, and decoding Agent Communication Language (ACL) messages. Furthermore, these methods, together with initiator and responder classes for submitting and receiving messages, ensure that the agent communication is strictly compliant to the FIPA interaction protocols and communicative acts.

28.3 Pluggable Behaviors

One of the major goals of the CHIL agent infrastructure is to provide a mechanism that allows a distributed development of services and an easy integration and configuration of multiple services in the CHIL system. A simple service can easily be integrated by creating an agent to handle the framework tasks and control the service, and then integrating this agent in the CHIL system. In this way, the agent acts as a wrapper for the service. In simple cases, the agent could also embed the service logic itself.

Usually, a service is more complex and requires new functionality in other agents, too. The Memory Jog service (see Section 18.1), for example, uses the *personal agents* to interface the users with the CHIL system, the *situation modeling agent* to retrieve information about the current status and location of each participant in the smart room, and the *smart room agent* to present information and messages on the various output devices. Therefore, the Memory Jog service needs to implement message exchange functionality in several agents. Implementing this functionality in the agents themselves would mean that several service providers have to modify the agent's code. This method would quickly raise significant problems in coordinating the implementation and configuration of software components. Agents dealing with multiple services, resulting in service-specific code inside the agent, show the critical points of distributed development.

Hence, a plug-in mechanism has been designed that allows an easy integration of agent handlers from different services developed by several partners. By using this mechanism, all service-specific code will be moved to these pluggable handlers, and the agent itself becomes service-independent. It contains only the common methods and attributes all concerned partners have agreed on and thus becomes a stable module. Three types of pluggable handlers are considered necessary:

1. **setup handler**: handles service specific initialization in the setup phase of an agent;
2. **event handler**: is registered for certain events from outside the agent world, e.g., the user's GUI, a perceptual component, the situation model, or a Web service;
3. **pluggable responder**: is triggered by incoming ontology-based messages from other agents.

Example of a Pluggable Behavior

Table 28.1 illustrates an example of a pluggable behavior, a responder that informs a user (i.e., the user's personal agent) about connection requests from other users. This responder (exclusively) accepts ACL request messages using the agent action *InformAboutConnect*. More precisely, the responder uses ontology-based behavior registration and reacts on ACL messages of the type *REQUEST*, which contain an instance of the ontology concept *InformAboutConnect*.

In order to be handled correctly by the plug-in mechanism, the *InformAboutConnectResponder* has to implement the *getAcceptedMessages* and *getBehavior* methods of the *PluggableResponder* interface. The *getAcceptedMessages* method returns

```
public class InformAboutConnectResponder extends CHILSimpleAchieveREResponder
   implements PluggableResponder {

   public InformAboutConnectResponder(Agent agent) {
      super(agent, ((PersonalAgent)agent).matchAbsOntoRequestAction
                   (new AbsAgentAction("InformAboutConnect")));
   }

   public MessageTemplate getAcceptedMessages() {
      return getPA().matchAbsOntoRequestAction
             (new AbsAgentAction("InformAboutConnect"));
   }

   public Behavior getBehavior() {
      return this;
   }

   protected ACLMessage prepareResponse(ACLMessage request)
      throws CHILFipaRefuseException {
      ...
      return response;
   }

   protected ACLMessage prepareResultNotification
      (ACLMessage request, ACLMessage response)
      throws CHILFipaFailureException {
      ...
      return reply;
   }
}
```

Table 28.1. An example of a pluggable responder for the personal agent accepting ontology-based messages.

a JADE *MessageTemplate* to indicate the type of message the responder accepts. The behavior uses the CHILAgent's *matchAbsOntoRequestAction* method to generate an ontology-based message template that causes the responder to react on agent actions of the type *InformAboutConnect*. The *getBehavior* method returns the concrete behavior object, which realizes the responder's functionality. In this case, it is the responder itself, but the developer may also create and return a separate object.

The *prepareResponse* and *prepareResultNotification* methods implement the agree/refuse and failure/inform paths of the FIPA Request Interaction Protocol, thus ensuring FIPA compliance. *PrepareResponse* returns an agree or a refuse message or, alternatively, throws a *CHILFipaRefuseException* to refuse the request. *PrepareResultNotification* returns an inform or a failure message; throwing a *CHILFipaFailureException* results in the return of a failure message.

28.4 Scalable Services

The *Pluggable Behaviors* mechanism allows a service provider to plug new functionality into multiple agents without the need to recompile. Moreover, it supports the facility of the CHIL agent framework for interagent communication on a semantic level, based on the CHIL communication ontology. However, to fully exploit the features of the CHIL system and other services and to go beyond the limits of the predefined CHIL communication concepts, a service must be able to define new messages that are understood and can be compiled by other services. These new messages must be handled in the same manner as the pluggable behaviors: They must be able to be plugged into the system without the need for recompiling the source code in the integration phase.

To this end, the *Pluggable Behaviors* mechanism has been extended to *Pluggable Services* by realizing a plug-in method for service-specific communication ontologies; each service provider may create his own service communication ontology. An example can be seen in Table 28.2.

```
<?xml version="1.0"?>
<rdf:RDF
    xmlns:rdf="http://www.w3.org/1999/02/22-rdf-syntax-ns#"
    xmlns:xsd="http://www.w3.org/2001/XMLSchema#"
    xmlns="http://www.owl-ontologies.com/YourServiceOntology.owl#"
    xmlns:rdfs="http://www.w3.org/2000/01/rdf-schema#"
    xmlns:owl="http://www.w3.org/2002/07/owl#"
    xmlns:ca="http://chil-svn.ira.uka.de/ontologies/chil/chil-ca.rdf#"
    xmlns:dc="http://purl.org/dc/elements/1.1/"
    xml:base="http://www.owl-ontologies.com/YourServiceOntology.owl">
   <owl:Ontology rdf:about="">
      <owl:imports
       rdf:resource="http://chil-svn.ira.uka.de/ontologies/
                     chil/chil-ca.rdf"/>
   </owl:Ontology>
   <owl:Class rdf:ID="YourServiceRootClass">
      <rdfs:subClassOf
       rdf:resource="http://chil-svn.ira.uka.de/ontologies/
                     chil/chil-ca.rdf#CommConcept"/>
   </owl:Class>
   <owl:Class rdf:ID="YourServiceAgentAction">
      <rdfs:subClassOf rdf:resource="#YourServiceRootClass"/>
   </owl:Class>
   <owl:Class rdf:ID="YourServiceConcept">
      <rdfs:subClassOf rdf:resource="#YourServiceRootClass"/>
   </owl:Class>
</rdf:RDF>
```

Table 28.2. Sample service ontology defined using OWL.

After generating an ontology class (preferably using the CHIL tool *JadeOntologyGenerator*) and implementing the required agents and agent behaviors, the service provider only has to specify her new service in an XML-based configuration file to integrate it in the CHIL system. This file defines the agents, which participate in the service, their behaviors, and the service ontology. Each pluggable behavior is defined by its type (responder, event, setup) and its classname. A priority value assigned to each behavior can be used to determine the order of execution, which may particularly be important for setup behaviors. The service ontology is specified by a name, the namespace, the location, and the classname of the generated ontology class.

Furthermore, the configuration file provides an additional feature to system developers and administrators: It is possible to disable or enable certain functionality just by removing or adding, respectively, the appropriate elements in the configuration file without having to recompile the source code. Table 28.3 shows an example of a service configuration file. In this example, *YourService* adds to two agents: a responder to the travel agent and an event handler to *YourAgent*.

```
<?xml version="1.0" encoding="UTF-8"?>
<serviceconfig version="1.1"
 xmlns:xsi="http://www.w3.org/2001/XMLSchema-instance"
 xsi:noNamespaceSchemaLocation="../xmlschema/CHIL_ServiceConfig_1.1.xsd">
   <service name="YourService">
      <agent name="YourAgent">
         <handler type="event"
                  priority="1"
                  classname="de.yourNamespace.service.handler.yourAgent.
                             YourServiceEventHandler"/>
      </agent>
      <agent name="TravelAgent">
         <handler type="responder"
                  priority="1"
                  classname="de.yourNamespace.service.handler.travelAgent.
                             YourServiceResponder"/>
      </agent>
   </service>
   <ontology
      name="YourOntology"
      namespace="http://www.owl-ontologies.com/YourServiceOntology.owl"
      locationPath="/$ChilHome/lib/yourService.jar"
      className="de.yourNamespace.ontology.yours.YourOntology">
   </ontology>
</serviceconfig>
```

Table 28.3. Sample service configuration file using multiple agents and a service-specific communication ontology.

At startup, each agent parses the service configuration files to determine which behaviors have to be instantiated and added to the agent's behaviors queue and the ontology class to be loaded. Since the necessary code for this mechanism is concentrated in the basic *CHILAgent* and a few helper classes, the source of the configuration data can be changed easily. For example, by using the *CKBSAgent*, which interfaces with the knowledge base, it is possible to get this data from a knowledge base instead of the XML file. Furthermore, the plug-in mechanism is available to all agents derived from *CHILAgent* without extra work for the agent developers.

In the same way a service is informed about all participating agents, the CHIL system is informed about all participating services: A master configuration file, also based on XML, specifies the services that are activated upon system startup, by their names and configuration files. A sample master configuration file is shown in Table 28.4.

28.5 Autonomy

Fault tolerance has been discussed excessively in the literature, as in [2, 7, 10]. Our approach in providing autonomous behavior of both the context acquisition layer and

```
<?xml version="1.0" encoding="UTF-8"?>
<services
   version="1.0"
   xmlns:xsi="http://www.w3.org/2001/XMLSchema-instance"
   xsi:noNamespaceSchemaLocation="../xmlschema/CHIL_Services_1.0.xsd">

   <service name="CoreService"      configFile="CoreService.xml"/>
   <service name="MeetingService"   configFile="MeetingService.xml"/>
   <service name="ConnectorService" configFile="ConnectorService.xml"/>
   <service name="TravelService"    configFile="TravelService.xml"/>
   <service name="SmartroomService" configFile="SmartroomService.xml"/>
   <service name="SitWatchService"  configFile="SitWatchService.xml"/>
   <service name="AutonomicService" configFile="AutonomicService.xml"/>
   <service name="GatewayService"   configFile="GatewayService.xml"/>
</services>
```

Table 28.4. The master configuration file for services.

the end-service delivery has been presented in [1]. In this section, we will outline the structural components that provide this features, and we will focus mostly on the flexibility of the introduced mechanism.

Service Autonomy

In our implementation, some dedicated agents play a very specialized role. The self-healing control is managed by two of the core agents, the *AutonomicAgent* and the *AutonomicAgentManager*. The *AutonomicAgent* is the base class for all agents requiring self-healing handling by the architectural framework. Exploiting the benefits of the JADE framework for interagent communication at regular intervals the *AutonomicAgentManager* queries JADE's Agent Management System for the status of specific agents. The result of this request reflects the current status of all registered agents participating in the framework. In the case of a dead agent, the *AutonomicAgentManager* initiates the regeneration sequence. This sequence is dependent on the type of self-healing the agent has requested during registration. We consider two types of registration, which signify a different handling method: stateless and stateful handling. The agents specify their registration type during their registration to the *AutonomicAgentManager* as soon as they boot up.

Stateless Handling

Stateless handling implies that the agent does not maintain any state during its execution. The regeneration sequence for this type of agent is a straightforward procedure based on restarting the agent, if necessary on a different platform. The agent registers again and participates in the framework by processing incoming messages. We follow this approach in the cases when agents are assigned to control actuating devices, such as the targeted audio device or projectors. The reason is that such agents wait for incoming requests and do not maintain a state during the execution, as they actually act as message receptors.

Stateful Handling

When an agent needs a state to be maintained during its execution, the *AutonomicAgentManager* behaves differently. At regular intervals, each *AutonomicAgent* stores a serialized object in a database followed by a timestamp. This procedure is done using the Hibernate library [4], which enables such data manipulation. As soon as the agent dies and the *AutonomicAgentManager* is aware of this fact, it attempts to restart the corresponding agent, which in turn reads the last entry of its serialized object from the database. As soon as the old state is reloaded, the agent is able to participate again in the framework.

28.6 Directory Service

Sophisticated context-aware services require the presence of a directory service mechanism for registration and later lookup. In our implementation, we have developed a directory service leveraging a knowledge base that services requests and handles registration of any component in the architectural framework [9].

The managing agents are the *CKBSAgent* and *KBAgent*, which wrap around the knowledge base and database, respectively. The two agents interpret agent requests into ontology and SQL queries and are appropriately diverting the request to the corresponding service. In the case of the database, the resulting reply is compiled to be a member of the communication ontology, described in Section 28.2. This enables the bundling of the information to be transmitted as a single ontology class, such as "Camera", which contains information about the camera type, the location of the camera, etc.

28.7 Conclusion

In this chapter, we have presented a distributed agent framework allowing developers of different services to concentrate on their service logic, while exploiting existing infrastructures for perceptual processing, information fusion, and sensors and actuators control. The core concept of this framework is to decouple service logic from context-aware and sensor/actuator control middleware. Hence, service logic can be "plugged" into a specific placeholder based on well-defined interfaces. The agent framework has been implemented based on the JADE environment, and accordingly instantiated within real-life smart rooms comprising a wide range of sensors and context-aware middleware components. The benefits of this framework have been manifested in the development of different applications.

References

1. N. Dimakis, J. Soldatos, L. Polymenakos, M. Schenk, U. Pfirrmann, and A. Bürkle. Perceptive middleware and intelligent agents enhancing service autonomy in smart spaces.

In *IEEE/WIC/ACM International Conference on Web Intelligence and Intelligent Agent Technology*, pages 276–283, Hong Kong, Dec. 2006.
2. N. Faci, Z. Guessoum, and O. Marin. Dimax: a fault-tolerant multi-agent platform. In *SELMAS '06: Proceedings of the 2006 International Workshop on Software Engineering for Large-Scale Multi-Agent Systems*, pages 13–20, New York, NY, 2006.
3. FIPA. The foundation for intelligent physical agents. `http://www.fipa.org`.
4. Hibernate. Relational persistence for java. `http://www.hibernate.org`.
5. JADE. Java Agent DEvelopent Framework. `http://jade.tilab.com`.
6. N. R. Jennings, K. Sycara, and M. Wooldridge. A roadmap of agent research and development. *Journal of Autonomous Agents and Multi-Agent Systems*, 1(1):7–38, 1998.
7. S. Kumar and P. R. Cohen. Towards a fault-tolerant multi-agent system architecture. In *AGENTS '00: Proceedings of the Fourth International Conference on Autonomous Agents*, pages 459–466, New York, NY, 2000.
8. OWL. World Wide Web Consortium (W3C), `http://www.w3.org/OWL/`, 2004.
9. I. Pandis, J. Soldatos, A. Paar, J. Reuter, M. Carras, and L. Polymenakos. An ontology-based framework for dynamic resource management in ubiquitous computing environments. In *2nd International Conference on Embedded Software and Systems*, Dec. 2005.
10. A. F. Zorzo and F. R. Meneguzzi. An agent model for fault-tolerant systems. In *SAC '05: Proceedings of the 2005 ACM Symposium on Applied Computing*, pages 60–65, New York, NY, 2005.

29

CHIL Integration Tools and Middleware

Jan Cuřín, Jan Kleindienst, Pascal Fleury

IBM Czech Republic, Praha, Czech Republic

In this chapter, we introduce two integration tools for smart environments: SITCOM and *CHiLiX*. *CHiLiX* is an eventing middleware component for receiving messages from remote perceptual components and for controlling their status from services and situation modeling components. SITCOM is a platform for developing context-aware services, editing and deploying context situation models, and simulating perceptual components. SITCOM, which stands for Situation Composer, is constructed as an open and extensible framework with rich graphical rendering capabilities, including 3D visualization. One of SITCOM's main goals is to simulate interactions among people and objects in various settings such as presentation rooms, meetings halls, social places, automobiles, etc. Moreover, SITCOM captures and models such environmental context and delivers the abstracted information to context-aware user services. Contextual models are designed as pluggable modules. They can be configured to build hierarchies of contextual models, and deployed and reused across applications. SITCOM allows the plugging-in of various multimodal sensing components such as body trackers, face detectors, speech recognition engines, gesture recognizers, etc. Therefore, switching from simulated to real components with SITCOM is seamless, typically involving no change in the contextual models and the services.

During the process of architecture design and implementation in the CHIL project [1], we had to decide whether it is better to use an existing pervasive computing framework (ROS, Gaia, Aura, Solar, UbiREAL, Context toolkit) or to build a new system fitting exactly our requirements, i.e., to be OS-independent and -distributable, having logging and simulation capability, and supporting the integration of components in Java and C++ at least. We have identified three reliable pervasive computing systems satisfying a majority of our criteria: UbiREAL [9] and Gaia [10]. But in contrast to these two systems, which integrate contextual information directly from various sensors, we needed a system that utilizes information provided by more complex perceptual components, i.e., more complex context acquisition components such as person trackers and speech recognizers. A body tracker might be simultaneously capable of detecting the location, heading, and posture of persons, identifying them, and tracking subjects of their interest. Such scenarios led us to separate the perceptual components layer from the layer that deals with higher abstraction – the situation modeling. Defining and modeling situations based on a wide range of context acquisition components was not supported by other environments such as UniREAL and Context Toolkit [5], so we decided to implement a new framework based on two tools: SITCOM and *CHiLiX*. Our system is used as an integrator for the perceptual components providers and is capable of dynamically exchanging perceptual components of the same kind as well as enabling the seamless replacement of these components by simulators, a requirement from service developers.

SITCOM is a tool and run-time application that deals with editing, configuring, implementing, and deploying context-aware applications [6]. As a complement, *CHiLiX* deals with the communication of components spanning a typical pervasive computing system. Using SITCOM and *CHiLiX*, developers can simulate and test perceptual and situation modeling components in complex scenarios.

29.1 SITCOM: Situation Composer

SITCOM (Situation Composer) is a simulation tool and run-time application for the development of context-aware applications and services. Context-aware applications draw data from the surrounding environment (such as an ongoing meeting in the room, a person's location, body posture, etc.), and their behavior depends on the respective situation (e.g., while in meeting, mute the phone). In SITCOM, the environmental characteristics are captured by situation models that receive input events from real sensor inputs (cameras, microphones, proximity sensors, etc.), simulated data, or a combination of real and simulated inputs. SITCOM allows for the composition of situation models into hierarchies to provide event filtering, aggregation, and combination to construct higher-level meaning. Through the IDE controls, SITCOM also facilitates the capture and creation of situations (e.g., a 10-minute sequence of several people meeting in a conference room) and their subsequent realistic rendering as 3D scenarios. These scenarios can be replayed to invoke situations relevant for the application behavior, and thus provide a mechanism for the systematic testing of the context-aware application under different environmental conditions.

29.1.1 SITCOM Input

SITCOM operates with both real and simulated input. Simulated input includes recorded data, which can be either synthetic or real. Simulated input is particularly important at the beginning of a pervasive systems project, when very sparse data are usually available about most parts of a complex system. In such cases, working with synthetic data is the best way to bootstrap the development of situation modeling and service design. SITCOM supports creation of synthetic data, while also allowing service testing and simulation with such data. Thus, simulated input can be split into two aspects: input data generation and input data use.

Along with generating data, SITCOM can also record and/or store data using the same data formats. Data recordings can be useful in recording services in operation, and subsequently replaying or postprocessing them. Using SITCOM, recorded events and services can be viewed multiple times based on different views (e.g., cameras) and situation models. SITCOM is oblivious to the origin of the events that it receives. This provides support for interactive simulated input, i.e., input that mixes real-time live information with recorded information. Interactive simulated input is handy in cases where part of the system has to be tightly controlled (e.g., for recording or demonstration purposes). Information mixing in SITCOM can be based on the following methods:

- Manual triggering: This involves the lower level of interactivity, where a button is used to start a particular scenario. Manual triggering is most useful in a demonstration mode, as one can simulate any event that might be of interest for a scenario.
- Manually inserting entities: This allows SITCOM users to interact with the world model as perceived by the situation model. For example, in a meeting simulation, one could manually add a person or two, and observe how the situation model or service reacts to this new event stream.
- Interacting with a mock-up: This provides the highest level of interactivity by having the models use a mock-up of the scenario. The mock-up will make entities behave in some programmed way. An example would be to simulate an incoming call to a (live) participant, and have the participant decide if she would answer. The mock-up would then have to deal with that fact, and may be calling back later if the call has not been answered, or playing a voice and not calling back if the participant has answered.

All three methods have been proved to be very valuable tools for the verification of situation models, demonstrations, as well as service development.

29.1.2 Internal Structures

SITCOM relies on data representation structures, which are specially designed to bridge context information and pervasive services in smart spaces. The basic unit visible at the context acquisition level is an entity. An *entity* is characterized by a type (e.g., person, whiteboard, room) and its access to a set of property streams. A *property stream* represents a set of events for a particular aspect of the entity. As an example, properties for an entity of the type "person" are location (indicating the physical location of the person), heading (tracking direction the person is looking at), and identity (i.e., a stream sending events about the identification of the person). At the *situation modeling* layer, entities are stored in a container called the entity repository. The *entity repository* is capable of triggering events upon creation and registration of new entities. Accordingly, situation models and services can be notified about changes for a particular *entity* or *stream*, based on an appropriately designed subscription mechanism.

For processing the contextual information, wherin the concept of a situation network [2] is used, we introduce abstraction of *situation machines*. *Situation machines*

(SM) interpret information inferred from observing entities in the entity repository and current states of other situations. As SMs may use different techniques, such as a rule-based or statistical approach, SITCOM can accommodate a mix of models at the same time. More details about the actual implementation are given in Section 29.3.

SITCOM itself plugs into the layer stack at the situation modeling level. So it sees the service layer as its upper layer, and the perceptual components as its lower layer. Also, as SITCOM is a framework for modeling systems enabling the plugging-in of diverse situation modeling technologies, it is split into two distinct parts: the simulation framework, named SITCOM, and the actual situation modeling being developed, named SITMOD. In Fig. 29.1, the two parts are clearly separated.

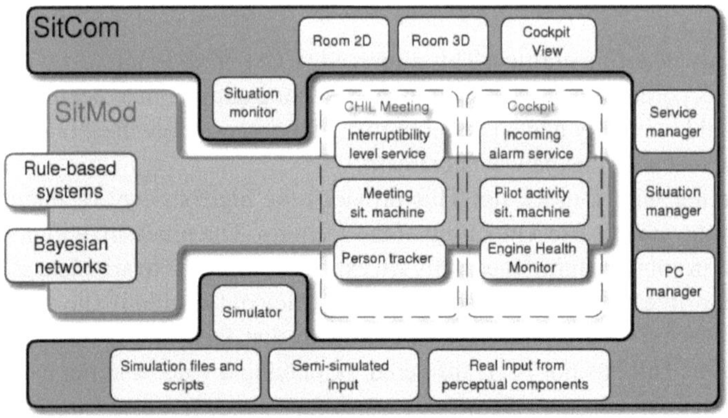

Fig. 29.1. Architecture diagram overview of SITCOM, the Situation Composer, and SITMOD, the situation model framework.

SITMOD (and modules plugged into it) is the portion of the system that will eventually be deployed in the field. During simulation, but also once deployed, nothing prevents the SITMOD from running situation models for multiple (possibly unrelated) scenarios. In Fig. 29.1, a meeting scenario and a cockpit scenario are running at the same time. SITMOD has modules for all the mentioned layers we model: a *person tracker* and an *engine health monitor* at the perceptual component level, which present their input to the situation model; a *meeting situation machine* and a *pilot activity situation machine* at the situation modeling level, which are scenario-specific; and an *interruptibility service* and an *incoming alarm service* at the service level, presenting the visible output of the system.

Through its GUI, SITCOM can enable the easy setup of the experiment in loading the situations, perceptual components, and services through its set of manager components for service, situation, and PCs. It can then feed SITCOM with its simulator components, sending either synthetic or recorded data from script files, having semisimulated components, or handling the activity of real components. On the upper side, the output of the situation model and the services can be displayed in raw

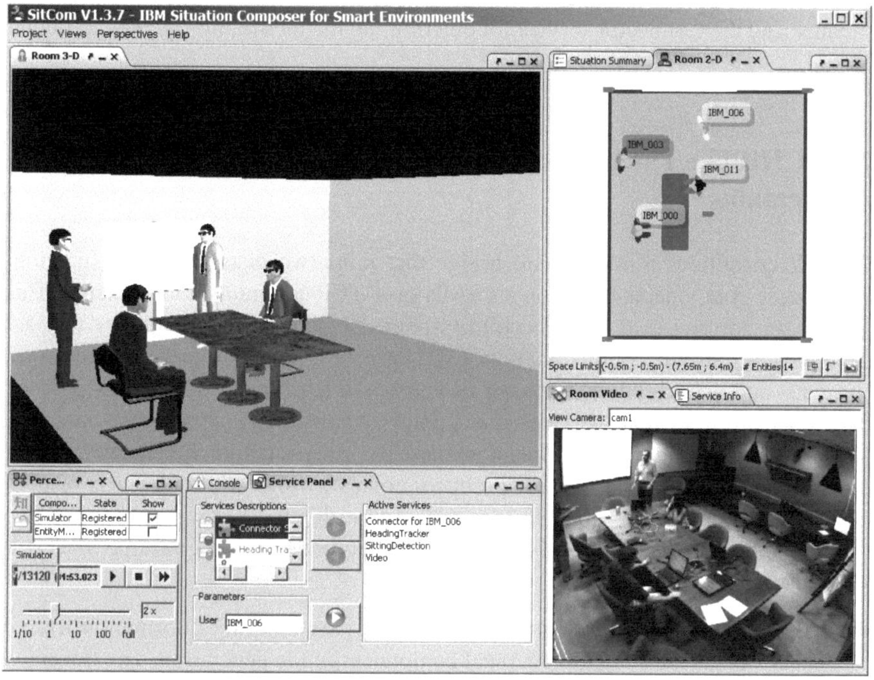

Fig. 29.2. Screenshot of SITCOM running the services on meeting data recorded in the IBM smart room.

form, with its Situation Summary, or in a more scenario-specific display, as the Room 2D for the meeting scenario or the Cockpit view for the cockpit scenario.

29.1.3 SITCOM Output

SITCOM users, typically developers of context acquisition components or services, are provided with a rich GUI. The user can allocate, start, and stop real or simulated perceptual components, load services and situation machines, and observe the current state of entities and streams in various SITCOM views (including 3D visualization). It is possible to add user-defined renderers for displaying scenario-specific streams of entities, and to add special views for showing the current state of situation machines.

The tool, specifically the inner module called SITMOD provides both pulling and pushing APIs to services. The pulling API of a *situation model* is done by direct calls to Java methods, and the pushing API is done by an event and listeners mechanism, which can be accessed remotely via *CHiLiX*.

Figure 29.2 shows the SITCOM graphical environment with both 2D and 3D visualizations and one camera view corresponding to a real situation in the IBM smart room. Three services are running in this particular configuration: the Occupancy service tracking participants of the meeting (see the output of the Attendance SM in the Situation Summary panel), the Video service allowing the online display of camera streams, and the Connector service for determining the interruptibility of a particular person.

29.2 *CHiLiX*: The Eventing Middleware for Context-aware Applications

CHiLiX constitutes a middleware bridge that joins two distributed and functionally diverse components based on an XML-over-TCP communication schema. This bridge can be seen as a flexible point-to-point message exchange library that accommodates multiple communication XML formats while also supporting both synchronous and asynchronous message exchanges. *CHiLiX* can flexibly support the communication between services and context acquisition components, based on XML message structures that implement message encapsulation. Specifically, a layer of context acquisition components produces context, while another layer is responsible for realizing the application logic. Based on this layered separation, the two communicating endpoints can be classified into producers and consumers, as follows:

- The producer is the endpoint responsible for providing the information from a context acquisition (e.g., perceptual component) upon generation. The transmission can be either synchronous or asynchronous.
- The consumer is the endpoint waiting to receive information from a producer. The consumer is also responsible for knowing where the component that generates this information is located. To help this discovery, producers may advertise their existence by registering themselves with a directory service.

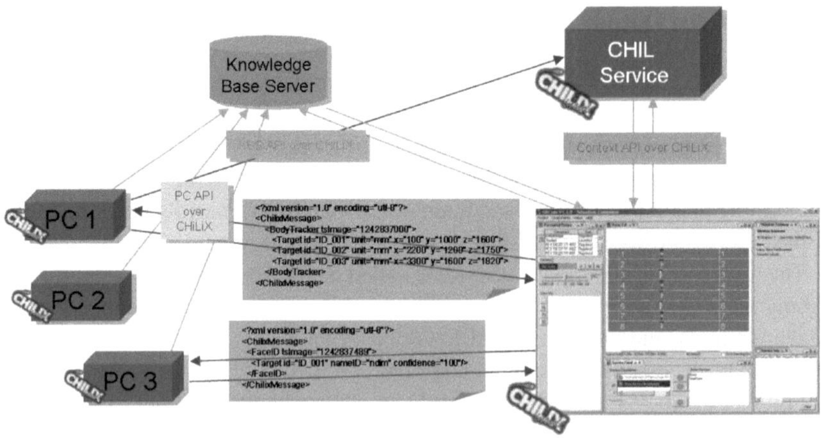

Fig. 29.3. Schema of *CHiLiX* communication channels between various CHIL components and the SITCOM application.

Our implementation of *CHiLiX* offered a number of benefits, as measured from external usage feedback. First and foremost, it is lightweight and can run on embedded hardware as on some mobile phones, which is a major requirement for a variety of pervasive and ubiquitous computing applications. It is a single package, yet it supports both synchronous and asynchronous messaging, as well as a publish-subscribe mechanism. It also allows simple reconnection in the case of service or network failure and has an application feedback mechanism to indicate when the connection is lost. Thus, it is more versatile than most conventional technologies.

Its portability was in the center of its development, and hence it can run on multiple platforms (Linux, Windows, MacOS, etc.) and provides bindings for a variety of programming languages (Java, C/C++, Python, Perl, Tcl). Its portability is similar to Web services, without assuming familiarities with standards like WSDL, SOAP, or Web containers. This low learning curve helped its adoption by technology providers, who have to deal with near real-time issues and are less inclined to include huge libraries and frameworks in their systems. *CHiLiX* has also benefited from a non-intrusive API, so that existing perceptual components can be *CHiLiX*-enabled with minimal code changes. The use of different *CHiLiX*-based APIs within the reference architecture is shown in Fig. 29.3. The use of *CHiLiX* in the multimodal perceptual systems is further described in [7].

29.3 Designing with SITCOM: Connector Service Scenario

Let us describe the SITCOM framework on a Connector scenario proposed and exploited in [3, 4]. The Connector service is responsible for detecting acceptable interruptions (phone call, SMS, targeted audio, etc.) of a particular person in the smart room. During the meeting, for example, a member of the audience might be interrupted by a message during the presentation, whereas the service blocks any calls for the meeting presenter. It therefore uses the context-aware service outlined in Section 26.2.

29.3.1 Perceptual Input

Our smart room is equipped with multiple cameras and microphones on the *sensor* level. The audio and video data are streamed into the following *perceptual components*, as depicted in Fig. 29.4:

- **Body tracker** is a video-based tracker providing 3D coordinates of the head centroid for each person in the room;
- **Facial features tracker** is a video-based face visibility detector, providing nose visibility for each participant from each camera;
- **Automatic speech recognition** (ASR) provides speech transcription for each participant in the room.

SITCOM will present these components in its GUI, if they already exist, and will use artificial data as well as data recorded on other sites to start designing the context

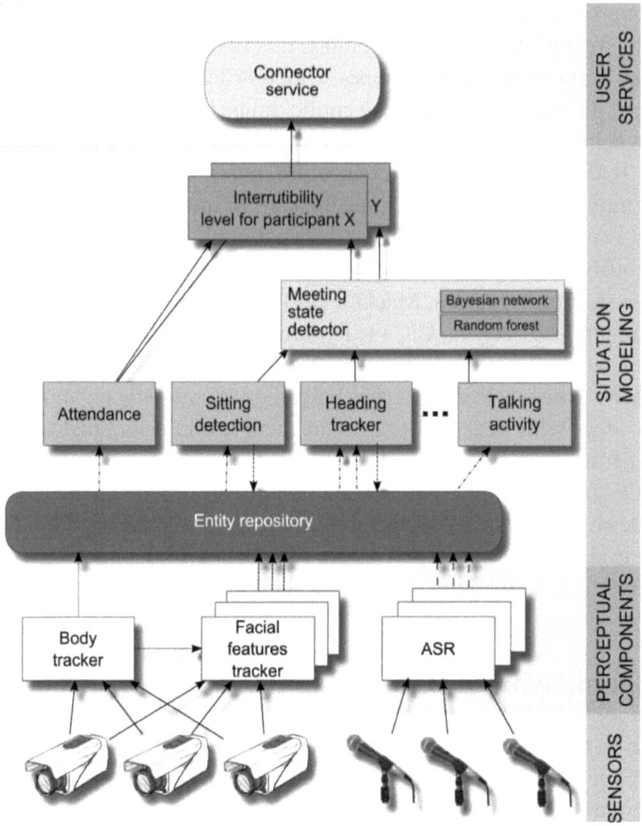

Fig. 29.4. Schema of data flow in the meeting scenario.

models. For room data, Fig. 29.2 shows the rendering of the aggregated information in a *2D* fashion for people's locations, headings and speech activity as well as indicating if they are sitting or not. The *3D* is useful when annotating information, as it's a more natural view to assess whether or not the meeting is going on. Note also the view indicating the output of the sensors (in this case, the cameras), which is useful for perceptual component developers to assess the quality of their output.

A key aspect of the perceptual components is that they implement a specific API. This API is defined along *functional* aspects of the perceptual components, e.g., *body location tracker API*, so that a body tracker may do its tracking based on video signals, audio signals, or even a combination of other monomodal trackers. The important part is that the API represents the type of information that we can get from the component.

Such an API also presents the advantage that certain components that may deliver multiple types of information (e.g., speaker ID and audio transcription) will implement multiple simple APIs (e.g., a speaker ID API and a transcription API) instead of one single-purpose combined API. This simplifies the situation model, as it can then request information from only one aspect of such a component. It also reduces the set of APIs we have to deal with, directly benefiting the integration and deployment of the system.

29.3.2 Situation Model

As depicted in Chapter 26, the goal of the situation model is to extract semantically higher information from these facts. This is achieved through *situation machines*. A situation machine models a single situation that consists of a set of *states*; the state of the situation machine is detected by observing entities in the entity repository or current states of other situation machines.

The Connector scenario uses the following situation machines:

- **Attendance** tracks the number of participants in the room;
- **Motion detection** reports how many participants have moved over a certain distance threshold in a given time period;
- **Heading tracker** infers the head orientation for each person from her position and face visibility for all cameras;
- **Attention direction** tracks the number of participants looking at the same spot, using their heading information;
- **Sitting detection** infers whether or not a particular person is sitting from the Z-(height) coordinate;
- **Crowd detection** searches for groups of at least three people whose relative distance does not exceed a threshold;
- **Talking activity** tracks the duration of speech activity and the number of speaker changes in a given period;
- **Meeting state detector** infers the current state of the meeting in the room;
- **Interruptibility level** selects the level of *interruptibility* of a particular participant according to the current meeting state and the current speaker.

Again, through its GUI, we can see all the situation machines and their state, for any particular time in the simulation data. In Fig. 29.2, we can see the center lower panel showing the currently active set of situation machines, along with their state information. Later, we could use the same status information to follow live what our system is doing, all while recording the additional real data.

29.3.3 SitCom GUI

The SitCom user, which is typically either a perceptual component or service developer, is provided with a graphical user interface (GUI). The user can allocate, start, and stop real or simulated perceptual components, load services and situation machines, and observe the current state of entities and streams in SitCom views, such as the Room 2D or Room 3D view reflecting the entity repository or the Room Video view displaying pictures captured by one of the cameras at the corresponding time. Again, it is possible to add user-defined renderers for displaying scenario-specific streams of entities, and to add special views for showing the current state of situation machines.

Through the IDE controls, SITCOM also facilitates the capture and creation of situation machines and subsequent realistic rendering of the scenarios in a 3D visualization module.

29.3.4 Services

SITCOM's architecture allows for many situation machines (SMs) to be registered with the situation model, but only the SMs that are needed are actually loaded. In Fig. 29.2, we run the Connector service for a participant named *IBM_006* that will pull in the necessary set of situation machines. It can be seen that the participant is presenting at this time, so his interruptibility level is set to *no interruptions*. The service will use this information and match it to the importance of the incoming interruption request, so that only very important requests get through at this particular time.

SITCOM provides a good infrastructure for plugging components, so that we have used it for other helper services, like the *Video* service, that open a view showing the sensor information from a relevant camera for the current simulated time of the meeting.

29.4 Conclusion

We have introduced a development environment that has already been exercised by several developers. SITCOM works on the division-of-concerns principle, supports different developer roles, and thus fosters the shorter and more efficient development of contextual services.

Both SITCOM and *CHiLiX* help in all phases of the development of nontrivial life cycle of context-aware services. We equipped SITCOM with a set of functionalities that we found beneficial in the development of perceptual applications:

- simulates the environment and the perceptual input (manually created or recorded scenarios);
- provides 2D and 3D visualization of the scenes, situations, and scenarios;
- works as the middleware between user services and the layer of perception (situation modeling);
- serves as IDE for repetitive testing of context-aware services and applications by replaying recorded or simulated scenarios;
- provides portability between virtual and real devices;
- serves as a tool for annotation of recorded data.

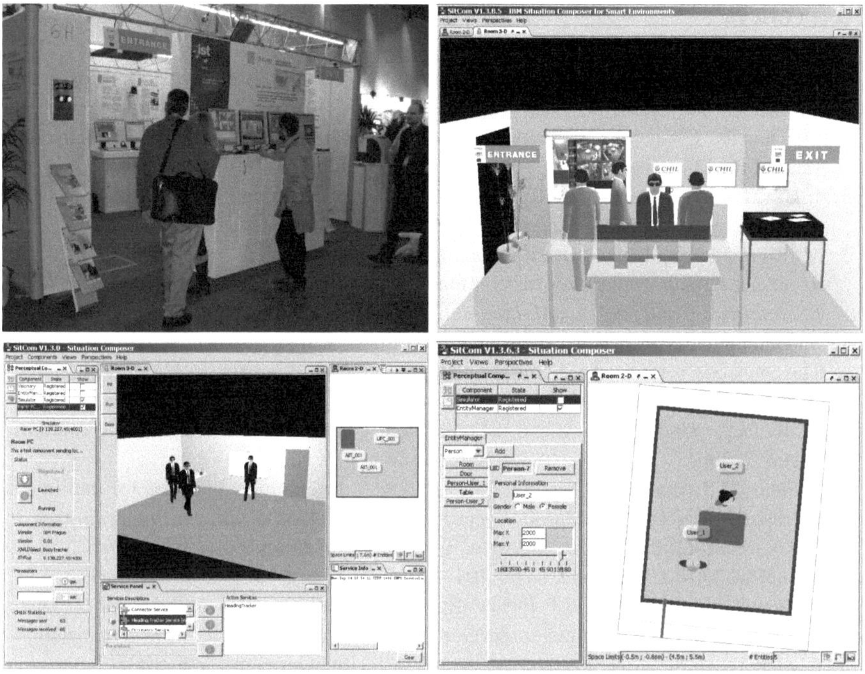

Fig. 29.5. Some of the deployments of SITCOM: The top row shows the booth at IST'06 in Helsinki along with its model representation; the bottom left shows SITCOM at the first attempt to exchange perceptual components across sites; and the bottom right shows the manual editing of a scene with the entity manager GUI.

We have successfully used the tool in meeting room detection and car cockpit situational analysis. It was also used in the IST'06 EU projects showcase in Helsinki to display tracked people in the exhibition booth (Fig. 29.5).

In many ways, having an integrated tool specialized for context-aware applications has been helpful to identify necessary pieces for the application, like the needed room information, the specific sensor information, or the definition of the participants' roles. We have also used SITCOM's rendering capabilities for annotation purposes, for example, in the CHIL project when there was no available corpus of annotated meeting data containing multimodal nonmeeting parts. Our tool is in use in more than 10 sites and has seen plug-ins written by multiple developers, who have integrated a rule-based situation modeling, as well as multiple perceptual components. SITCOM is also used to provide contextual information and a GUI to an agent-based service architecture and is used as a manual trigger for events in scenario-based experiments [8].

References

1. A. Bürkle, J. Crowley, J. Cuřín, J. Edlund, P. Fleury, J. Kleindienst, J. Mossgraber, W. Müller, M. Okon, A. Paar, U. Pfirrmann, J. Reuter, J. Soldatos, G. Szeder, and M. Thomas. Cooperative information system software design, 2004. Public Deliverable D2.1 of the EC's project CHIL (Computers in the Human Interaction Loop).
2. J. L. Crowley, J. Coutaz, G. Rey, and P. Reignier. Perceptual components for context aware computing. In *Proceedings of UbiComp*, pages 117–134, London, 2002.
3. M. Danninger, T. Kluge, and R. Stiefelhagen. Myconnector: Analysis of context cues to predict human availability for communication. In *ICMI '06: Proceedings of the 8th International Conference on Multimodal Interfaces*, pages 12–19, New York, NY, 2006.
4. M. Danninger, E. Robles, L. Takayama, Q. Wang, T. Kluge, R. Stiefelhagen, and C. Nass. The connector service - predicting availability in mobile contexts. In *MLMI*, LNCS 4299, pages 129–141. Springer, 2006.
5. A. K. Dey. Understanding and using context. *Personal and Ubiquitous Computing*, 5(1):4–7, 2001.
6. P. Fleury, J. Cuřín, and J. Kleindienst. SITCOM – development platform for multimodal perceptual services. In V. Marik, V. Vyatkin, and A. Colombo, editors, *Proceedings of 3rd International Conference on Industrial Applications of Holonic and Multi-Agent Systems, HoloMAS 2007*, LNAI 4659, pages 104–113, Regensburg, Germany, Sept. 2007.
7. J. Kleindienst, J. Cuřín, and P. Fleury. Reference architecture for multi-modal perceptual systems: Tooling for application development. In *Proceedings of 3rd IET International Conference on Intelligent Environments (IE'07)*, Ulm, Germany, Sept. 2007.
8. Z. Míkovec, I. Malý, P. Slavík, and J. Cuřín. Visualization of user activities in specific environment. In S. G. Henderson, B. Biller, M.-H. Hsieh, J. Shortle, J. D. Tew, and R. R. Barton, editors, *Proceedings of the 2007 Winter Simulation Conference*, Washington, D.C., US, Dec. 2007.
9. H. Nishikawa, S. Yamamoto, M. Tamai, K. Nishigaki, T. Kitani, N. Shibata, K. Yasumoto, and M. Ito. UbiREAL: Realistic smartspace simulator for systematic testing. In *Proceedings of the 8th International Conference on Ubiquitous Computing (UbiComp2006)*, LNCS 4206, pages 459–476, 2006.
10. M. Roman, C. K. Hess, R. Cerqueira, A. Ranganathan, R. H. Campbell, and K. Nahrstedt. Gaia: A middleware infrastructure to enable active spaces. *IEEE Pervasive Computing*, pages 74–83, Oct.-Dec. 2002.

Part V

Beyond CHIL

30

Beyond CHIL

Alex Waibel

Universität Karlsruhe (TH), Interactive Systems Labs, Fakultät für Informatik, Karlsruhe, Germany

Despite tremendous progress, the CHIL project represents a milestone rather than the final vision. Whilst many questions have been answered, the project has also opened up new challenges and directions for further exploration, and more work remains to be done.

Three main areas of concern have been explored in CHIL and are presented in this book: Perceptual Technologies, Services and Infrastructure. In CHIL, the project, of these, Perceptual Technologies has received the greatest attention primarily because when CHIL was initially proposed the perceptual technologies available at the outset of the program simply did not offer the robustness necessary to permit the development of flexible, perceptually informed CHIL Services. Nevertheless, the CHIL program managed to propose and showcase a set of initial CHIL services within its short lifetime and managed to evaluate and examine their effectiveness for real users. This was made possible via architectural and organizational tools and processes that were designed explicitly for the purpose of rapid prototyping and exploration.

At the conclusion of the CHIL program, it is clear that research will continue on all fronts, perceptual components, infrastructure services, and that further advances in perceptual technologies will also lead to more advanced and more daring new CHIL services in the future. The following comments speculate on future directions for CHIL computing, based on ideas and insights learned in the course of the three year CHIL effort.

30.1 Perceptual Technologies

Among the areas of concern in the CHIL program, Perceptual Technologies has received the most attention, and tremendous improvements have resulted from the effort, thanks in part to concerted, worldwide benchmarking efforts carried out in each of the perceptual processing technologies considered. The benchmarking has proven to be extremely beneficial for community building and to achieve a focused intense effort around common databases, leading to rapid progress, delivering high quality results and greater robustness in a short time frame. Robustness has benefited from the concerted technical effort in each of the processing technologies as well as innovative work on fusing them across modalities.

- Robustness in Individual components: CHIL broke new ground in building all its benchmarks around real data collected in real environment, with real people carrying out real tasks. The program did – on purpose – not resort to artificial scenarios, intrusive sensors, or slow expensive processing technology. Everything had to work in real-time, and all the data came from real meeting and seminar rooms in real organization. The data was also collected at multiple (5) sites to ensure generalization of the results across idiosyncrasies of each local environment. Yet, numerous challenges still exist:
  - Environments: Despite the generalization across spaces explored in CHIL, all of our environments were still meeting rooms, and did not cover the full range of human spaces, such as corridors, lobbies, airports, railway stations, shops, offices, restaurants, streets, the outdoors, and many more. To generalize perceptual processing to all human experience, new and different environments and transitions between them need to be included.
  - Sensor Positioning: At the conclusion of CHIL good speech recognition and speaker identification accuracies have been achieved over lecture and meeting data, with close speaking or lapel microphones. Significant advances were also observed with remote microphones and microphone arrays. Nevertheless, even though error rates have dropped, sometimes from 70% to 20%-30%, the remaining errors are still too high for certain applications, where error rates below 10% are necessary. Further research will be required.
  - Interferences: In both acoustic and visual sensing, different environments bring greater interference. In speech, the most well known is the so-called cocktail party phenomenon, the apparent ability of humans to follow a conversation at a cocktail party, despite an overwhelming level of jamming noises and jamming human conversations, and despite the distant and variable positioning of a listener's sensors. Toward the end of CHIL, such questions were raised and corresponding challenges formulated, but the problem remains largely open and unaddressed. In vision, analogous interferences exist. Here too, our environment (meetings and lectures) provided a real but comparatively benign version of these. Railway stations and large crowded places still present greater challenges in resolution, noise and occlusion. Moreover, they may require a combination of different techniques aimed at dealing with long range vs. short range perception, and with the smooth integration of entirely different techniques available at high resolution vs. low resolution (for example, in the case of person identification: face ID vs. gait or color histograms of clothing).
  - Sensor Coordination: A surprising discovery in CHIL was the challenges and opportunities emerging from the coordination of multiple sensors. When perceptual processing was done only in directed, well positioned human-machine tasks, there was typically one well positioned sensor from which data was collected such as a camera straight ahead or a close-speaking, head-

mounted microphone. Similarly, the phases of recording were well defined, by way of an on/off switch, or a shutter release at the right moment. CHIL removed this artificial constraint, and instead placed multiple sensors in a space. With multiple cameras observing the same scene, and multiple microphones listening to the same acoustic events, the question of coordination and integration became paramount. In CHIL, several calibration techniques were successfully tried and the signals collected from multiple sensors combined for the benefit of several of the perceptual processors (e.g., speech recognition, focus of attention tracking, people tracking, etc.). Despite these advances, more is clearly going to be required such as how do sensors know where they are in any environment and how do they communicate and merge their results more effectively? How do sensors know when the collected signal is unreliable how does the perceptual processor identify the most informative signals from multiple sensors and varying reliability?

- Robustness through Fusion: CHIL examined a number of perceptual tasks, where multiple modalities cooperate to describe human communicative events. Speaker localization, or person identification, for example can be done based on the acoustic signal as well as the visual signal. A number of results from such multimodal processing have been reported in this book, and will continue to attract research interest moving forward. In addition to work on the mere combination of multimodal signals, two new research directions have emerged:
  - Opportunistic Multimodal Fusion: With multiple signal streams from different modalities and multiple sensors, the same event cannot always be detected at the same time and may vary in robustness and reliability. A speaker may, for example, speak or be silent, a face may be temporarily occluded and different cameras or microphones may yield better and more reliable signals at different times. Fusion must therefore be selective and accumulate evidence over time. Identifying such moments of high robustness, better confidence measures and better integration across time, will continue to drive research as we aim for increasingly natural interactions and environments.
  - Self-Calibration: In addition to opportunistic fusion, we must also consider a more adaptive approach to achieving perceptual robustness. For a start, sensors must be able to self-calibrate better, and identify their own positioning and their own role in performing a perceptual task, vis a vis the other sensors. This is particularly important if we hope to build flexible, and general, perceptual components for practical commercially relevant deployments. For such deployments, the sensors and their precise positioning will vary and the arrangement cannot be redesigned on site in a cost-effective manner: The sensors must therefore arrange and determine their cooperation by themselves.
  - Active Fusion: Beyond Opportunistic Fusion and Self-Calibration across fixed sensors, we may also consider the possibility of sensors that perform more active perceptual processing by moving into position. This is of particular interest in applications that offer the possibility of moving the sensors during processing, such as humanoid robots, vehicles, or transportation systems. A moving platform could thus be positioned to take a "better look" or

turn to better listen in from different angles and distances. Such multimodal perceptual processing will require considerably more complex models of the perceptual systems and their environments.

30.2 CHIL, A Family of Services

CHIL computing is not and was never meant to be limited to the four CHIL services, considered in this book. Rather it represents a vision for numerous proactive services that aim to support human interaction without necessitating direct commands or human-machine interaction. Numerous additional services are possible. Even during the project, several such "surprise" services emerged at the participating laboratories: The "hummer", a system observing and modeling human conversational speech and turn-taking, the "Lecture Translator" a simultaneous speech translation system for seminar speakers with selective audio presentation for select subgroups in an audience, or a meeting coaching system for consultants, are among the surprising outcomes of some of the work in CHIL that has already occurred during the life-time of the project. Further advances are likely, and are facilitated by the CHIL architecture and the availability of interchangeable perceptual modules among the partners. Security applications, advertising, coaching, and assistance to the elderly, are further applications that appear now possible and that are already being considered beyond CHIL. And architecture for assembling components into CHIL services, the methodology for evaluating perceptual components and evaluating usability of the resulting systems permit a rapid prototyping to explore CHIL services beyond the ones discussed in this book. Indeed, we very much hope that we have only just scratched the surface.

30.3 From CHIL to CHHIL Services

In considering CHIL systems, we began with the rather extreme position of exploring the "disappearing" computer and the technologies that could be necessary to make this reality. We felt that such an extreme position was necessary to drive progress and to lay the groundwork for implicit computing capabilities. Robust perceptual technologies, implicit computing services, and flexible architectures, have all been advanced and benefited considerably from taking this view, as they are key elements of truly flexible computer systems in natural human environments.

Nevertheless, for many practical systems, there is no need to make a hard decision between implicit CHIL systems and services and more traditional human-machine interaction and dialog. In fact, it is useful to think of human-machine interaction as part of CHIL services or of Computers and Humans in the Human Interaction Loop (CHHIL). Already in CHIL, we have carried out work on human-machine dialog, suggesting a reappearing computer that not only observes but also listens and occasionally "comments" on the interactions and communications between humans. Indeed, a truly useful autonomous device of the future may take a more balanced approach between pure CHIL and Human-Machine interaction and seek a middle ground between implicit and explicit interaction. Why should a perceptually well-informed, proactive, and intelligent social agent not engage in an occasional direct dialog with its human partner, while also at the same time proactively taking the initiative? There are numerous scenarios that would make this an attractive proposition:

- Humanoid robots that interact occasionally with their masters, but that are capable of doing their work autonomously
- Smart rooms with avatars occasionally speaking up, or engaging in an occasional directed dialog
- CHIL services, that are occasionally addressed explicitly by humans in the room
- Learning CHIL services that occasionally request clarification or instructions from humans

All this, will necessitate further exploration of study of the interplay and trade-offs between direct/explicit and indirect/implicit interaction, and between autonomy and explicit command & control. Attaining a balance between these will indeed be challenging. Not only will it involve further advances in CHIL Computing, Human-Machine Interaction, and their technical integration, but also raise new social and philosophical issues. How is a balanced integration achieved? When should a computer system speak up, when to interrupt and when to patiently observe? What social norms are to be applied and how? How does the system assess relevance and urgency in interacting with or in the presence of humans? How does a system decide whether to take on the initiative and proceed proactively and when to await instructions? And finally, how much autonomy would we want and how much would we be willing to yield to a computer artefact and under what circumstances and tasks?

For the moment, these questions are only suggestive for potentially profitable ongoing research, whilst work on more mundane but no less challenging tasks, remains. Despite the impressive advances so far, it is clear that further work is needed to advance our understanding of perception, cognition and human interaction in order to achieve computer systems that will blend in with humans and as gracefully as humans, in a community and a social partnership.

Index

MIX
Papier aus verantwortungsvollen Quellen
Paper from responsible sources
FSC® C105338

If you have any concerns about our products,
you can contact us on
ProductSafety@springernature.com

In case Publisher is established outside the EU,
the EU authorized representative is:
Springer Nature Customer Service Center GmbH
Europaplatz 3, 69115 Heidelberg, Germany

Printed by Libri Plureos GmbH
in Hamburg, Germany